T0177695

Quantum Mechanics for Beginners

Quantum Mechanics for Beginners

with applications to quantum communication and quantum computing

M. Suhail Zubairy

Texas A&M University

OXFORD

UNIVERSITY PRESS

OXFORD
UNIVERSITY PRESS

Great Clarendon Street, Oxford, OX2 6DP,
United Kingdom

Oxford University Press is a department of the University of Oxford.
It furthers the University's objective of excellence in research, scholarship,
and education by publishing worldwide. Oxford is a registered trade mark of
Oxford University Press in the UK and in certain other countries

First Edition published in 2020

Impression: 3

Published in the United States of America by Oxford University Press
198 Madison Avenue, New York, NY 10016, United States of America

British Library Cataloguing in Publication Data
Data available

Library of Congress Control Number: 2020934341

ISBN 978-0-19-885422-7 (hbk.)
ISBN 978-0-19-885423-4 (pbk.)

DOI: 10.1093/oso/9780198854227.001.0001

Printed and bound by
CPI Group (UK) Ltd, Croydon, CR0 4YY

Dedicated to my beloved

Zoya, Aliya, Qasim, Sameer, and Khalid

Preface

The laws of quantum mechanics were formulated about a hundred years ago, replacing the classical laws of Newton and Maxwell. Since then, quantum mechanics has been applied remarkably successfully to understand a very wide range of observations and systems. The success of the laws of quantum mechanics in predicting and explaining essentially all the known physical phenomena is astounding. However, in spite of the great success, it remains a mysterious theory and the concepts of wave–particle duality, complementarity, the probabilistic nature of measurement, quantum interference, and quantum entanglement are still hotly discussed. However, it is not just the remarkable success in explaining all the known phenomena that makes quantum mechanics a fascinating subject. It is truly amazing that, even today, a mere knowledge of the basic postulates can lead to startling new ideas and devices. For example, just the knowledge of the principle of complementarity can lead to perfectly secure communication systems, or the understanding of a beam splitter for a single photon can lead to a highly counterintuitive communication protocol with no particle present in the transmission channel, or the resource of quantum entanglement can lead to novel quantum computing algorithms. Therefore it becomes possible to convey not only the foundations of quantum mechanics but also some mind-boggling applications, such as in quantum communication and quantum computing, with just elementary knowledge of basic physics and mathematics.

With this background it is interesting to ask whether it is possible to convey the basic concepts of quantum mechanics and its amazing applications to someone with a limited knowledge of physics and mathematics. In the fall of 2018, I offered a course on Quantum Mechanics to incoming freshman students at Texas A&M University. These students, just out of high school, took this course before they took the usual Mechanics and Electricity/Magnetism courses. This book grew out of the lecture notes of this course. The main objective of this book is to present an introduction to quantum mechanics in an almost self-contained way for someone with a high school physics and mathematics background.

The book challenges the common perception that quantum mechanics is a highly mathematical and abstract subject that is inaccessible to anyone without an advanced knowledge of mathematics. This book, except the last chapter on the Schrödinger equation, is entirely algebra-based. An effort is made to derive some amazing results from very simple ideas and elementary mathematical tools. Ideally every chapter offers results that are highly counterintuitive and interesting. This book can be used as a text for a course on quantum mechanics or quantum informatics at the undergraduate level. However it can also be a useful and accessible book for those who are not familiar with but want to learn some of the fascinating recent and ongoing developments in areas related to the foundation of quantum mechanics and its applications to areas such as quantum communication and quantum computing.

The book is divided into four parts. After an introductory chapter, some basic mathematical tools such as complex numbers, vector analysis, and introduction to probability as well as a classical description of particles and waves are introduced in the next three chapters. In the next eight chapters, basic concepts of quantum mechanics are discussed, such as wave–particle duality, complementarity, the Heisenberg uncertainty relation, quantum interference and entanglement, no-cloning theorem, as well as issues at the foundations of quantum mechanics such as the delayed-choice quantum eraser, the Schrödinger's cat and EPR paradoxes, and Bell theorem. In the following chapters, these fundamentals of quantum mechanics are applied to applications in areas such as secure quantum communication, quantum teleportation, counterfactual communication, and quantum computation. In the last part, the Schrödinger equation is introduced with its relation to Newtonian dynamics and its applications for a particle inside a box and the hydrogen atom.

Each chapter is followed by a short bibliography, guiding an interested reader to some relevant books and papers. In some instances, the original papers are included in the list. No attempt has, however, been made to give an exhaustive list of references. A number of problems are also given at the end of each chapter for the students in case the book is followed as a text for a course.

I owe my gratitude to several individuals for their support and encouragement in the preparation of this book. First and foremost, I thank Marlan Scully for his long friendship and many fruitful collaborations that helped shape my thinking about aspects of the foundations of quantum mechanics. A person most responsible for this book is David Lee who first proposed the idea and remained an enthusiastic supporter and inspiration throughout the writing of this book. I am also grateful to Peter McIntyre whose enthusiastic support, as the Head of the Department, for teaching an unprecedented course on quantum mechanics to freshmen, was vital to this project. Robert Brick and Wenchao Ge graciously read parts of the book and gave me their unvarnished, but extremely helpful, comments. I also thank Jiru Liu and Chaofan Zhou for their help with proofreading the manuscript. Special thanks are due to Sonke Adlung and Harriet Konishi of the Oxford University Press and Cheryl Brant of SPi Global for all their help during the publication process.

Finally I wish to acknowledge the loving support of my family members, Sarah, Neo, Sahar, Shani, Raheel, and Reema. My deepest gratitude is however reserved for my wife, Parveen. She has been relentless in her support not just during the writing of this book, but for all the projects, big and small, during my life.

M. Suhail Zubairy
College Station, Texas
October 9, 2019

Contents

1 What is this Book About?   1

   1.1 From Classical to Quantum Mechanics   2

   1.2 Outline of the Book   5

PART I: INTRODUCTORY TOPICS   11

2 Mathematical Background   13

   2.1 Complex Numbers   13

   2.2 Trigonometry   16

   2.3 Vector and Scalar Quantities   20

   2.4 Elements of Probability Theory   25

3 Particle Dynamics   32

   3.1 Classical Trajectory   32

   3.2 Linear Momentum   35

   3.3 Kinetic and Potential Energy   37

   3.4 Inelastic and Elastic Collisions   39

   3.5 Angular Motion   39

   3.6 Angular Momentum   44

   3.7 Motion of an Electron in Electric and Magnetic Fields   45

4 Wave Theory   50

   4.1 Wave Motion   50

   4.2 Young's Double-slit Experiment   57

   4.3 Diffraction   61

   4.4 Rayleigh Criterion   66

PART II: FUNDAMENTALS OF QUANTUM MECHANICS   71

5 Fundamentals of Quantum Mechanics   73

5.1 Quantization of Energy   73

5.2 Wave–Particle Duality   74

5.3 End of Certainty—Probabilistic Description   75

5.4 Heisenberg Uncertainty Relations and Bohr's Principle of Complementarity   76

5.5 Coherent Superposition and Quantum Entanglement   78

6 Birth of Quantum Mechanics—Planck, Einstein, Bohr   81

6.1 Brief History of Light   81

6.2 Radiation Emitted by Heated Objects   84

6.3 Einstein and the Photoelectric Effect   88

6.4 History of the Atom till the Dawn of the Twentieth Century   90

6.5 The Rutherford Atom   92

6.6 The Hydrogen Spectrum   93

6.7 Quantum Theory of the Atom: Bohr's Model   94

7 De Broglie Waves: Are Electrons Waves or Particles?   100

7.1 De Broglie waves   100

7.2 Wave–Particle Duality—A Wavefunction Approach   105

7.3 Bose–Einstein Condensation   108

7.4 Heisenberg Microscope   110

7.5 Compton Scattering   114

8 Quantum Interference: Wave-Particle Duality   121

8.1 Young's Double-slit Experiment for Electrons   121

8.2 Einstein–Bohr Debate on Complementarity   127

8.3 Delayed Choice   130

8.4 Quantum Eraser   131

9 Simplest Quantum Devices: Polarizers and Beam Splitters   137

9.1 Polarization of Light   137

9.2 Malus' Law for a Single Photon—Dirac's ket-bra Notation   142

9.3 Input-Output Relation for a Classical Beam Splitter   148

9.4   Beam Splitter for a Single-photon State   149

9.5   Polarization Beam Splitter and Pockel Cell   150

10 Quantum Superposition and Entanglement   154

10.1   Coherent Superposition of States   154

10.2   Quantum Entanglement and the Bell Basis   158

10.3   Schrödinger's Cat Paradox   162

10.4   Quantum Teleportation   164

10.5   Entanglement Swapping   167

11 No-cloning Theorem and Quantum Copying   172

11.1   Cloning and Superluminal Communication   172

11.2   No-cloning Theorem   175

11.3   Quantum Copier   176

12 EPR and Bell Theorem   182

12.1   Hidden Variables   182

12.2   The Einstein–Podolsky–Rosen (EPR) Paradox   183

12.3   Bohr's Reply   186

12.4   Bell's Inequality   187

12.5   Quantum Mechanical Prediction   190

12.6   Experiments to Test Bell's Inequality   192

12.7   Bell–CHSH Inequality   193

PART III: QUANTUM COMMUNICATION   199

13 Quantum Secure Communication   201

13.1   Binary Numbers   202

13.2   Public Key Distribution, RSA   203

13.3   Bennett–Brassard 84 (BB-84) Protocol   207

13.4   Bennett-92 (B-92) Protocol   211

13.5   Quantum Money   214

14 Optical Communication with *Invisible* Photons   217

14.1   Mach–Zehnder Interferometer   218

14.2 Interaction-free Measurement   220

14.3 An Array of *N* Mach–Zehnder Interferometers   221

14.4 Counterfactual Communication   223

PART IV: QUANTUM COMPUTING   227

15 Quantum Computing I   229

15.1 Introduction to Quantum Computing   229

15.2 Quantum Logic Gates   233

15.3 The Deutsch Problem   237

15.4 Quantum Teleportation Revisited   240

15.5 Quantum Dense Coding   241

16 Quantum Computing II   245

16.1 How to Factorize *N*?   245

16.2 Discrete Quantum Fourier Transform   248

16.3 Shor's Algorithm   251

16.4 Quantum Shell Game   254

16.5 Searching an Unsorted Database   257

PART V: THE SCHRÖDINGER EQUATION   263

17 The Schrödinger Equation   265

17.1 The Schrödinger Equation in One Dimension   265

17.2 Kinematics in Classical and Quantum Mechanics—Newton vs. Schrödinger   270

17.3 Particle Inside a Box   275

17.4 Tunneling Through a Barrier   278

17.5 The Schrödinger Equation in Three Dimensions and the Hydrogen Atom   283

Index   289

1 What is this Book About?

Our common sense is based on what we observe in the world around us. The laws that govern this world appear to be completely deterministic. If we apply a force, any kind of force, on an object, we can predict the response quite accurately. For us light is a wave and a ball is a particle—there is no doubt about it. Light cannot behave like a particle and a ball cannot behave like a wave. This seems to be the world we live in. The laws of physics that explain such behavior were formulated during several centuries leading up to the beginning of the twentieth century and form the core of what we call classical physics.

What was found at the dawn of the twentieth century was that these laws are good only for big objects and intense light. For small objects like electrons, atoms, and very weak light signals, the laws of classical physics fail miserably. For example, light can behave both like wave and particle. Similarly an atom can also behave as both particle and wave. Soon, it was realized that a new set of laws were needed to explain the observations related to atoms and molecules. It took about 25 years to formulate a theory that could explain all the known observations up to that time. This theory is called quantum mechanics. Quantum mechanics is the fundamental theory of physics and classical mechanics is an approximation when considering macroscopic objects. This book presents an introduction to quantum mechanics and some of its interesting applications.

Quantum mechanics is one of the two most successful theories in human history, the other being Einstein's theory of relativity. The justification for this remarkable claim is that, after the passage of almost 100 years, no physical phenomenon has been found to be in violation of the predictions of quantum mechanics. This is true in spite of the tremendous advances in the precision with which the measurements can be made. For example, time can be measured with an accuracy of a billionth of a billionth of a second, distance to a trillionth of a meter, temperature to a millionth of a Kelvin, and mass to a billionth of a gram. We can see and manipulate a single atom and cool a gas to an extent that atoms and molecules lose their identity. We can carry out experiments where light consists of a single "photon" and even manipulate the interaction of a single "photon" with a single atom. In all such experiments, the results are dramatically different from what classical physics predicts but they are remarkably in full agreement with the predictions of quantum mechanics.

In this chapter, we first give a brief history of how classical mechanics evolved into quantum mechanics. We then give a bird's eye view of the basic aspects of quantum mechanics and its applications. In subsequent chapters, these topics will be discussed with reasonable completeness, with a minimum of mathematical background. Except for the last chapter, the entire book is only algebra-based. Most of the mathematical tools needed are discussed in Chapter 2.

Quantum Mechanics for Beginners: With Applications to Quantum Communication and Quantum Computing. M. Suhail Zubairy.
© M. Suhail Zubairy 2020. Published in 2020 by Oxford University Press. DOI: 10.1093/oso/9780198854227.001.0001

1.1 **From Classical to Quantum Mechanics**

We trace the beginning of the modern era of science to the year 1543, when Nicholaus Copernicus published his book "*De Revolutionibus Orbium Coelestium*" (On the Revolutions of the Heavenly Spheres). He proposed a heliocentric model of the solar system, a system in which the sun was held at rest and all the planets including earth circled around it, replacing the long held Ptolemaic geocentric model in which earth was at rest and at the center of the cosmos. Without the benefit of the knowledge of the law of gravitation, it was hard to believe how earth could be moving around the sun still maintaining the stability of all objects including the humans on its surface. The hostility to a model that took away the centrality of earth in a solar system was so great that Copernicus could not publish his heliocentric theory till the end of his life. According to a legend, Copernicus received the published copy of his book *De Revolutionibus* on the very last day of his life, thus dying without knowing that his work heralded a new era of human history. Copernicus was followed by Johannes Keppler (1571–1631) and Galileo Galilei (1564–1642) who studied the motion of planets within the framework of the heliocentric theory.

Isaac Newton (1642–1727) is the next defining figure in the history of science. His "*Principia*" laid the foundation of classical mechanics. His law of gravitation is a bright example of the nature of scientific law—a law that applies equally well to all objects, big and small. His contributions in mathematics, particularly his co-discovery of calculus (with Wilhelm Leibniz) provided tools that would be vital for almost all the subsequent major discoveries in physics and many other branches of science. They played key roles in shaping the physics of the coming centuries. Newton's laws of motion, discussed in Chapter 3, are the corner stone of classical mechanics as they provided both physical and mathematical tools to make scientific predictions. If we knew all the forces acting on an object and we knew the position and velocity of the particle at an initial time, we could trace the trajectory of the particle for all subsequent time.

Newton's description of light as consisting of particles was however not able to explain phenomena such as interference and diffraction. The work of Thomas Young (1773–1829) and Augustin Jean Fresnel (1788–1827) showed unambiguously (as we discuss in Chapter 4) that light consisted of waves, instead. Young's double-slit experiment was not only decisive in debunking Newton's corpuscular theory of light, but it also continued to play a crucial role in our understanding of the nature of light and matter even in the twentieth century as we see in Chapter 8. It was left to James Clerk Maxwell (1831–1879) to complete the classical picture of light as consisting of electric and magnetic waves. This was a truly remarkable outcome of his efforts to unify the two known forces of nature: electric force and magnetic force.

This was the situation that existed at the end of the nineteenth century. So much was the satisfaction with the existing laws of physics that a very eminent British scientist, Lord Kelvin, is quoted as saying in an address to the British Association for the Advancement of Science in 1900, "*There is nothing new to be discovered in physics now. All that remains is more and more precise measurement*". The classical theories of mechanics, electromagnetics, thermodynamics, and, of course, light, were firmly in place and it was justified in feeling that the basic laws of nature were fully understood.

There were however a small number of unresolved problems at the dawn of the twentieth century that could not be explained on the basis of existing theories. A resolution of these problems led to the birth of quantum mechanics. The development of quantum mechanics, which replaced the classical mechanics of Newton and Maxwell, took place in two distinct eras.

The first era began in December 1899, when Max Planck introduced the notion of the quantization of energy to explain the frequency spectrum radiated by hot bodies. This problem had remained unresolved for almost 40 years. The two other big heroes of this era are Albert Einstein and Niels Bohr. In 1905, Einstein used Planck's hypothesis to explain some observations about the emission of electrons when light shines on metals that could not be explained on the basis of the theories that existed at that time. In the process, as we learn in Chapter 6, Einstein introduced the notion of light quanta which were subsequently called photons. This was, in some sense, going back to Newton's corpuscular theory to explain certain phenomena. Another stunning success of Planck's quantum postulate came in 1913 when Niels Bohr used these ideas to describe a model of the atom that could explain the discrete frequencies emitted by hydrogen atoms.

These successful attempts to explain some unresolved phenomena based on the quantization hypothesis of Planck led to the realization that the old classical theory, as formulated by Newton, Young, Maxwell, and others, may not be valid when we try to understand phenomena at the atomic level. Planck, Einstein, and Bohr could explain some unresolved phenomena based on postulates that involved quantization of energy that had no basis in classical theories. Despite these successes, there was however no theory that could explain these and all other phenomena in a unified manner. The period between 1913 and 1925 was a period of unprecedented crisis. It was becoming apparent with the difficulties faced in explaining new emerging results at the microscopic level that a full-fledged theory was needed that should replace Newtonian mechanics.

The second era began with the breakthrough that came in the summer of 1925 when the 24-year-old Werner Heisenberg took the first major step in formulating a quantum theory, making a clean break with the past. In January 1926, Erwin Schrödinger independently formulated the quantum theory and wrote down a dynamical equation that is called the Schrödinger equation in his honor. Later it was shown that the theories of Heisenberg and Schrödinger were two different but completely equivalent formulations of quantum mechanics. Schrödinger's equation, like Newton's equation $F = ma$ and Maxwell's equations for electromagnetic fields, is one of the most famous equations in physics. We introduce it in the last chapter of this book.

Quantum mechanics, as formulated by Heisenberg and Schrödinger (along with other founding fathers including Max Born, Pascual Jordan, Paul Dirac, and Wolfgang Pauli), could not only explain all the existing phenomena at the microscopic and macroscopic levels but also predict new phenomena that could then be observed experimentally. Despite these stunning successes of the new theory, the conceptual foundations of the theory became a major point of discussion. What we see is that, at the level of a single atom or an electron or a photon, quantum mechanics makes predictions that are startling. They are dramatically different from the corresponding results for our everyday objects that can be described very successfully using Newtonian mechanics. The mind-boggling aspect of quantum mechanics was not lost on the founding fathers. Indeed, in spite of the great successes in explaining and predicting novel

Fig. 1.1 One of the most famous photographs in the history of physics taken at the Solvay conference in 1927. Almost all the founders of the quantum mechanics attended the conference with 17 of the 29 attendees were or to become Noble Laureates. **Back:** Auguste Piccard, Émile Henriot, Paul Ehrenfest, Édouard Herzen, Théophile de Donder, Erwin Schrödinger, J.E. Verschaffelt, Wolfgang Pauli, Werner Heisenberg, Ralph Fowler, Léon Brillouin. **Middle:** Peter Debye, Martin Knudsen, William Lawrence Bragg, Hendrik Anthony Kramers, Paul Dirac, Arthur Compton, Louis de Broglie, Max Born, Niels Bohr. **Front:** Irving Langmuir, Max Planck, Marie Curie, Hendrik Lorentz, Albert Einstein, Paul Langevin, Charles-Eugène Guye, C.T.R. Wilson, Owen Richardson. Photograph by Benjamin Couprie, Institut International de Physique Solvay, Brussels, Belgium.

phenomena, the conceptual foundation of quantum mechanics remains a hotly debated issue. Some of these discussions will be topics of later chapters.

The major milestones in the formulation of the quantum theory are as follows:

- 1899 – Max Planck introduces first ideas about quanta to explain blackbody radiation
- 1905 – Albert Einstein explains the photoelectric effect by treating light as consisting of particles
- 1913 – Niels Bohr presents a planetary model for the atom based on a quantum postulate
- 1924 – Louis de Broglie postulates that particles behave like waves
- 1925 – Werner Heisenberg invents quantum mechanics
- 1926 – Erwin Schrödinger presents the Schrödinger wave equation
- 1927 – Max Born introduces the probabilistic nature of quantum mechanics
- 1927 – Werner Heisenberg derives Heisenberg uncertainty relations
- 1927 – Niels Bohr formulates the principle of complementarity
- 1928 – Paul Dirac unifies the wave–particle description of light

The era spanning over almost 30 years (from 1900 till 1930) when the foundations of quantum mechanics were laid is perhaps the most remarkable in the history of science. A whole new way of thinking and doing physics emerged. New effects and phenomena were predicted and observed based on quantum mechanics that led to the birth of many new fields of study. Just as the development of classical mechanics in the seventeenth, eighteenth, and nineteenth centuries ushered in an industrial revolution, an understanding of the quantum mechanical laws have led to amazing technological developments. The electronics industry, the communication revolution, the computer technology, the sources of energy, nanotechnology devices, lasers, and numerous other products and outcomes would not have been possible without an understanding of the laws of quantum mechanics. Our world would have been rather old-fashioned and simple in most respects if we still existed in an era of classical physics of the nineteenth century.

1.2 Outline of the Book

The objectives of this book are two-fold. On one hand, we discuss the foundation of quantum mechanics and the laws of quantum theory as were formulated by the founding fathers in the first quarter of the twentieth century. On the other, we discuss some novel applications of these ideas to modern and evolving fields of quantum communication and quantum computing. In the following, some questions (that would appear bizarre, even crazy, within the framework of classical physics) are listed. These are discussed and answered in this book with as little mathematics as possible. The amazing nature of these questions and their answers within the context of quantum mechanics should indicate what this book is all about. However, first a warning! For someone not familiar, certain terms and expressions used in this section may be new and incomprehensible. There is no need to be concerned as the premise of this section is only to give a flavor and not the full explanation. These terms and more are explained when they appear in later chapters of the book.

Can light behave like particles? The studies on the nature of light started in the seventeenth century. One of the earliest theories was advanced by Newton when he postulated that light consists of small particles. As we study in Chapter 4, this theory was completely discredited through the work of Thomas Young on interference and Augustine Fresnel's work on diffraction, who showed that light behaves like a wave. However Albert Einstein, while explaining the photoelectric effect (that we discuss in Chapter 6) showed in 1905 that the experimental results on this effect can only be explained if we treat light as consisting of quanta of energy called photons. Thus we have a paradoxical situation with which we have lived for well over a hundred years: light can behave like waves in some experiments and particles in some others. A milestone experiment was performed in 1923 by Arthur Compton, who showed that the results of the scattering of light by an electron can be explained only if we treat the light as consisting of particles with well-defined momentum (a particle concept). The Compton effect is discussed in Chapter 7.

Can electrons behave like a wave? Almost twenty years after Einstein showed that light can behave like particles, Louis de Broglie postulated that the converse should also be true— particles should also be expected to behave like a wave. His predictions, as discussed in Chapter 7, were verified in experiments involving electrons incident on crystals. The results of the experiments done independently by G. P. Thomson and by Clinton J. Davisson and Lester H. Germer could only be explained based on de Broglie's conjecture. More recently, in 1961, it was shown in a landmark experiment by Claus Jönsson that Young's double-slit experiment done with electrons (instead of light) leads to interference fringes, which is a hallmark of waves. This amazing experiment and its role in the study of the foundations of quantum mechanics are discussed in Chapter 8.

Delayed-choice and quantum eraser: An analysis of the double-slit experiment continues to amaze us with new paradoxical results. If somehow we are able to obtain the information about which slit the photon or electron passed through in the double-slit experiment, the interference fringes disappear and the particle nature is exhibited. John Wheeler argued that we can make a delayed choice as to whether the wave nature is exhibited and interference fringes are formed or the particle nature is exhibited by finding the which-path information, long after the photon (or electron) passed through the double slit. In 1982, Marlan Scully and Kai Drühl went one step further and came up with the ingenious but highly counterintuitive notion of a quantum eraser in which the which-path information is 'erased' and the interference pattern is recovered *after* the photon has passed through the slits and is detected on the screen. All this and more is discussed in Chapter 8.

Can atoms and molecules in a gas lose their identity and become one? In a gas, atoms and molecules are tiny particles that are moving around randomly in all directions, occasionally colliding with each other. This is the picture of a gas we have had since the laws of thermodynamics were formulated in the nineteenth century. This picture gives a quite accurate description of the properties of the gas such as pressure and temperature in our everyday life. However it was observed by Nathan Bose and Albert Einstein in 1925, that, when a gas is cooled to very low temperatures, almost at the nano-Kelvin level, the atoms shed their particle nature and act like waves—the de Broglie waves—losing their identity in the process and becoming a big massive object. This is a new state of matter—neither solid nor liquid nor gas—a state we call a Bose–Einstein condensate. We study this amazing effect that was experimentally observed in 1995—some 70 years after its prediction—in Chapter 7.

Can we measure position and velocity with arbitrary accuracy? When Newton formulated the laws of mechanics, the underlying principle was that, if we know all the forces acting on an object, we can predict both its location and velocity with arbitrary precision. The limitation on the measurements came only from the quality of the measuring apparatus. Thus when we throw a ball, we can predict with absolute certainty where it will be and how fast it will be moving at a given time. We can thus trace a trajectory of the ball. But is this really true? Heisenberg showed in 1927 that, no matter how accurate and precise our measurements are, we cannot measure two complementary variables such as position and momentum with as much accuracy as we would like. We have inevitable uncertainties in measuring both quantities such that the product of these uncertainties is above a minimum value, extremely tiny but nevertheless nonzero. This uncertainty relation called the Heisenberg uncertainty relation is "derived" in Chapter 7 and is one of the foundation principles of quantum mechanics. Heisenberg was a research assistant to Niels Bohr when he came up with this most amazing result. Bohr came up with his own principle of complementarity at the same time according to which two observables are *complementary* if precise knowledge of one of them implies that all possible outcomes of measuring the other one are equally probable. According to Bohr's principle of complementary, we cannot see both particle and wave nature in an experiment simultaneously. This principle is discussed within the context of Young's double-slit experiment in Chapter 8.

Can we predict anything with certainty? An important consequence of the wave–particle duality is that there is no determinism in quantum mechanics. Using Newtonian mechanics, we can predict with certainty where a ball thrown will hit the wall. The same cannot be said about an electron—quantum mechanics only allows us to calculate the probability that it hits the screen at a given point but does not allow a definite prediction where it will hit. Even after the electron is detected at a certain location, quantum mechanics does not allow us to talk about the trajectory it may have followed except in some very special cases. In the quantum mechanical description, a particle like an electron is not described as a point object that follows a definite trajectory. Instead it is described by a wave packet or wave function that spreads as it propagates. The wave function is NOT a real object representing the object—it (or rather its modulus squared) represents the probability that the object is found at a certain location. This inherent probabilistic nature of quantum mechanics makes it very distinct from classical mechanics and this aspect is discussed in different contexts at various places in the book, most notably in Chapters 9, 10, and 17.

Einstein–Bohr debate: One of the major events in the evolution of the quantum theory is the debate that took place between the two giants: Albert Einstein and Niels Bohr. Through conference presentations, letters, and research papers, these two iconic figures argued vigorously about the fundamental issues concerning quantum mechanics. This debate has left a lasting imprint on any discussion on the foundation of quantum mechanics. Einstein was one of the founding fathers of quantum mechanics through his work on the photoelectric effect and introduction of light quanta (Chapter 6). Yet he never reconciled with the final product that emerged with the theories of Heisenberg and Schrödinger. What he appeared to be most uncomfortable with was the lack of the determinism that we cherish in the classical theories of Newton and Maxwell. He could not reconcile with Bohr's principle of complementarity that implied the mutual incompatibility between the wave and particle natures of, for example, light and electrons. He came up with many clever thought experiments that demonstrated how we could see both particle and wave aspects in any given experiment. Bohr, however, defended his principle of complementarity each time, sometimes by invoking Heisenberg uncertainty relation. We discuss aspects of this debate in Chapters 8 and 12.

Does the moon exist when we do not look at it? This rhetorical question was posed by no ordinary mortal but by Albert Einstein. There are two fundamental principles that we cherish as almost "self-evident truth". These are reality and locality. Reality means that an object is real when it exists even in the absence of an observer. Paraphrasing Einstein, we have no doubt that the moon exists even when none of us is looking at it, i.e., it does not cease to exist when nobody is looking at it. Locality means that no information can be sent faster than the speed of light in vacuum. This means that if two objects are separated by a distance that light takes one hour to go from one to the other then what happens to one object (destroyed, divided, rotated etc.) cannot be influenced at all by the other object in any way whatsoever for one hour. One of the most startling results is that quantum mechanics violates any theory that is based only on these self-evident truths of reality and locality, and the remarkable result is that experiments are all in agreement with the predictions of quantum mechanics. Thus reality and locality cannot co-exist. We discuss this amazing result with far reaching consequences in Chapter 12.

Is perfect cloning possible? It is our common experience that, given the expertise and resources, we can make identical copies of any object. An expert carpenter can make such a copy of an object like a chair that the copy is indistinguishable from the original. A document can similarly be copied such that the original and the copy are identical. But is this true even at the microscopic level? Can we build a cloning machine that can make an identical copy of a single photon or an electron? The answer is, surprisingly, no. We discuss the no-cloning theorem of quantum mechanics in Chapter 11. An immediate question is if we cannot make identical copies then what is the best copy we can make? The answer to this question is also presented in the same chapter.

Can we accomplish teleportation like it is done in "Star Trek"? In other words can we make someone disappear at one place and recreate them at another place? This is still a far-fetched dream. What has, however, been possible is that the exact state of an atom or a photon can be destroyed in one place and generated at another. This is done by first preparing an entangled state of two particles at the two locations. Through some joint measurements and sending the measurement results from one location to another, the state of the atom or photon can be "teleported". Quantum teleportation is discussed in Chapter 10.

Can we have absolutely secure communication and detect an eavesdropper with certainty? Exchanging secret information between two parties at a very large distance from each other is a problem addressed for thousands of years. A field called cryptography emerged to address this problem. The basic idea has been to exchange a random key between the two parties through secure and reliable channels such that the key is known only to the two parties and is completely inaccessible to a potential eavesdropper. Then a message encoded with the key is sent from one party to the other. Even if some eavesdropper intercepts the encoded message, the actual message cannot be deciphered without the knowledge of the key. Only the receiving party can decipher the message with the key. A question of interest in the modern world of e-commerce and heightened concerns about security is whether it is possible to exchange a key on a public channel (which is accessible to everyone) and if an eavesdropper attempts to listen to the transmitted massage they can be detected by both the sender and the receiver. In Chapter 13, we show that this impossible task can be accomplished by using quantum mechanical systems. We first discuss the RSA algorithm that is used in the present-day public key distribution. We then present quantum mechanical algorithms that can ensure absolutely secure key distribution on a public channel but also ensures that an eavesdropper can be detected with certainty.

Is psychic communication possible? It is a paradigm that an exchange of information between two parties requires an exchange of particles and objects that carry the information. In ordinary conversation, atoms and molecules of the atmosphere are the carrier of our words and speech (no verbal communication will be possible if we go to outer space where there are no particles present). In present-day optical communication, the carrier of information is a light beam or photons. Is it possible for two parties to communicate with no particle existings between them? If this becomes possible, it would be akin to something like psychic communication. In 2013, the author and his colleagues showed that it is indeed possible to communicate with no particles in the transmission channel. This highly counterintuitive result is discussed in Chapter 14.

Can we build a quantum computer? All the computers that we see around ourselves use materials such as semiconductors whose properties are understood only by quantum mechanical analysis. Every computer down to its basic building blocks such as transistors is a quantum device. However the basic processing unit for the purpose of doing any manipulation, called a bit, is a classical object. A bit can take two possible values: "0" or "1". A quantum computer is a computing device in which the processing unit is a quantum-bit (or qubit). A qubit has highly non-classical properties—it can exist in a state where it is simultaneously in the states "0" and "1". Similarly two qubits can exist in an entangled state—a state in which a manipulation of one qubit can influence the other qubit no matter how far away they are from each other. A quantum computer can solve some problems at a speed much faster than we can imagine in comparison with our classical or conventional computer. The basics of quantum computing and some potential applications are discussed in Chapters 15 and 16.

Can we factorize a very large number in few million steps? Multiplication is easy—we can multiply two arbitrarily large numbers in a relatively very small time. The converse, finding the factors of a number, is a notoriously difficult problem. The finding of the prime factors of a number is not a mundane problem of pure mathematics—it lies at the heart of present-day e-commerce (Chapter 13). The reason we can send our credit card number and other personal information on the internet without any concerns about the information falling into hostile hands is the difficulty of finding prime factors of a number. For example, it may take several decades to factorize a 256-digit number (which is typically used in the RSA algorithm for communication safety) on the fastest computer today. One of the major potential successes of a quantum computer is that it can solve this problem in few million steps and thus giving an answer in a remarkably short time. The protocol for factoring a number, called Shor's algorithm, is discussed in Chapter 16.

Can we find a needle in a haystack? A game we may have played in our childhood is the shell game. A pea is hidden underneath one of the four shells and we have to guess the shell that is hiding the pea. If we are lucky, we can find the pea in just one guess but the probability of that happening is only 25%. Can we come up with a trick such that we can find the pea with certainty the first time every time? In classical systems, this seems impossible. In Chapter 16, we show that, in a quantum version of the shell game, we can indeed find the "pea" in only one attempt every time. We also show that this simple procedure can be applied to the search for a marked item (a needle) in an unsorted database (haystack) much faster than we can imagine using the tricks of quantum computing.

Can a particle tunnel through a barrier even when it does not have sufficient energy? It is a common observance that if we want to cross a hurdle or a barrier we need to have enough energy to surpass the barrier. For example, in a pole vault, an athlete acquires enough kinetic energy by running very fast that is at least equal to the potential energy corresponding to the height of the barrier. A smaller incident energy will not allow the athlete to jump over the barrier. An amazing consequence of quantum mechanics, first observed by Friedrich Hund in 1927, is that a particle such as an electron can "tunnel" through a barrier even when its energy is less than the minimum energy required to jump over the barrier. This effect, which is extensively used in transistors and microscopes, is discussed as a consequence of the Schrödinger equation in Chapter 17.

What does an atom look like? In the nineteenth century, an atom, which is the basic building block of matter, was thought of as an indivisible small object. With the discovery that an atom consists of both positive and negative electric charges, it was thought that negative charges were embedded in a sea of positive charges. A major breakthrough came in the early twentieth century when a model of the atom emerged through quantum mechanical postulates in which positive charges are concentrated in the central part of the atom called the nucleus and the electrons revolve around the nucleus in fixed orbits like planets revolve around the sun. This picture, though having serious problems within the classical electromagnetic theory, is at least easy to visualize. This evolution of the picture of the atom is discussed in Chapter 6. The final understanding of how the electrons are distributed within an atom developed with the advent of full-fledged quantum mechanics, and the picture of how electronic charge is distributed inside the atom is nothing that we can visualize with our classical intuition. This picture is discussed in the last chapter of this book when we consider the Schrödinger equation and its solution for the hydrogen atom.

 BIBLIOGRAPHY

George Gamow, *Thirty Years that Shook Physics: The Story of Quantum Theory* (Dover Publications 1985).

Steven Weinberg, *To Explain the World: The Discovery of Modern Science* (HarperCollins Publisher, New York 2015).

R. J. Scully and M. O. Scully, *The Demon and the Quantum* (John-Wiley-VCH 2010).

J. Baggott, *The Quantum Story: A History in 40 Moments* (Oxford University Press 2011).

A. Ray, *Quantum Physics: A Beginner's Guide* (Oneworld 2005).

PART

Introductory Topics

2 Mathematical Background

Mathematics is the language of modern physics. Mathematics provides the tools to solve physics problems. Much of the progress of science, and particularly physics, owes to the discovery of algebra and calculus. In the spirit of making this book reasonably self-contained, certain topics are discussed that may be required in understanding the foundation and the applications of quantum mechanics. Foremost are the definition and properties of complex numbers. The central quantity of quantum mechanics is the wave function which is, in general, a complex quantity. Trigonometry and vector analysis are necessary topics for almost any discussion of physical phenomena. In this chapter we discuss these topics to the extent that makes their use in subsequent chapters quite natural and normal. Another topic that will reverberate throughout this book due to the nature of quantum mechanics is probability theory. A clear understanding of the concept of probability is essential for the study of quantum mechanical predictions and phenomena. The theory of probability is a vast topic. Here the main ideas of probability theory are presented that should be sufficient for an understanding of the topics discussed in this book.

2.1 Complex Numbers

There are two classes of numbers we are very familiar. The first one consists of integers:

$$\cdots -3, -2, -1, 0, 1, 2, 3 \cdots \tag{2.1}$$

The integers can both be positive and negative. The other class of numbers is more general. It consists of what are called real numbers.

$$\cdots, -45.346, -1.872, 0.236, 2.000, 3.458, 2.639, \cdots \tag{2.2}$$

All integers are also real numbers. These are solutions of some algebraic equations. As examples, the solutions of the quadratic equation[1]

$$x^2 - 3x + 2 = 0 \tag{2.3}$$

are

$$x = 1, 2. \tag{2.4}$$

---

[1] The solutions of a general quadratic equation, $ax^2 + bx + c = 0$, where a, b, and c are arbitrary real or complex numbers, are given by $x = \left(-b \pm \sqrt{b^2 - 4ac}\right)/2a$.

Quantum Mechanics for Beginners: With Applications to Quantum Communication and Quantum Computing. M. Suhail Zubairy.
© M. Suhail Zubairy 2020. Published in 2020 by Oxford University Press. DOI: 10.1093/oso/9780198854227.001.0001

We can verify that these values of x are indeed the solutions of Eq. (2.3) by substituting them in the equation and seeing that the left hand side is equal to zero. For example,

$$1^2 - 3 \times 1 + 2 = 0. \tag{2.5}$$

Similarly the solutions of the quadratic equation

$$x^2 - 5x + 6 = 0 \tag{2.6}$$

are

$$x = 2, 3. \tag{2.7}$$

Real numbers can be represented on a line as shown in Fig. 2.1. We can choose a point on the line representing '0'. All the points to the right of '0' are positive numbers and the points to the left are negative numbers.

Next we consider another simple quadratic equation:

$$x^2 + 1 = 0. \tag{2.8}$$

The solutions of this equation are

$$x = +\sqrt{-1} \text{ and } -\sqrt{-1}. \tag{2.9}$$

These numbers are neither integers nor real numbers. They are what we call *imaginary numbers*. For a real number x, the square x^2 is always positive even when x is negative. For example, $(-2.5)^2 = +6.25$. However the square of an imaginary number is negative, i.e., $\left(\sqrt{-1}\right)^2 = -1$.

Let us consider another quadratic equation

$$(x + 1)^2 + 9 = 0. \tag{2.10}$$

The solutions of this equation are

$$x = -1 \pm 3\sqrt{-1}. \tag{2.11}$$

These solutions are not real numbers. They are of the form

$$a + bi,$$

where a and b are real numbers. Here we used the designation

$$i = \sqrt{-1}. \tag{2.12}$$

Such numbers which have a real part a and an imaginary part bi are called *complex numbers*. The solution of the equation $(x + 1)^2 + 9 = 0$ can therefore be written as $x = -1 \pm 3i$.

Fig. 2.1 Real numbers, both positive and negative, are represented on a line. (The number e is defined in Eq. (2.40).)

Geometrically, complex numbers extend the concept of the one-dimensional line to the two-dimensional complex plane by using the horizontal axis for the real part and the vertical axis for the imaginary part. This is shown in Fig. 2.2. The complex number $a + bi$ can be identified with the point (a, b) in the complex plane. A complex number whose real part is zero is said to be purely imaginary; these numbers lie on the vertical axis of the complex plane. A complex number whose imaginary part is zero can be viewed as a real number, lying on the horizontal axis of the complex plane.

According to the fundamental theorem of algebra, all polynomial equations with real or complex coefficients in a single variable have a solution in complex numbers. In the following we discuss some properties of complex numbers.

Complex conjugate: Complex conjugate of the complex number

$$z = x + iy \tag{2.13}$$

is defined as

$$z^* = x - iy. \tag{2.14}$$

The complex conjugate can be obtained by replacing i by $-i$. The geometric representation of z and its conjugate z^* in the complex plane can be seen in Fig. 2.3. The complex conjugate z^* is the reflection of z about the real axis. It can be verified that conjugating twice gives the original complex number:

$$\left(z^*\right)^* = z. \tag{2.15}$$

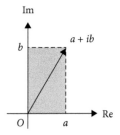

Fig. 2.2 A complex number is represented on a plane. The x-component gives the real part and the y-component gives the imaginary part.

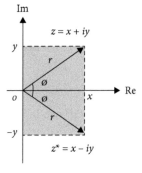

Fig. 2.3 The complex conjugate of a complex number $z = x + iy$ is the reflection of z about the x-axis.

The real and imaginary parts of a complex number z can be extracted using its complex conjugate via

$$x = \frac{1}{2}(z + z^*) \,;\, y = \frac{1}{2i}(z - z^*).$$ (2.16)

A complex number is real if and only if it equals its own conjugate: $z = z^*$.

Addition and subtraction: Complex numbers are added by separately adding the real and imaginary parts of the summands. That is to say:

$$(a + ib) + (c + id) = (a + c) + i(b + d).$$ (2.17)

Here the real part of the sum of the two complex numbers is the sum of the real parts and the imaginary part is the sum of the imaginary parts. Similarly, subtraction is defined by

$$(a + ib) - (c + id) = (a - c) + i(b - d).$$ (2.18)

Multiplication and division: The multiplication of two complex numbers is defined by the following formula:

$$(a + ib)(c + id) = (ac - bd) + i(bc + ad).$$ (2.19)

In deriving this equation, we used $i^2 = -1$. In particular,

$$(a + ib)(a - ib) = a^2 + b^2.$$ (2.20)

The real and imaginary parts of the ratio of two complex numbers are obtained by multiplying both numerator and denominator by the complex conjugate of the denominator. This makes the denominator a real number and the numerator becomes the product of two complex numbers, which can be separated into real and imaginary parts. Thus

$$\frac{a + ib}{c + id} = \frac{(a + ib)(c - id)}{(c + id)(c - id)} = \frac{(ac + bd)}{(c^2 + d^2)} + i\frac{(bc - ad)}{(c^2 + d^2)}.$$ (2.21)

Modulus: The modulus of a complex number z is defined as follows:

$$|z| = \sqrt{zz^*} = \sqrt{(x + iy)(x - iy)} = \sqrt{(x^2 + y^2)}.$$ (2.22)

Its properties are:

(i) $|z|$ is real.

(ii) $|z| \geq 0$.

(iii) $|z| = 0$ if and only if $x = y = 0$.

2.2 Trigonometry

Let us consider a right angle triangle ABC as shown in Fig. 2.4. For such a triangle, according to the Pythagoras theorem,

$$a^2 + b^2 = c^2,$$ (2.23)

where a is the perpendicular, b is the base, and c is the hypotenuse.

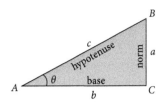

Fig. 2.4 A right angle triangle.

We can define the trigonometric functions as follows:

$$\sin\theta = \frac{\text{Norm}}{\text{Hypotenuse}} = \frac{a}{c}, \tag{2.24}$$

$$\cos\theta = \frac{\text{Base}}{\text{Hypotenuse}} = \frac{b}{c}, \tag{2.25}$$

$$\tan\theta = \frac{\text{Norm}}{\text{Base}} = \frac{a}{b}. \tag{2.26}$$

We can verify that:

$$\sin^2\theta + \cos^2\theta = 1, \tag{2.27}$$

$$\tan\theta = \frac{\sin\theta}{\cos\theta} = \frac{a}{b}. \tag{2.28}$$

Some useful trigonometric formulae are given as follows:

$$\sin(\theta_1 \pm \theta_2) = \sin\theta_1 \cos\theta_2 \pm \cos\theta_1 \sin\theta_2, \tag{2.29}$$

$$\cos(\theta_1 \pm \theta_2) = \cos\theta_1 \cos\theta_2 \mp \sin\theta_1 \sin\theta_2, \tag{2.30}$$

$$\tan(\theta_1 \pm \theta_2) = \frac{\tan\theta_1 \pm \tan\theta_2}{1 \mp \tan\theta_1 \tan\theta_2}, \tag{2.31}$$

$$\sin(2\theta) = 2\sin\theta\cos\theta, \tag{2.32}$$

$$\cos(2\theta) = \cos^2\theta - \sin^2\theta. \tag{2.33}$$

The functions $\sin\theta$ and $\cos\theta$ are oscillating functions of θ as shown in Fig. 2.5. However, these functions have series expansions,

$$\sin\theta = \theta - \frac{\theta^3}{3!} + \frac{\theta^5}{5!} - \cdots \tag{2.34}$$

$$\cos\theta = 1 - \frac{\theta^2}{2!} + \frac{\theta^4}{4!} - \frac{\theta^6}{6!} + \cdots, \tag{2.35}$$

where θ is given in radians and

$$n! = n(n-1)(n-2)\cdots 3\cdot 2\cdot 1. \tag{2.36}$$

As examples, $3! = 3\cdot 2\cdot 1 = 6$ and $5! = 5\cdot 4\cdot 3\cdot 2\cdot 1 = 120$.

A condition that we encounter frequently in the book is $\theta \ll 1$ where θ is given in radians. When this happens, we can see from the series expansions (2.34) and (2.35), that

$$\sin\theta \approx \theta, \tag{2.37}$$

$$\cos\theta \approx 1 - \frac{\theta^2}{2}, \tag{2.38}$$

$$\tan\theta = \frac{\sin\theta}{\cos\theta} \approx \theta \tag{2.39}$$

We can verify the series expansions (2.34) and (2.35) and the small θ limits of $\sin\theta$, $\cos\theta$, and $\tan\theta$ as given in Eqs. (2.37), (2.38), and (2.39), respectively, from Table 2.1.

Next we define the quantity e called the exponent which is defined by the following series

$$e = 1 + \frac{1}{1!} + \frac{1}{2!} + \frac{1}{3!} + \frac{1}{4!} + \cdots = 2.71828\cdots. \tag{2.40}$$

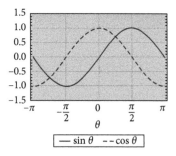

Fig. 2.5 The trigonometric functions $\sin\theta$ and $\cos\theta$ are plotted as a function of θ.

Table 2.1 Selected values of trigonometric quantities (θ is in radians)

θ	$\sin\theta$	$\cos\theta$	$\tan\theta$
0.00	0.0000	1.0000	0.0000
0.01	0.0100	1.0000	0.0100
0.05	0.0500	0.9988	0.0500
0.10	0.0998	0.9950	0.1003
0.14	0.1395	0.9902	0.1409
1.05	0.8674	0.4976	1.7433
2.10	0.8632	−0.5048	−1.7098

It can be shown that

$$e^x = 1 + x + \frac{x^2}{2!} + \frac{x^3}{3!} + \cdots \tag{2.41}$$

Sometimes e^x is expressed as exp x (an abbreviation of exponential of x). An important property of e is that, for two numbers x and y (real or complex),

$$e^x e^y = e^{x+y}. \tag{2.42}$$

This can be proven by expressing the functions e^x and e^y in their respective series expansions, then multiplying term by term and keeping terms of the same orders together.

An important formula is the Euler's formula:

$$e^{i\theta} = \cos\theta + i\sin\theta. \tag{2.43}$$

We can verify this formula by comparing the series expansions of $e^{i\theta}$, $\cos\theta$, and $\sin\theta$ as given in Eqs. (2.41), (2.34), and (2.35), respectively. In addition, the property of i, namely $i^2 = -1$, is used. On taking the complex conjugate,

$$e^{-i\theta} = \cos\theta - i\sin\theta. \tag{2.44}$$

On adding and subtracting Eqs. (2.43) and (2.44) we can rewrite $\cos\theta$ and $\sin\theta$ as

$$\cos\theta = \frac{e^{i\theta} + e^{-i\theta}}{2}, \tag{2.45}$$

$$\sin\theta = \frac{e^{i\theta} - e^{-i\theta}}{2i}. \tag{2.46}$$

As a special case of Euler's formula

$$e^{i\pi} = -1. \tag{2.47}$$

Here we used $\cos\pi = -1$ and $\sin\pi = 0$. One of the most famous theoretical physicists of our time, Richard Feynman, described this relation as "the most remarkable formula in mathematics." This is also called Euler's identity. As an example

$$e^{i3\pi/2} = e^{i\pi/2}e^{i\pi} = -e^{i\pi/2} = -(\cos(\pi/2) + i\sin(\pi/2)) = -i. \tag{2.48}$$

A pictorial representation of $e^{i\theta}$ is given in Fig. 2.6. Here $e^{i\theta}$ is a phasor whose magnitude is 1 and which is rotated by an amount θ in the complex plane. We can see that the projection on the x-axis representing the real part is $\cos\theta$ and the projection on the y-axis representing the imaginary part is $\sin\theta$. Thus

$$e^{i\theta} = \cos\theta + i\sin\theta. \tag{2.49}$$

The Euler's theorem helps us in proving that

$$(\cos\theta + i\sin\theta)^n = e^{in\theta} = \cos(n\theta) + i\sin(n\theta). \tag{2.50}$$

This is called De Moivre's theorem.

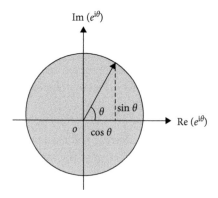

Fig. 2.6 A pictorial representation of $e^{i\theta}$.

These relations help us in writing complex numbers in the polar coordinates (r, θ) as follows. The complex number

$$z = x + iy \tag{2.51}$$

can be transformed to the polar coordinates via

$$x = r\cos\theta; \quad y = r\sin\theta. \tag{2.52}$$

We thus have

$$z = r\cos\theta + ir\sin\theta = re^{i\theta}. \tag{2.53}$$

The polar coordinate r is the magnitude of z as can be seen by first noting that

$$|z| = \sqrt{zz^*} = \sqrt{re^{i\theta}re^{-i\theta}} = r. \tag{2.54}$$

We thus have

$$r = \sqrt{x^2 + y^2}. \tag{2.55}$$

2.3 Vector and Scalar Quantities

All physical quantities can be classified into two kinds. The scalar quantities are described by a single number (including any units). Examples are temperature, volume, and time. The vector quantities, on the other hand, require both magnitude and direction for their complete description. Examples of vector quantities include displacement, velocity, and force. We designate a scalar quantity by a normal font, such as the temperature is represented by T and the time by t. To describe vectors we use the bold font. For example, velocity is $\mathbf{v}$ and force is $\mathbf{F}$. To describe the magnitude of a vector we use the absolute value sign: $|\mathbf{A}|$ or just A. The magnitude is always positive and is equal to the length of the vector.

Here we mention some of the properties of vectors:

Equality of two vectors: Two vectors are equal if they have the same magnitude and the same direction.

Movement of vectors: Any vector can be moved parallel to itself without being affected, i.e., two parallel vectors of the same magnitude are the same vectors.

Negative vectors: Two vectors are negative if they have the same magnitude but are 180° apart (opposite directions):

$$A = -B; A + B = A + (-A) = 0. \tag{2.56}$$

Adding Vectors: While adding vectors, we can use both geometric and algebraic methods. We first describe the geometric method.

In the geometric method for the addition of vectors, we use scale drawings. As we noted, a vector can be moved parallel to itself without changing itself. Thus if we want to add two vectors A and B, we first draw the vector A and then draw the vector B with the tail of B coinciding with the tip of A by appropriate parallel movement of the vector B. The resultant vector $R = A + B$ is drawn from the tail of the vector A to the tip of vector B as shown in Fig. 2.7. This method is known as the triangle method of addition. When we add two vectors, the sum is independent of the order, i.e.,

$$A + B = B + A. \tag{2.57}$$

This same general approach can also be used to add more than two vectors. For example, the vector sum of four vectors A, B, C, and D,

$$R = A + B + C + D \tag{2.58}$$

is as shown in Fig. 2.8.

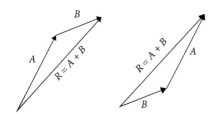

Fig. 2.7 A geometric method for adding two vectors **A** and **B**.

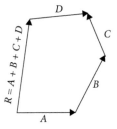

Fig. 2.8 A geometric addition of four vectors **A**, **B**, **C**, and **D**.

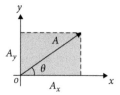

Fig. 2.9 A vector **A** is decomposed in its x- and y-components.

In many situations, it is inconvenient to use the geometric method for the addition of vectors. A more convenient method is the algebraic method, adding the components of each vector separately and then finding the magnitude and the direction of the vector.

Before we discuss this method, we show how the components of a vector are obtained. We only consider the vectors in two dimensions but a generalization to three dimensions is straightforward.

In Fig. 2.9, we consider a vector **A** of length A and making an angle θ with the x-axis. At this point, we note that the coordinate system (the orientations of mutually perpendicular x- and y-axes) is completely arbitrary. The only condition is that x- and y- axes are mutually perpendicular. As seen in Fig. 2.9, the x-component of the vector **A** is obtained by drawing a perpendicular from the tip of the vector onto the x-axis. The x-component A_x is then equal to the distance from the origin to the point where the perpendicular intersects the x-axis. This distance is equal to $A \cos \theta$, i.e.,

$$A_x = A \cos \theta. \tag{2.59}$$

Similarly, the y-component, A_y, of the vector **A** is obtained by drawing a perpendicular from the tip of the vector **A** to the y-axis. The y-component is then equal to the distance $A \sin \theta$ from the origin to the intersection of the perpendicular on the y-axis, i.e.,

$$A_y = A \sin \theta. \tag{2.60}$$

From the Pythagoras theorem (2.23) as well as Eq. (2.27), it can be seen that the magnitude of the vector **A** in terms of the components A_x and A_y is

$$A = \sqrt{A_x^2 + A_y^2}. \tag{2.61}$$

It follows from Fig. 2.9 (or by dividing A_y and A_x in Eqs. (2.60) and (2.59)) that

$$\tan \theta = \frac{A_y}{A_x}. \tag{2.62}$$

The angle θ is obtained as

$$\theta = \tan^{-1} \left(\frac{A_y}{A_x} \right). \tag{2.63}$$

With this description of vector components, we now discuss a component method to obtain the vector sum of two vectors.

Component method: In this method two or more vectors can be added algebraically by adding the x- and y-components of all the vectors.

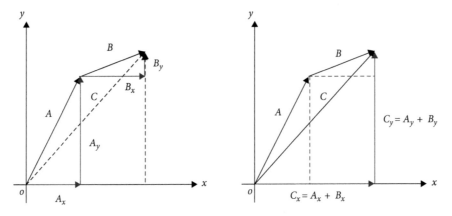

Fig. 2.10 Sum of two vectors **A** and **B** via component method.

Let us consider two vectors **A** and **B** as shown in Fig. 2.10. We can decompose each vector into its x- and y- components:

$$A = A_x\hat{x} + A_y\hat{y}, \tag{2.64}$$

$$B = B_x\hat{x} + B_y\hat{y}. \tag{2.65}$$

Here $\hat{x}$ and $\hat{y}$ are unit vectors along x- and y-axis, respectively. These unit vectors are used to specify direction and have magnitude equal to 1. Any vector **A** (and **B**) in the xy-plane can be written in the form (2.64) (and (2.65)).

Then the sum of the two vectors is

$$\begin{aligned} A + B &= \left(A_x\hat{x} + A_y\hat{y}\right) + \left(B_x\hat{x} + B_y\hat{y}\right) \\ &= (A_x + B_x)\hat{x} + \left(A_y + B_y\right)\hat{y}. \end{aligned} \tag{2.66}$$

Thus the resultant vector **C** can be written as

$$C = A + B = (A_x + B_x)\hat{x} + \left(A_y + B_y\right)\hat{y}. \tag{2.67}$$

The x- and y-components of the vector **C** are therefore given as

$$C_x = A_x + B_x, C_y = A_y + B_y \tag{2.68}$$

and the resulting magnitude of the vector C is

$$C = \sqrt{(A_x + B_x)^2 + \left(A_y + B_y\right)^2}. \tag{2.69}$$

Scalar or dot product of two vectors: The scalar product of two vectors **A** and **B** is written as $A \cdot B$. It is also called the dot product.

The dot product of two vectors can be thought of as the projection of one onto the direction of the other and is defined via

$$A \cdot B = AB\cos\theta \tag{2.70}$$

where θ is the angle between A and B. This is shown in Fig. 2.11. The dot product says something about how parallel two vectors are. For example, it has maximum value when two vectors are parallel. In this case $\theta = 0$ and $A \cdot B = AB$. On the other hand, if the two vectors are perpendicular to each other, $\theta = \pi/2$, and the dot product $A \cdot B = 0$.

The dot products of the unit vectors $\hat{x}$ and $\hat{y}$ can then be given as follows:

$$\hat{x} \cdot \hat{y} = 0; \hat{x} \cdot \hat{x} = 1; \hat{y} \cdot \hat{y} = 1. \tag{2.71}$$

Therefore, for $A = A_x \hat{x} + A_y \hat{j}$,

$$A \cdot \hat{x} = A \cos \theta = A_x. \tag{2.72}$$

In terms of the components, the dot product can be written as

$$A \cdot B = \left(A_x \hat{x} + A_y \hat{y}\right) \cdot \left(B_x \hat{x} + B_y \hat{y}\right) = A_x B_x + A_y B_y. \tag{2.73}$$

For vectors in three dimensions, we have

$$A \cdot B = A_x B_x + A_y B_y + A_z B_z \tag{2.74}$$

Cross Product: The cross product between two vectors is denoted as

$$C = A \times B. \tag{2.75}$$

Unlike the dot product, the cross product is a vector quantity. Its magnitude is

$$|C| = |A \times B| = AB \sin \theta, \tag{2.76}$$

where θ is smaller angle between the vectors and its direction is perpendicular to the plane containing the vectors A and B as shown in Fig. 2.12.

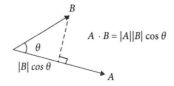

Fig. 2.11 The dot product of two vectors A and B.

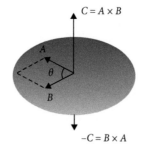

Fig. 2.12 The cross product of two vectors A and B.

The cross product of two vectors says something about how perpendicular they are: Cross product of any parallel vectors ($\theta = 0$) is zero and the cross product is maximum for perpendicular vectors ($\theta = \pi/2$).

The cross products of Cartesian unit vectors are:

$$\hat{x} \times \hat{y} = \hat{z}; \hat{x} \times \hat{z} = -\hat{y}; \hat{y} \times \hat{z} = \hat{x}, \tag{2.77}$$

$$\hat{x} \times \hat{x} = 0; \hat{y} \times \hat{y} = 0; \hat{z} \times \hat{z} = 0. \tag{2.78}$$

Using these results, we can show that

$$\boldsymbol{A} \times \boldsymbol{B} = \left(A_y B_z - A_z B_y\right)\hat{x} + (A_z B_x - A_x B_z)\hat{y} + \left(A_x B_y - A_y B_x\right)\hat{z}. \tag{2.79}$$

2.4 Elements of Probability Theory

Let us first give formal description of some simple properties of probability and then illustrate with some examples.

Suppose that $\{a_1, a_2, \cdots, a_N\}$ is the set of possible outcomes of an event A. The probability that a_i occurs is denoted as $P(a_i)$. In the case where the event never happens, the probability of that event is equal to 0. However if the event is definitely going to happen then the probability of that event is 1. In general, the probability $P(a_i)$ is greater than or equal to zero and is less than or equal to one. The sum of all the probabilities should be equal to 1, i.e.,

$$\sum_{i=1}^{N} P(a_i) = P(a_1) + P(a_2) + \cdots P(a_N) = 1. \tag{2.80}$$

For a single random event A, the set of probabilities $\{P(a_1), P(a_2), \cdots, P(a_N)\}$ provide a complete description.

The simplest example of a probabilistic event is the toss of a coin. The set of possible outcome is {Head, Tail}. For a fair coin toss, the probabilities of getting a Head or a Tail are both equal to 1/2, i.e.,

$$P(\text{Head}) = P(\text{Tail}) = 1/2. \tag{2.81}$$

It is easy to see that

$$P(\text{Head}) + P(\text{Tail}) = 1. \tag{2.82}$$

A question of interest is: Can we make a definite statement about the outcome of the tossed coin for a single event (that is a single toss) before the coin lands? The answer is no. We cannot make a definite statement about the outcome before the toss is completed. For a single toss we can get either a Head or a Tail. Then what does it mean to say that $P(\text{Head}) = P(\text{Tail}) = 1/2$? These probabilities can be determined by tossing the coin a large number of times, say N times. Let the number of times we get a Head be n_H and the number of times we get a Tail be n_T. The probability of getting a head is thus

$$P(\text{Head}) = \frac{n_H}{N}. \tag{2.83}$$

Similarly

$$P(\text{Tail}) = \frac{n_T}{N}. \tag{2.84}$$

For a small number of tosses, we may get random values of $P(\text{Head})$ and $P(\text{Tail})$. However, as the number of tosses increases, $n_H \approx n_T = N/2$, and both, $P(\text{Head})$ and $P(\text{Tail})$, approach the value 1/2. We can also verify that $P(\text{Head}) + P(\text{Tail}) = 1$.

We next consider a somewhat more complicated example. We consider a class of 200 students with the scores on a 10-point quiz as shown in Table 2.2.

The distribution is such that the numbers of students scoring very low and very high are relatively small. The maximum number of students score 7. The probability that a student has a particular score is then obtained by dividing the number of students with the same score by the total number of students. For example, the probability that a student has a score 6, $P(6)$, is given by $36/200 = 0.18$. The probabilities $P(n)$ are given in the third column of the above table. It is easy to verify that

$$\sum_{n=1}^{10} P(n) = P(1) + P(2) + \cdots\cdots + P(9) + P(10) = 1, \tag{2.85}$$

in agreement with Eq. (2.80) showing that the sum of the probabilities of all the possible outcomes is equal to one. The various moments of the variable n (in the above example, the score of the students) are defined as the following

$$\langle n^r \rangle = \sum_{n=1}^{n_{max}} n^r P(n), \tag{2.86}$$

Table 2.2 Distribution of scores of 200 students.

Score	Number of students	Probability
1	04	$P(1) = 04/200 = 0.02$
2	04	$P(2) = 04/200 = 0.02$
3	04	$P(3) = 04/200 = 0.02$
4	08	$P(4) = 08/200 = 0.04$
5	12	$P(5) = 12/200 = 0.06$
6	36	$P(6) = 36/200 = 0.18$
7	52	$P(7) = 52/200 = 0.26$
8	34	$P(8) = 34/200 = 0.17$
9	28	$P(9) = 28/200 = 0.14$
10	18	$P(10) = 18/200 = 0.09$

where $r = 1, 2, \cdots$ and n_{max} is the maximum value of the variable n. There are two quantities that are of particular interest: the average (or the mean) and the root-mean-square (*rms*) deviation.

The average or mean is defined as

$$\langle n \rangle = \sum_{n=1}^{n_{max}} n P(n). \tag{2.87}$$

For example, in the above example, the average score of the class is

$$
\begin{aligned}
\langle n \rangle &= \sum_{n=1}^{10} n P(n) \\
&= 1 \times P(1) + 2 \times P(2) + \cdots \cdots + 10 \times P(10) \\
&= 1 \times 0.02 + 2 \times 0.02 + 3 \times 0.02 + \cdots \cdots + 9 \times 0.14 + 10 \times 0.09 \\
&= 7.
\end{aligned}
\tag{2.88}
$$

A high average means that larger number of students scored high. An average is thus a measure of how good a class is.

The second important quantity is the *rms* (root-mean-square) deviation or fluctuation Δn defined by

$$
\begin{aligned}
\Delta n &= \sqrt{\langle (n - \langle n \rangle)^2 \rangle} \\
&= \sqrt{\langle n \rangle^2 - \langle n^2 \rangle}.
\end{aligned}
\tag{2.89}
$$

Here

$$\langle n^2 \rangle = \sum_{n=1}^{n_{max}} n^2 P(n). \tag{2.90}$$

In the above example

$$
\begin{aligned}
\langle n^2 \rangle &= \sum_{n=1}^{10} n^2 P(n) \\
&= 1^2 \times P(1) + 2^2 \times P(2) + \cdots \cdots + 10^2 \times P(10) \\
&= 1^2 \times 0.02 + 2^2 \times 0.02 + 3^2 \times 0.02 + \cdots \cdots + 9^2 \times 0.14 + 10^2 \times 0.09 \\
&= 52.86.
\end{aligned}
\tag{2.91}
$$

It follows from Eqs. (2.88) and (2.91) that

$$\Delta n = \sqrt{\langle n^2 \rangle - \langle n \rangle^2} = \sqrt{(52.86 - 49)} = 1.96 \approx 2. \tag{2.92}$$

The *rms* deviation or fluctuation is a measure of how spread is the score distribution. In the above example, most of the students scored between $\langle n \rangle - \Delta n$ and $\langle n \rangle + \Delta n$, i.e., between 5 and 9. We can verify that 162 students out of a total of 200 or 81% fall in this range. If, in a class, everyone scored 7, the mean would still be equal to 7 but the *rms* deviation would be zero. The *rms* deviation Δn is a measure of how broad a probability distribution is. It is also a measure of uncertainty when we discuss the concept of fluctuations in quantum mechanics.

So far we considered only one variable and the corresponding probabilities. Next we consider two variables and introduce the concept of joint probability. Let A and B be two events with the set of outcomes $\{a_i\}$ and $\{b_i\}$. In this case the individual probability distributions $P(a_i)$ and

$P(b_i)$ do not give complete information. For a complete probabilistic description we also need joint probabilities $\{P(a_i, b_j)\}$. The definition of $P(a_i, b_j)$ is that it is the joint probability that the event A has the value a_i and the event B has the value b_j.

First we consider the case when the events A and B are independent of each other. We then have

$$P(a_i, b_j) = P(a_i) P(b_j). \tag{2.93}$$

An example of such events is a throw of two coins. The outcome of each coin is independent of the outcome of the other. Each coin has a 50% probability of getting a Head or a Tail. For the two coins the possible set of events is {Head$_1$ Head$_2$, Head$_1$ Tail$_2$, Tail$_1$ Head$_2$, Tail$_1$ Tail$_2$}, i.e., both coins have Heads, the first coin has a Head and the second has a Tail, the first coin has a Tail and second has a Head, and both have Tails. Since the two coins are independent, the joint probability of getting two Heads is

$$P(\text{Head}_1, \text{Head}_2) = P(\text{Head}_1) P(\text{Head}_2) = \frac{1}{2} \cdot \frac{1}{2} = \frac{1}{4}. \tag{2.94}$$

Similarly

$$P(\text{Head}_1, \text{Tail}_2) = P(\text{Head}_1) P(\text{Tail}_2) = \frac{1}{2} \cdot \frac{1}{2} = \frac{1}{4}, \tag{2.95}$$

$$P(\text{Tail}_1, \text{Head}_2) = P(\text{Tail}_1) P(\text{Head}_2) = \frac{1}{2} \cdot \frac{1}{2} = \frac{1}{4}, \tag{2.96}$$

$$P(\text{Tail}_1, \text{Tail}_2) = P(\text{Tail}_1) P(\text{Tail}_2) = \frac{1}{2} \cdot \frac{1}{2} = \frac{1}{4}. \tag{2.97}$$

The sum of all the joint probabilities is equal to one, i.e.,

$$P(\text{Head}_1, \text{Head}_2) + P(\text{Head}_1, \text{Tail}_2) + P(\text{Tail}_1, \text{Head}_2) + P(\text{Tail}_1, \text{Tail}_2) = 1. \tag{2.98}$$

Next we consider events A and B that are not independent but are 'correlated'. In this case

$$P(a_i, b_j) \neq P(a_i) P(b_j). \tag{2.99}$$

In order to illustrate how various probabilities are related, we consider the same example as discussed above. Again we have a class of 200 students whose score distribution is given by Table 2.3. However this time, in addition to the scores, we have another variable, gender. Let us assume that 80 students are girls and 120 students are boys. We now have the probability distributions with two variables, the score and the gender. For example, we can ask the question: What is the joint probability that a girl scored 8 points? We denote it by $P(8, G)$. Or what is the probability that a boy scored 6 points, denoted by $P(6, B)$?

According to the Table 2.3, $P(8, G)$ can be calculated as follows. There are a total of 200 possibilities of all kinds (scores ranging from 1 to 10 and the student being a Girl or a Boy). Out of these 200 possibilities, there are only 18 instances when the student is a Girl and the score is 8. Therefore the joint probability $P(8, G)$ is given by

$$P(8, G) = \frac{18}{200} = 0.09.$$

Similarly there are only 26 instances when the student is a Boy and has a score of 6. Therefore the corresponding joint probability is

Table 2.3 Distribution of scores of 120 girls and 80 boys

Score	Number of students	Probability	Girls	Boys	Joint probabilities
1	04	$P(1) = 0.02$	2	2	$P(1, G) = 0.01; P(1, B) = 0.01$
2	04	$P(2) = 0.02$	0	4	$P(2, G) = 0.00; P(2, B) = 0.02$
3	04	$P(3) = 0.02$	4	0	$P(3, G) = 0.02; P(3, B) = 0.00$
4	08	$P(4) = 0.04$	2	6	$P(4, G) = 0.01; P(4, B) = 0.03$
5	12	$P(5) = 0.06$	4	8	$P(5, G) = 0.02; P(5, B) = 0.04$
6	36	$P(6) = 0.18$	10	26	$P(6, G) = 0.05; P(6, B) = 0.13$
7	52	$P(7) = 0.28$	18	34	$P(7, G) = 0.09; P(7, B) = 0.17$
8	34	$P(8) = 0.17$	18	16	$P(8, G) = 0.09; P(8, B) = 0.08$
9	28	$P(9) = 0.14$	14	14	$P(9, G) = 0.07; P(9, B) = 0.07$
10	18	$P(10) = 0.09$	8	10	$P(10, G) = 0.04; P(10, B) = 0.05$

$$P(6, B) = \frac{26}{200} = 0.13.$$

In Table 2.3, all the joint probabilities $P(n, G)$ and $P(n, B)$ are given.

At this point we note that the probability of a student scoring 8 points is

$$P(8) = 0.17$$

and the probability of a student being a Girl is

$$P(G) = \frac{80}{200} = 0.40.$$

It is clear that $P(8)P(G) = 0.17 \times 0.40 = 0.068$ and $P(8, G) = 0.090$. Therefore

$$P(8, G) \neq P(8)P(G).$$

Thus, for correlated events, the probability of the joint event is not the product of the probability of individual variables.

An important question is: How can we calculate single variable probabilities $P(a_i)$ and $P(b_j)$ from the joint probabilities $P(a_i, b_j)$? In order to see how this can be done, we again consider the example of the quiz score of the students. Suppose we want to find the probability that a student scored 8 points $P(8)$ from the joint probabilities $P(n, G)$ and $P(n, B)$. Since we do not care about whether the student is a Girl or a Boy, the probability that 'a student' scored 8 points is the sum of the joint probabilities that a Girl scored 8 points and a Boy scored 8 points, i.e.,

$$P(8) = P(8, G) + P(8, B).$$

This equation can be verified from Table 2.3.

In general, the single-event probabilities are given by

$$P(a_i) = \sum_j P\left(a_i, b_j\right),$$

(2.100)

$$P(b_j) = \sum_i P\left(a_i, b_j\right).$$

(2.101)

Problems

2.1 Find the real and imaginary parts of

$$z = \frac{2 + 3i}{5 - 7i}.$$

What are the real and imaginary parts of z^2?

2.2 What is the result of the multiplication of $3 + 4i$ and its complex conjugate? What are the real and the imaginary parts of $3 + 4i$ divided by its complex conjugate?

2.3 Express the complex number $z = 4 + 3i$ in polar coordinates r and φ.

2.4 Show that

$$1 + e^{i\frac{2\pi}{4}} + e^{i\frac{4\pi}{4}} + e^{i\frac{6\pi}{4}} = 0,$$

$$e^{i\frac{5\pi}{4}} + e^{i\frac{7\pi}{4}} + e^{i\frac{\pi}{4}} + e^{i\frac{3\pi}{4}} = 0,$$

$$e^{i\frac{6\pi}{4}} + e^{i\frac{2\pi}{4}} + e^{i\frac{6\pi}{4}} + e^{i\frac{2\pi}{4}} = 0.$$

2.5 A vector A has a magnitude of 8 units and an angle of 60° with the x-axis. What are the x- and y-components of A?

2.6 Find the dot product of two vectors $A = \hat{x} + \hat{y}$ and $B = \hat{x} + 2\hat{y}$. What is the angle between the two vectors?

2.7 Find out the cross product of two vectors $A = \hat{x} + \hat{y} + \hat{z}$ and $B = \hat{x} - \hat{y} + \hat{z}$.

2.8 Prove that

$$1 + \cos\theta = 2\cos^2\frac{\theta}{2},$$

$$1 - \cos\theta = 2\sin^2\frac{\theta}{2}.$$

2.9 Using the series expansion

$$e^x = 1 + x + \frac{x^2}{2!} + \frac{x^3}{3!} + \cdots$$

show that, for two numbers x and y (real or complex),

$$e^x e^y = e^{x+y}.$$

2.10 From Table 2.3, show that $P(6, B) \neq P(6)P(B)$.

BIBLIOGRAPHY

L. Susskind and G. Hrabovsky, *The Theoretical Minimum: What you Need to Know to Start Doing Physics* (Basic Books 2013).

R. Larson, *Algebra and Trigonometry* (CENAGE Learning 2014).

M. Spiegel, S. Lischutz, and D. Spellman, *Vector Analysis* (McGraw-Hill 2010).

D. Stirzaker, *Elementary Probability* (Cambridge University Press 2003).

3 Particle Dynamics

In physics, a particle is described as an object that is characterized by certain properties. The most important characteristics of a particle are its mass, position, velocity, and acceleration. A particle can be a microscopic object like an electron or an atom or a macroscopic object like a tennis ball or a stone. A particle is fundamentally different from a wave whose main characteristics are amplitude, frequency, wavelength, and phase. In this chapter we present the main characteristics of the dynamics of particles whereas in the next chapter we introduce the properties of the waves and the associated effects like interference and diffraction. An understanding of these effects is essential in understanding and appreciating the laws of quantum mechanics. In quantum mechanics, the particles and waves lose their distinctive behaviors and both of them carry each other's characteristics.

3.1 Classical Trajectory

All the physics before the advent of quantum mechanics in 1900 is called classical mechanics. The basic laws of classical mechanics were formulated by Isaac Newton in the late seventeenth century. The bedrock of classical mechanics are the three laws of motion. For simplicity's sake, we describe them only for one-dimensional motion along the x-axis.

First law: A particle in a state of uniform motion tends to remain in that state of motion unless an external force is applied to it.

Second law: When a force F is applied to a particle of mass m, it experiences an acceleration a. The acceleration is directly proportional to the applied force and is in the same direction as the force. The resulting equation is

$$F = ma. \tag{3.1}$$

Third law: When one body exerts a force on a second body, the second body simultaneously exerts a force equal in magnitude and opposite in direction on the first body.

An important consequence of Newton's laws of motion is that the motion of a particle is deterministic: If we know the initial position and velocity of the particle as well as all the forces acting on it, then we can predict with certainty its location and velocity at a subsequent time with arbitrary precision. In other words, the trajectory of the particle can be traced in advance.

The simplest example of the dynamics of a particle is a particle located at position x_i at rest (initial velocity, $v_i = 0$) and no force acts on it ($F = 0$). According to the first law of motion, it will remain at rest at the same position for all times. Thus, at time t, the position

$$x(t) = x_i \tag{3.2}$$

Quantum Mechanics for Beginners: With Applications to Quantum Communication and Quantum Computing. M. Suhail Zubairy.
© M. Suhail Zubairy 2020. Published in 2020 by Oxford University Press. DOI: 10.1093/oso/9780198854227.001.0001

and the velocity

$$v(t) = 0. \tag{3.3}$$

However if the particle is initially moving with a velocity v_i, then the position and the velocity of the particle at a later time are given by

$$x(t) = x_i + v_i t, \tag{3.4}$$

$$v(t) = v_i. \tag{3.5}$$

The particle continues to move with constant velocity in the absence of a force.

Next we consider the dynamical equations of the particle in the presence of a constant force. According to the second law of motion, the particle moves with a constant acceleration. An important example of such a motion is the movement of a particle under the action of gravity. When we throw a particle in the upward direction, it experiences a force of gravity that pulls it towards the earth given by

$$F = mg, \tag{3.6}$$

where $g = 9.81 \text{ m/s}^2$ is the acceleration in the downward direction. Before we discuss such a motion, we first derive the equations of motion for a particle moving with a constant acceleration a. These equations relate the position $x(t)$ and the velocity $v(t)$ of a particle at time t if we know the position x_i and velocity v_i at the initial time $t = 0$.

The first equation is obtained by realizing that, for constant acceleration a, the acceleration is given by the change in velocity divided by the elapsed time, i.e.,

$$a = \frac{v(t) - v_i}{t}. \tag{3.7}$$

We can rewrite this equation as

$$v(t) = v_i + at. \tag{3.8}$$

The second dynamical equation can be derived by noting that the velocity changes uniformly in time when a particle is moving with constant acceleration. Thus the displacement $x(t) - x_i$ during time t is given by the average velocity $(v(t) + v_i)/2$ times the elapsed time t, i.e.,

$$x(t) - x_i = \frac{(v(t) + v_i)}{2} t. \tag{3.9}$$

On substituting for $v(t)$ from Eq. (3.8) into Eq. (3.9) and making some rearrangements, we get

$$x(t) = x_i + v_i t + \frac{1}{2} a t^2. \tag{3.10}$$

Another equation that does not involve time t can be obtained by solving for t from Eq. (3.8), i.e., $t = (v(t) - v_i)/a$, and substituting in Eq. (3.9):

$$v^2(t) = v_i^2 + 2a\,(x(t) - x_i). \tag{3.11}$$

Equations (3.8) – (3.11) provide the full dynamics of a one-dimensional motion in the presence of a constant force (or constant acceleration). They are the consequence of the Newton's second law of motion.

As an example, we consider the motion of the car that starts ($v_i = 0$) at a location that we designate as the origin ($x_i = 0$). Let the car accelerate with $a = 10\ m/s^2$. We can calculate the location of the car at time t from Eq. (3.10),

$$x(t) = 5t^2. \tag{3.12}$$

Thus after 1 second, the car will be 5 m from the starting point and, after 2 seconds, it will be 20 m from the starting point. Similarly we see, from Eq. (3.8) that the velocity of the car after time t will be

$$v(t) = 10t. \tag{3.13}$$

Thus the car will be moving with velocity 10 m/s after 1 second and 20 m/s after 2 seconds. We therefore have precise knowledge about both the position and velocity at any subsequent time. This is the hallmark of the Newtonian or classical mechanics.

An immediate consequence of the classical mechanics is determinism! If we know the initial position and velocity as well as all the forces acting on an object, we can describe its trajectory to arbitrary accuracy. At a given time we can determine both the position and the velocity of the object precisely.

As another example of the deterministic nature of classical mechanics, we consider the two-dimensional motion. In the two-dimensional motion, the x- and y-components of motion are independent of each other. Let us consider that a particle, such as a ball, is thrown at the origin of the coordinate system ($x_i = 0, y_i = 0$) with a velocity $\mathbf{v}_i$ making an angle θ_0. The only force acting on the ball is the force of gravity. The classical dynamical equations can predict with certainty both the position and velocity at any point in the two-dimensional space.

At the initial time ($t = 0$), we can write the x- and y-components of the position

$$x_i = 0, y_i = 0 \tag{3.14}$$

and the velocity

$$v_{ix} = v_0 \cos \theta_0 \text{ and } v_{iy} = v_0 \sin \theta_0. \tag{3.15}$$

For the motion in the horizontal direction, i.e., x-direction, there is no force. Consequently there is no acceleration ($a = 0$). It then follows from Eqs. (3.10) and (3.8) that

$$x(t) = v_{ix}t = v_0 \cos \theta_0 t, \tag{3.16}$$

$$v_x(t) = v_0 \cos \theta_0. \tag{3.17}$$

In the vertical direction, i.e., y-direction, there is a constant force of gravitation leading to the acceleration due to gravity acting in the downward direction. We thus have $a_y = -g$. We then obtain from Eqs. (3.10) and (3.8) that

$$y(t) = v_{iy}t - \frac{1}{2}gt^2 = v_0 \sin \theta_0 t - \frac{1}{2}gt^2, \tag{3.18}$$

$$v_y(t) = v_{iy} - gt = v_0 \sin \theta_0 - gt. \tag{3.19}$$

Thus, at a time t, we can find $x(t)$, $v_x(t)$, $y(t)$, and $v_y(t)$ from Eqs (3.16) – (3.18). The location of the ball at time t is at a distance

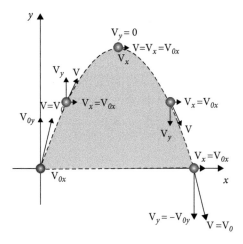

Fig. 3.1 In projectile motion, the location and the velocity of the particle is known precisely at all times.

$$r(t) = \sqrt{x^2(t) + y^2(t)} = \sqrt{\left(v_0 \cos \theta_0 t\right)^2 + \left(v_0 \sin \theta_0 t - \frac{1}{2}gt^2\right)^2} \qquad (3.20)$$

from the starting point in the direction

$$\theta_r = \tan^{-1}\left(\frac{v_0 \sin \theta_0 t - \frac{1}{2}gt^2}{v_0 \cos \theta_0 t}\right). \qquad (3.21)$$

Similarly we can calculate the velocity at time t. The magnitude and the direction of the velocity are

$$v(t) = \sqrt{v_x^2(t) + v_y^2(t)} = \sqrt{\left(v_0 \cos \theta_0\right)^2 + \left(v_0 \sin \theta_0 - gt\right)^2}, \qquad (3.22)$$

$$\theta_v = \tan^{-1}\left(\frac{v_0 \sin \theta_0 - gt}{v_0 \cos \theta_0}\right). \qquad (3.23)$$

Thus, at each moment, we know precisely both the location and the velocity of the particle. This is shown in Fig. 3.1.

In later chapters, we see that the quantum behavior is drastically different—quantum mechanics does not allow a description in terms of a trajectory of the particle and it does not allow a deterministic description of the motion of the particle. Contrary to the classical mechanics, position and velocity of an object cannot be simultaneously known to arbitrary precision according to the quantum mechanical description.

3.2 Linear Momentum

We now introduce some concepts that are basic to classical mechanics. One such concept is linear momentum. This is a fundamental quantity associated with a particle. The linear momentum p of an object of mass m moving with a velocity $\mathbf{v}$ is defined to be the product of the mass and velocity:

$$p = m\mathbf{v}. \tag{3.24}$$

In this book, we use the terms momentum and linear momentum interchangeably. Momentum is a vector quantity, the direction being the direction of velocity. We can therefore define the x- and y-components of the momentum as

$$p_x = mv_x, \tag{3.25}$$

$$p_y = mv_y. \tag{3.26}$$

The physical importance of this quantity is its close link with the force. A force on a particle is defined as rate of the change of momentum, i.e., if the momentum changes from an initial momentum $\mathbf{p}_i$ at time t_i to the final momentum $\mathbf{p}_f$ at time t_f, then the applied force F is given by

$$F = \frac{\mathbf{p}_f - \mathbf{p}_i}{t_f - t_i}. \tag{3.27}$$

This relationship follows from the Newton's second law. For example, for a body moving with constant acceleration,

$$F = m\mathbf{a} = m\frac{\mathbf{v}_f - \mathbf{v}_i}{t_f - t_i} = \frac{m\mathbf{v}_f - m\mathbf{v}_i}{t_f - t_i} = \frac{\mathbf{p}_f - \mathbf{p}_i}{t_f - t_i}. \tag{3.28}$$

Here we used the definition of acceleration from Eq. (3.6) and the definition of momentum from Eq. (3.24). Momentum is conserved if the net force, F, on an object is zero. In this case the particle's momentum does not change and the final momentum is the same as the initial momentum, i.e.,

$$\mathbf{p}_f = \mathbf{p}_i. \tag{3.29}$$

The conservation of momentum in the absence of an external force is a more general concept and can be applied when two objects collide (Fig. 3.2).

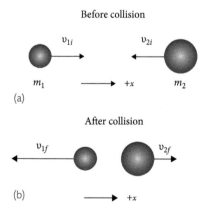

Fig. 3.2 Collision between two particles in one dimension.

In any collision, linear momentum is always conserved and it remains constant both in magnitude and in direction. The momenta of the individual objects in the system may change, but the vector sum of all the momenta does not change. Total momentum of all the particles before the collision is equal to the total momentum of all the particles after the collision.

As an example, we consider a one-dimensional collision between two particles, like balls, as shown in Fig. 3.2. The particle of mass m_1 moving with an initial velocity v_{1i} collides with another particle of mass m_2 moving with an initial velocity v_{2i}. After the collision, particles 1 and 2 move with velocities v_{1f} and v_{2f}, respectively. Since there is no external force on either particle 1 or particle 2, the total momentum is conserved, i.e., the sum of the momenta of the two particles is equal to the sum of momenta after the collision. We thus have

$$\boldsymbol{p}_{1i} + \boldsymbol{p}_{2i} = \boldsymbol{p}_{1f} + \boldsymbol{p}_{2f} \tag{3.30}$$

In terms of mass and velocities,

$$m_1\mathbf{v}_{1i} + m_2\mathbf{v}_{2i} = m_1\mathbf{v}_{1f} + m_2\mathbf{v}_{2f}. \tag{3.31}$$

The conservation of momentum is a manifestation of Newton's third law of motion: When one body exerts a force on a second body, the second body simultaneously exerts a force equal in magnitude and opposite in direction on the first body. Therefore if $\boldsymbol{F}_{12}$ is the force exerted by the particle 1 on particle 2 and $\boldsymbol{F}_{21}$ is the force exerted by particle 2 on particle 1, then

$$\boldsymbol{F}_{12} = -\boldsymbol{F}_{21}$$

or

$$\boldsymbol{F}_{12} + \boldsymbol{F}_{21} = 0. \tag{3.32}$$

This is a statement that there is no external source acting on the system of two particles. The conservation of momentum (3.30) follows if we substitute

$$\boldsymbol{F}_{21} = \frac{\boldsymbol{p}_{1f} - \boldsymbol{p}_{1i}}{t_f - t_i}, \tag{3.33}$$

$$\boldsymbol{F}_{12} = \frac{\boldsymbol{p}_{2f} - \boldsymbol{p}_{2i}}{t_f - t_i}, \tag{3.34}$$

in Eq. (3.32).

3.3 Kinetic and Potential Energy

Another important property of the particles is energy. When a particle of mass m is moving with a velocity $\mathbf{v}$, the kinetic energy is equal to

$$KE = \frac{1}{2}mv^2. \tag{3.35}$$

How does this expression for the kinetic energy come about? We first recall that energy is the ability to do work. Work, in scientific terms, is defined as follows: If a particle is displaced by a distance $(x_f - x_i)$ under the action of a force F, then the work done W is given by

$$W = F \cdot (x_f - x_i). \tag{3.36}$$

From Newton's second law, $F = ma$. We thus have

$$W = ma \cdot (x_f - x_i). \tag{3.37}$$

However, according to Eq. (3.11), $a \cdot (x_f - x_i) = (v^2(t) - v_i^2)/2$. On substituting this expression in Eq. (3.37), we obtain

$$W = \frac{1}{2}mv^2(t) - \frac{1}{2}mv_i^2. \tag{3.38}$$

Thus the work done is equal to the change in the kinetic energy where the expression of the kinetic energy is given by Eq. (3.35).

Next we consider the potential energy. Potential energy represents the 'potential' of doing work and we designate it as $V(x)$. For example, if we take a particle, such as a stone, to a height h above the ground, the potential to do work (not the actual work) increases. The potential is represented by the fact that if we drop the stone from the height h, a force of gravity equal to mg acts on it and if the stone can move through the distance h, then the potential of doing the work is equal to mgh. Thus the potential energy in the presence of the gravitational force is

$$PE = mgh. \tag{3.39}$$

Another example of potential energy is the harmonic oscillator. If we attach a mass m to a spring of spring constant k and compress the spring by an amount x from the equilibrium position, then, upon releasing the spring, it executes a simple harmonic motion between $+x$ and $-x$. At the displacement, x, the potential energy is

$$PE = \frac{1}{2}kx^2. \tag{3.40}$$

As a final example, we consider two equal and opposite charges, such as a proton and an electron, each carrying equal but opposite charge equal to $e = 1.6 \times 10^{-19}$C, separated by a distance r. The Coulomb force between a proton and an electron is attractive and is given by

$$F = -\frac{1}{4\pi\varepsilon_0}\frac{e^2}{r^2}, \tag{3.41}$$

where $\varepsilon_0 = 8.85 \times 10^{-12}$ Farad/m is the so-called free-space permittivity. The potential energy is given by

$$V(r) = -\frac{1}{4\pi\varepsilon_0}\frac{e^2}{r}. \tag{3.42}$$

The potential energy is equal to zero when the two particles are far apart, $r = +\infty$. As the two charges come closer, the potential energy decreases.

The kinetic energy and momentum are simply related. We recall that the kinetic energy of a particle of mass m moving with velocity v is given by

$$KE = \frac{1}{2}mv^2.$$

We also recall the definition of momentum:

$$p = mv.$$

Therefore kinetic energy is related to momentum via:

$$\mathrm{KE} = \frac{1}{2}mv^2 = \frac{1}{2m}(mv)^2 = \frac{p^2}{2m}. \tag{3.43}$$

3.4 Inelastic and Elastic Collisions

As we discussed, momentum is conserved in any collision. A collision between two particles where momentum is conserved but kinetic energy is not is described as an inelastic collision. In such a collision, part of the energy is lost as heat or some other form of energy. This is at the expense of the kinetic energy of the particles. An example of an inelastic collision is when two pieces of putty collide and stick together and move with some common velocity after the collision.

In an elastic collision, in addition to the conservation of momentum, the total kinetic energy of all the particles *before* the collision is equal to the total kinetic energy of all the particles *after* the collision. Thus, in an elastic collision between two particles of masses m_1 and m_2, conservation of energy and momentum lead to

$$\frac{1}{2}m_1 v_{1i}^2 + \frac{1}{2}m_2 v_{2i}^2 = \frac{1}{2}m_1 v_{1f}^2 + \frac{1}{2}m_2 v_{2f}^2, \tag{3.44}$$

$$m_1 \mathbf{v}_{1i} + m_2 \mathbf{v}_{2i} = m_1 \mathbf{v}_{1f} + m_2 \mathbf{v}_{2f}. \tag{3.45}$$

Here, as before, $\mathbf{v}_{1i}$ and $\mathbf{v}_{2i}$ are the initial velocities and $\mathbf{v}_{1f}$ and $\mathbf{v}_{2f}$ are the final velocities of particles of masses m_1 and m_2, respectively.

3.5 Angular Motion

So far we considered the motion of a particle along a straight line. Another class of motion is when a particle is moving in a circle. In this section, we consider the dynamics of a particle in a circular motion along the same lines as we discussed for a linear motion.

For a circular motion, the axis of rotation is the center of the circle. We choose a fixed reference line as shown in Fig. 3.3. This line is equivalent to the origin for the circular motion.

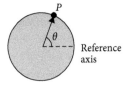

Fig. 3.3 Angular motion of a particle with respect to a reference axis.

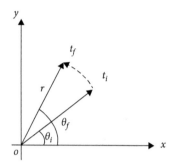

Fig. 3.4 A particle undergoes an angular displacement from $\theta = \theta_i$ at time t_i to $\theta = \theta_f$ at a later time t_f.

A point P on the circle is at a fixed distance r from the origin. As the particle moves, the only coordinate that changes is θ and as the particle moves through θ, it moves though an arc length s. The angle θ, measured in radians, is called the angular position.

Each quantity in the linear motion has an analogous quantity in the case of circular motion. The angular displacement is defined as the angle the object rotates through

$$\Delta\theta = \theta_f - \theta_i \tag{3.46}$$

during some time interval $t_f - t_i$ as shown in Fig. 3.4. This is the angle that the reference line of length r sweeps out. The analog of the velocity v is the angular velocity ω. If an angular displacement $\Delta\theta$ takes place in an infinitesimal time Δt, then the angular velocity is given by

$$\omega = \frac{\Delta\theta}{\Delta t}. \tag{3.47}$$

Similarly, if there is a change in the angular velocity $\Delta\omega$ during an infinitesimal time Δt, then angular acceleration α is defined as

$$\alpha = \frac{\Delta\omega}{\Delta t}. \tag{3.48}$$

We also note that the angular displacement θ can be related to the linear displacement x along the circular path of radius r via

$$\theta = \frac{x}{r}. \tag{3.49}$$

Similarly we can show that

$$\omega = \frac{v}{r} \tag{3.50}$$

$$\alpha = \frac{a}{r}. \tag{3.51}$$

The basic dynamical equations of the circular motion involving the angular displacement θ, the angular velocity ω, and angular acceleration α can be derived following a similar approach as that for the linear motion as done in Section 3.1. For a circular motion with constant angular acceleration α, the following equations of motion can be derived by replacing x by θ, v by ω, and a by α in Eqs. (3.8) – (3.11):

$$\omega(t) = \omega_i + \alpha t, \tag{3.52}$$

$$\theta(t) - \theta_i = \frac{(\omega(t) + \omega_i)}{2}t, \tag{3.53}$$

$$\theta(t) = \theta_i + \omega_i t + \frac{1}{2}\alpha t^2, \tag{3.54}$$

$$\omega^2(t) = \omega_i^2 + 2\alpha\,(\theta(t) - \theta_i). \tag{3.55}$$

Next we discuss the analogs for the force F and the mass m in the angular motion, which we refer as the torque τ and the moment of inertia I. Just as the force causes acceleration, a torque causes angular acceleration.

Torque: Torque is a measure of the effectiveness of a force in accelerating a rotation or changing the angular velocity over a period of time. The magnitude of torque is defined to be

$$\tau = rF \sin \theta, \tag{3.56}$$

where r is the distance from the origin (or the pivot point) to the point where the force is applied, F is the magnitude of the force, and θ is the angle between the force and the vector directed from the point of application to the pivot point (Fig. 3.5). The torque τ, like the force F, is a vector quantity, but its direction is perpendicular to both the direction of force F and the vector $\mathbf{r}$. It is formally defined as the cross product of $\mathbf{r}$ and F, i.e.,

$$\tau = r \times F. \tag{3.57}$$

The analog of force in a rotational motion can be understood most easily by considering the example of a door which can rotate around a hinge as shown in Fig. 3.5. We can see that, no matter how large a force we apply on the hinge, we cannot rotate the door as depicted in Fig. 3.5a. Similarly we cannot rotate the door if we apply a force, no matter how large, along the direction of the door or parallel to $\mathbf{r}$. The most effective location and direction of the door are the farthest point from the hinge (maximum value of the distance r) and the maximum angle, 90°, between the force and the vector $\mathbf{r}$ as shown in Fig. 3.5c. At those points torque is maximum according to Eq. (3.56).

Moment of Inertia: What about Newton's second law for circular motion? In analogy with the second law for linear motion, $F = ma$, we can write the equivalent law for rotational or circular motion as

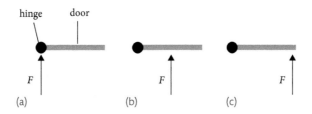

Fig. 3.5 A door rotates around a hinge that serves as the pivot point for angular motion.

Fig. 3.6 A point mass located at a distance r rotates around the pivot point in a circular motion when a force F is applied in a direction perpendicular to the position vector **r**.

$$\tau = I\alpha, \tag{3.58}$$

where we have replaced the force F by the torque τ and the linear acceleration a by angular acceleration α. Here I plays the role of mass in the rotational motion and is called the moment of inertia. The moment of inertia depends on the mass distribution as well as the axis of rotation of the rotating object.

A calculation of the moment of inertia for a general object is complicated. Here we consider a simple case where a rotating point mass is located at a distance r from a pivot point and a force F is applied on the point mass perpendicular to r as shown in Fig. 3.6. In this case, torque

$$\tau = rF. \tag{3.59}$$

It follows from Newton's second law $F = ma$ and the relation between the angular acceleration α and the linear acceleration a, $\alpha = a/r$, that

$$\alpha = \frac{a}{r} = \frac{F}{mr}, \tag{3.60}$$

and, according to Eq. (3.58),

$$\tau = I\alpha = \frac{I}{mr}F = rF. \tag{3.61}$$

Thus, $I/mr = r$, and the moment of inertia of a point mass m a distance r from the center of rotation is

$$I = mr^2. \tag{3.62}$$

This quantity is analogous to mass (or inertia).

The moment of inertia of an arbitrary object can be found by breaking up the object into little pieces, multiplying the mass of each little piece by the square of the distance it is from the axis of rotation, and adding all these products. This leads to

$$I = \sum mr^2. \tag{3.63}$$

Here the notation $\sum mr^2$ means summation over all the little pieces. To see the analogy of the moment of inertia with mass, we note that the mass represents the difficulty in moving an object – the larger the mass of an object the more difficult it is to move. Thus mass represents the property of inertia. In the same way, moment of inertia represents the difficulty in rotating an object. It becomes more difficult to rotate an object if it has a large mass and is located farther from the axis of rotation.

Centripetal force: We know that acceleration is a change in velocity, either in its magnitude or in its direction, or both. In uniform circular motion, the direction of the velocity changes

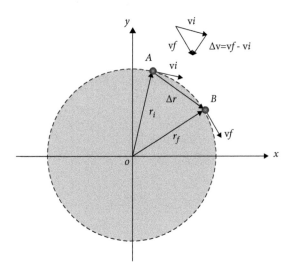

Fig. 3.7 In a uniform circular motion of a particle, the velocity changes from $\mathbf{v}_i$ to $\mathbf{v}_f$. The change in velocity $\Delta\mathbf{v} = \mathbf{v}_f - \mathbf{v}_i$ points towards the center of the circular path.

constantly. Therefore there is always an associated acceleration, even if the magnitude of the velocity is constant. We consider an object of mass m moving in a circular path of radius r with a constant speed v as shown in Fig. 3.7. We consider its change in location and the change in velocity during time Δt. The object always stays on the circular path and we designate the position vectors at the beginning and at the end of the time duration Δt by $\mathbf{r}_i$ and $\mathbf{r}_f$, respectively. The velocity is along the tangent to the circle and the velocity vectors (of equal magnitude for constant speed) are shown as $\mathbf{v}_i$ and $\mathbf{v}_f$. The change in displacement is given by

$$\Delta\mathbf{r} = \mathbf{r}_f - \mathbf{r}_i. \tag{3.64}$$

Acceleration is in the direction of the change in velocity:

$$\mathbf{a}_c = \frac{\Delta\mathbf{v}}{\Delta t} = \frac{\mathbf{v}_f - \mathbf{v}_i}{\Delta t}. \tag{3.65}$$

As shown in Fig. 3.7, $\Delta\mathbf{v}$ points directly toward the center of rotation O (the center of the circular path). This acceleration pointing towards the center is therefore called the centripetal acceleration.

Next we note that the triangle formed by the velocity vectors and the one formed by the displacement vectors are similar. Both the triangles are isosceles triangles (two equal sides). The two equal sides of the velocity vector triangle are the speeds $v_f = v_i = v$. Using the properties of two similar triangles, we obtain

$$\frac{\Delta v}{v} = \frac{\Delta r}{r} \tag{3.66}$$

yielding

$$\Delta v = \frac{v\Delta r}{r}. \tag{3.67}$$

The magnitude of the centripetal acceleration is therefore equal to

$$a_c = \frac{\Delta v}{\Delta t} = \frac{\Delta r}{\Delta t}\frac{v}{r} = \frac{v^2}{r}.$$ (3.68)

The corresponding centripetal force is obtained by multiplying the centripetal acceleration by the mass of the object, i.e.,

$$F_c = \frac{mv^2}{r},$$ (3.69)

and it points towards the center O.

3.6 Angular Momentum

We have defined various quantities like mass, velocity, acceleration, and force for linear motion. In analogy with these quantities, we defined moment of inertia, angular velocity, angular acceleration, and torque. One important quantity is momentum. We defined the linear momentum as the product of mass and velocity:

$$\boldsymbol{p} = m\mathbf{v}.$$ (3.70)

Momentum is a vector quantity whose direction is the same as the direction of the velocity.

What if the object is not linearly moving, but it is rotating? Then we can define the angular momentum

$$L = I\omega.$$ (3.71)

Here I is the moment of inertia of the object rotating about an axis and ω is the angular velocity and is perpendicular to the plane of rotation (or along the direction of the axis of rotation). The angular momentum L is positive when the object rotates in a counter-clockwise direction and is negative when the object rotates in a clockwise direction.

Let us consider a point particle of m moving around in an orbit of radius r. The moment of inertia is $I = mr^2$ and the angular frequency $\omega = v/r$. On substituting these expressions in the definition of the angular momentum, Eq. (3.71), we obtain the following expression for the magnitude of the angular momentum:

$$L = mrv.$$ (3.72)

The direction of the angular momentum is perpendicular to the plane of the circular motion. In the more general form,

$$\boldsymbol{L} = \boldsymbol{r} \times (m\mathbf{v}) = \boldsymbol{r} \times \boldsymbol{p},$$ (3.73)

where $\mathbf{r}$ is the particle's instantaneous position vector and $\boldsymbol{p}$ is its instantaneous linear momentum. Only tangential momentum components contribute to the angular momentum. As shown in Fig. 3.8, the vectors $\mathbf{r}$ and $\boldsymbol{p}$ form a plane and the angular momentum $\boldsymbol{L}$ is perpendicular to this plane.

Just as the linear momentum of a system of particles is conserved in the absence of an external force, the angular momentum is conserved in the absence of a torque.

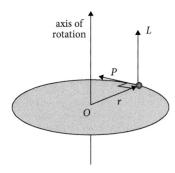

Fig. 3.8 Angular momentum **L** is perpendicular to the plane containing the linear momentum **p** and the position vector **r**.

3.7 Motion of an Electron in Electric and Magnetic Fields

The analysis of the motion of a charged particle, such as an electron, in the presence of an electric field or a magnetic field, or both, provides an interesting example of the particle dynamics studied in this chapter. Interestingly this analysis provides the basis of the landmark experiments done in the 1890s by John J. Thomson which led to the discovery of the electron as a subatomic particle.

First we introduce the basic concepts of electric and magnetic fields. The force on a charge q in the presence of a charge Q separated by a distance r is given by Coulomb's law:

$$F = \frac{1}{4\pi\epsilon_0}\frac{qQ}{r^2}. \tag{3.74}$$

The force is attractive if the charges have opposite sign and repulsive if they have the same sign. The direction of the force is along the line joining the two charges. We can define the electric field at a point r generated by the charge Q as the force a unit charge experiences due to the presence of the charge Q. Thus according to the Coulomb's law, the field generated by Q at a distance r is

$$E = \frac{1}{4\pi\epsilon_0}\frac{Q}{r^2} \tag{3.75}$$

and the force on the charge q in the presence of the field E is

$$F = qE. \tag{3.76}$$

It can be shown that a uniform electric field E can be generated by applying a voltage V between two parallel metallic plates separated by a distance d which is given by

$$E = \frac{V}{d}. \tag{3.77}$$

The direction of the field, and hence the force, on a negative charge q is in the direction of the positively charged plate and away from the negatively charged plate (Fig. 3.9). Thus a negatively charged electron entering a region of uniform field is deflected in the direction of the positively charged plate.

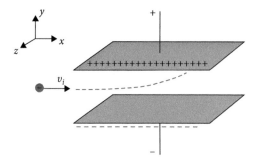

Fig. 3.9 An electron passing through a region of electric field is deflected towards the positively charged plate.

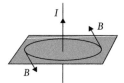

Fig. 3.10 A current carrying wire produces magnetic field **B**.

Just as a static charge creates an electric field, a current produces the magnetic field. A long current-carrying wire produces a magnetic field according to Ampere's law. The magnetic field at a distance r away from a wire carrying current I is given by

$$B = \frac{\mu_0}{2\pi} \frac{I}{r}. \tag{3.78}$$

where $\mu_0 = 4\pi \times 10^{-7}$ Henry/m is the permeability of the free space. Magnetic fields have both direction and magnitude. The field around a long straight wire is found to be in circular loops. According to the right hand rule, if we point the thumb in the direction of the current, the fingers curl in the direction of the magnetic field loops created by it (Fig. 3.10).

The force on a particle of charge q moving with a velocity $\mathbf{v}$ in the presence of the magnetic field $\mathbf{B}$ is given by the Lorentz formula

$$F = q\mathbf{v} \times \mathbf{B}. \tag{3.79}$$

The magnetic force is therefore in a direction perpendicular to both the direction of velocity of the charged particle, $\mathbf{v}$, and the magnetic field, $\mathbf{B}$. A uniform magnetic field in a region can be created via electromagnets. In Fig. 3.11, we show that an electron with charge $-e$ moving along the x-axis in a region of uniform magnetic field B in the $-z$-direction (into the page) is deflected in the $\hat{x} \times \hat{z} = -\hat{y}$ direction and the magnitude of the force is

$$F = evB. \tag{3.80}$$

With this introduction, we now turn to Thomson's experiment. The objective of the experiment was to prove the existence of negatively charged electrons and finding the charge to mass ratio of these particles. The experimental set up is shown in Fig. 3.12.

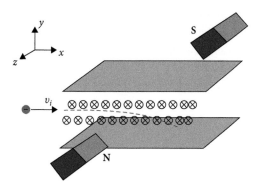

Fig. 3.11 An electron passing through a region of magnetic field is deflected in a direction perpendicular to both its velocity and the magnetic field.

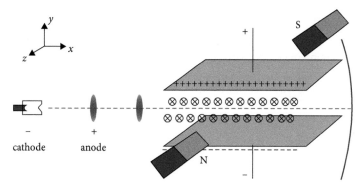

Fig. 3.12 A collimated electron beam passes through a region where both electric and magnetic fields are applied. The fields are set up in such a way that the electric field and the magnetic field forces on the electron are adjusted such that the net force on the electron in the y-direction is zero.

A heated metallic plate called a cathode generates electrons. Initially they move in random directions after being emitted from the metal. However they can be collimated in the x-direction by a positively charged plate called the anode. The electrons are attracted towards the anode with a small hole in the middle. The electrons passing through the anode form a beam moving with a velocity **v** whose magnitude is not known. We can find the velocity by passing the electrons through a region with both electric and magnetic fields. The fields are set up in such a way that the electric field forces the electron to be deflected in the $+y$-direction with the force eE and the magnetic field forces the electron to be deflected in the $-y$ direction with the force evB. The two forces can be adjusted such that the net force on the electron in the y-direction is zero and the electron keeps moving in the x-direction without any deflection. This happens when

$$eE = evB. \tag{3.81}$$

The velocity of the electron is then given by

$$v = \frac{E}{B}. \tag{3.82}$$

The role of the magnetic field is only to find how fast the electrons are moving.

In the next stage of the experiment, the magnetic field is switched off and only the electric field is applied. The electric field deflects the electron beam in the upward direction and gives a glow at a point y on a fluorescent screen. We can find the location of this point via the dynamical equations discussed in Section 3.1.

As there is no force in the x-direction the distance travelled by the electrons in time t is

$$x(t) = vt = \frac{E}{B}t, \tag{3.83}$$

where we substituted for v from Eq. (3.82). The vertical motion of the electron is however governed by the electric force $F_E = eE$ and the displacement in the y-direction is given by

$$y(t) = \frac{1}{2}a_y t^2 = \frac{1}{2}\frac{F_E}{m}t^2 = \frac{1}{2}\frac{eE}{m}t^2. \tag{3.84}$$

If the horizontal distance covered is $x(t) = L$ then

$$L = \frac{E}{B}t \rightarrow t = \frac{LB}{E}. \tag{3.85}$$

The transverse location of electrons on the fluorescent screen is

$$y = \frac{1}{2}\frac{eE}{m}\left(\frac{LB}{E}\right)^2 = \frac{eL^2B^2}{2mE}. \tag{3.86}$$

Thus the charge to mass ratio of an electron can be inferred from this experiment via the relation

$$\frac{e}{m} = \frac{2Ey}{L^2B^2}. \tag{3.87}$$

Thomson found the value for e/m to be $1.76 \times 10 \ \mathrm{C \ kg}^{-1}$.

Problems

3.1 An object is launched at a velocity of 25 m/s in a direction making an angle of 30° upward with the horizontal (assuming $g = 10$ m/s^2). (a) What is the maximum height reached by the object? (b) How much time does the object take to reach to the ground? (c) What is the horizontal displacement when the object reaches the ground? (d) What are the magnitude and direction of the velocity of the object just before it hits the ground?

3.2 Two masses m_1 and m_2 are in an elastic collision with initial velocities v_1 and v_2, and final velocities u_1 and u_2. Prove that after the collision

$$u_1 = \frac{(m_1 - m_2)v_1 + 2m_2v_2}{m_1 + m_2}$$

and

$$u_2 = \frac{(m_2 - m_1)v_2 + 2m_1v_1}{m_1 + m_2}$$

3.3 An object has a kinetic energy of 250 J and a momentum of magnitude 20.0 kg · m/s. Find the (a) speed and (b) mass of the object.

3.4 Consider two objects of masses m and $3m$ moving toward each other with equal speed v_0 along the x-axis with the mass $3m$ moving to the right and the mass m moving to the left. After an elastic collision, the mass m is moving downward at right angles from its initial direction. (a) Find the final speeds of the two objects. (b) What is the angle θ with respect to the x-axis at which the object with mass $3m$ is scattered?

3.5 A rotating wheel requires 4.00 s to rotate 48.0 revolutions. Its angular velocity at the end of the 4.00 s interval is 98.0 rad/s. What is the constant angular acceleration (in rad/s^2) of the wheel?

3.6 A model of the hydrogen atom is an electron of mass $m_e = 9 \times 10^{-31}$ kg circling around the massive proton with a velocity v. The attractive force between the electron and the proton, according to Coulomb's law, is given by

$$\frac{e^2}{4\pi\varepsilon_0 r^2}$$

where $e = 1.6 \times 10^{-19}$ C is the charge of the electron, ε_0 is the permittivity of the free space $(1/4\pi\varepsilon_0 = 9 \times 10^9$ N · m^2/C$^2)$, and r is the radius of the electron orbit. The angular momentum of the electron is assumed to be equal to a constant whose value is 1.05×10^{-34} J · s. What is the radius of the electron orbit? What is the velocity of the electron?

BIBLIOGRAPHY

R. P. Feynman, R. Leighton, and M. Sands, *The Feynman Lectures on Physics, Vol. I* (Addison-Wesley, Reading, MA 1965).

L. Susskind and G. Hrabovsky, *The Theoretical Minimum: What you Need to Know to Start Doing Physics* (Basic Books 2013).

D. C. Giancoli, *Physics: Principles with Applications* (Pearson 2013).

H. D. Young and R. A. Freedman, *University Physics* (Pearson 2015).

R. P. Feynman, *Six Easy Pieces: Essentials of Physics Explained by its Most Brilliant Teacher* (Basic Books 2011).

4 Wave Theory

One of the earliest and most important tenets of quantum mechanics is wave–particle duality: light behaves sometimes like a wave and at other times as a particle, and similarly an electron can also behave both like a particle and as a wave. When the formal laws of quantum mechanics are formulated, the central quantity that describes the particles is the wave function. All this and more lies ahead in the later chapters. However they point to the need for a good understanding of the properties of waves. This chapter introduces the concepts and most essential applications that are required to follow the discussion of quantum mechanical laws and systems.

4.1 Wave Motion

Waves are traveling disturbances and are found in an amazingly diverse range of physical systems. The concept of a wave is most clearly understood by looking at the periodic disturbances through the water when we drop a small object like a pebble in a body of still water. The disturbance produces water waves, which move away from the point where the pebble entered the water. There are numerous other kind of waves, such as a sound wave or a light wave. In this section, we present the basic characteristics of waves.

Waves can be classified under different criteria. One classification differentiates between two types of waves: longitudinal waves and transverse waves. In a longitudinal wave, the particles are disturbed in a direction parallel to the direction that the wave propagates. A longitudinal wave consists of "compressions" and "rarefactions" where particles are bunched together and spread out, respectively (see Fig. 4.1). Sound waves are an example of longitudinal waves where the molecules of the medium oscillate in the direction of propagation of the wave. In a transverse wave, the particles are disturbed in a direction perpendicular to the direction that the wave propagates (see Fig. 4.2). Light, as we will study later, is an example of a transverse wave in which electric and magnetic fields oscillate in the direction perpendicular to the direction of propagation. As most of the waves we are concerned about in this book are transverse waves, in discussing the properties of waves we will concentrate on transverse waves as examples.

Consider the wave propagating in the x-direction. We look at the wave at different times—this is done in Fig. 4.3. In Fig. 4.3a, we consider a point P on the wave and we look at its location at different times. Let it be at the point of maximum amplitude $y = A$ at time $t = 0$. At a later time (we label it $t = T/4$), the point P is found on the axis ($y = 0$). At the time $t = T/2$, point P is found at the lowest point $y = -A$. The point P then starts going up and after time $t = T$, it is back to where it started. Thus it takes time T for one cycle of P starting from $y = A$ and coming back to the same point. This cycle is repeated in time.

Quantum Mechanics for Beginners: With Applications to Quantum Communication and Quantum Computing. M. Suhail Zubairy.
© M. Suhail Zubairy 2020. Published in 2020 by Oxford University Press. DOI: 10.1093/oso/9780198854227.001.0001

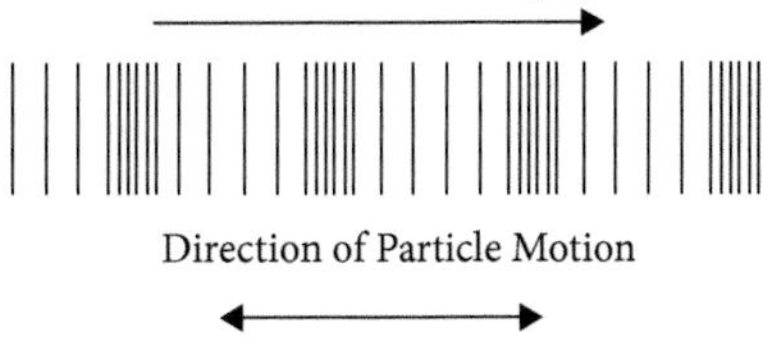

Direction of Wave Propagation

Direction of Particle Motion

Fig. 4.1 In a longitudinal wave, the disturbance moves in the direction of propagation of the wave.

Particle Motion

Fig. 4.2 In a transverse wave the disturbance is in the direction perpendicular to the direction of wave propagation.

We call A the amplitude of the wave (with the wave oscillating between $+A$ and $-A$) and call T the period of the wave (with the wave completing one cycle in time T). The period T is related to another important quantity called the frequency f which is the reciprocal of the period T, i.e.,

$$f = \frac{1}{T} \tag{4.1}$$

The frequency is the number of oscillations from $+A$ back to $+A$ in one second. The unit of frequency is Hertz. A frequency of $f = 10$ Hertz means that there are 10 oscillations per second and the time period is $T = 1/10 = 0.1$ second. Throughout this book, we use the angular frequencies ν (frequency in radians instead of Hertz) by multiplying f by 2π, i.e.,

$$\nu = 2\pi f. \tag{4.2}$$

We also note that, during the time T, the wave maximum at $t = 0$ at point P has advanced by a distance λ as shown in Fig. 4.3b. This distance is called the wavelength of the wave and is the distance between two maxima. Since the distance λ is travelled during the time T, the speed of the wave is

$$\text{v} = \frac{\lambda}{T} = \lambda f. \tag{4.3}$$

This is an important relation that applies to all kind of waves including both sound waves and light waves.

The periodic motion described above can be written in the following mathematical form

$$y(x = 0, t) = A \cos\left(2\pi f t\right), \tag{4.4}$$

where we have chosen the location of P (maximum amplitude A) as $x = 0$. This point oscillates back and forth between $+A$ and $-A$ at a frequency f. Next we note that, when the speed of the wave is v, the wave disturbance travels from $x = 0$ to some point x in time x/v. The motion of point x at time t is the same as the motion of point $x = 0$ at an earlier time $t - x/\text{v}$. We thus have

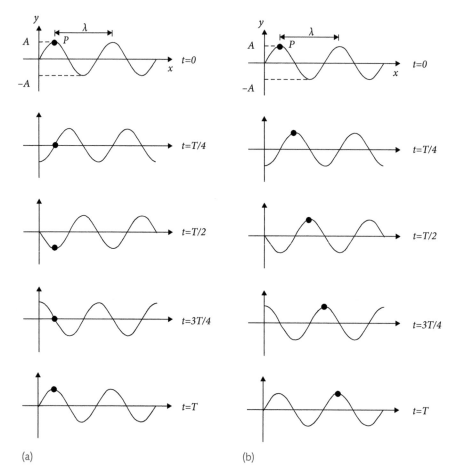

Fig. 4.3 A transverse wave of amplitude A and wavelength λ is depicted at different times. (a) A point on a wave undergoes a full oscillation during the time period T. (b) T is also the time period in which the wave propagates through a distance λ.

$$y(x,t) = A \cos\left(2\pi f\left(t - \frac{x}{\mathrm{v}}\right)\right) = A \cos\left(\nu\left(t - \frac{x}{\mathrm{v}}\right)\right). \tag{4.5}$$

An alternate form for the wave amplitude $y(x,t)$ is obtained from the relations $\nu = 2\pi f = 2\pi/T$ and $\mathrm{v} = \lambda f$:

$$y(x,t) = A \cos\left(2\pi\left(\frac{t}{T} - \frac{x}{\lambda}\right)\right). \tag{4.6}$$

Next we define the wave number:

$$k = \frac{2\pi}{\lambda}. \tag{4.7}$$

The speed v is then related to the angular frequency ν and the wave number k via $\mathrm{v} = \lambda/T = \lambda f = (\lambda/2\pi)\,2\pi f = \nu/k$, i.e.,

$$\nu = k\mathrm{v} \tag{4.8}$$

and

$$y(x,t) = A \cos(vt - kx). \tag{4.9}$$

This represents a wave propagating in $+x$ direction. If we replace x by $-x$, then

$$y(x,t) = A \cos(vt + kx), \tag{4.10}$$

which represents a wave propagating in $-x$ direction.

A convenient description of the wave is in term of a complex notation

$$y(x,t) = Ae^{-ivt+ikx}. \tag{4.11}$$

In this case the real part of $y(x,t)$ is $A \cos(vt - kx)$. This description is useful in many applications as we shall see later. The source of usefulness of the complex form of the wave is that we can depict it as a point in the complex plane instead of a full-fledged sinusoidal description as shown in Fig. 4.3. As an example, we show the wave at the location $x = 0$ and at time t by the point P on the complex plane in Fig. 4.4. Here the x-axis corresponds to the Real part of $y(0, t)$ and the y-axis corresponds to the Imaginary part of $y(0, t)$. The distance from the origin to the point P, A, is the amplitude of the wave and $\theta = vt$ is the phase of the wave. As time increases, the point P rotates in the complex plane.

Principle of superposition: An important property of any kind of wave is that, when two or more waves arrive at the same point, they superimpose themselves upon one another. This is called the principle of superposition. This important principle has many applications, particularly in understanding interference and diffraction. We study these phenomena in later sections of this chapter.

The principle of superposition for two waves can be stated mathematically as follows. If two waves are described by $y_1(x, t)$ and $y_2(x, t)$ at a point x at time t, then the resulting wave at this point is given by

$$y(x,t) = y_1(x,t) + y_2(x,t). \tag{4.12}$$

As simple examples, we consider two waves in Fig. 4.5a and Fig. 4.5b. In Fig. 4.5a, the two waves, $y_1(x, t)$ and $y_2(x, t)$, are shifted with respect to each other by a distance equal to half the wavelength. According to the principle of superposition, both waves cancel each other at every point and the resulting amplitude at each point is zero or the resulting wave is of zero amplitude. In Fig. 4.5b, the two waves have the same amplitude at each point and the resulting wave is identical to each wave with the only difference that the amplitude is twice of each

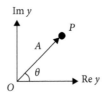

Fig. 4.4 A wave is represented by a point P on a complex plane. The distance A is the amplitude and θ is the phase.

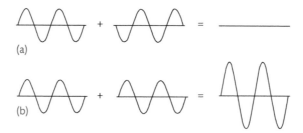

Fig. 4.5 Superposition of two waves. (*a*) The two waves are displaced by a distance equal to half the wavelength, $\lambda/2$, and the two waves cancel each other at every point, resulting in a wave of zero magnitude. (*b*) The two waves are displaced by a distance equal to the wavelength, λ, and the two waves superimpose on each other yielding a wave of twice the amplitude.

individual wave. We say that the two waves interfere destructively in Fig. 4.5*a* and constructively in Fig. 4.5*b*.

This description can also be used to introduce another important notion—the phase. A displacement by a distance equal to the wavelength λ corresponds to a phase shift of 2π. We can therefore say that the two waves in Fig. 4.5*a* have phase shift equal to

$$\phi = \frac{2\pi}{\lambda}\frac{\lambda}{2} = \pi \tag{4.13}$$

whereas the phase shift in Fig. 4.5*b* is

$$\phi = 0 \text{ or } \frac{2\pi}{\lambda}\lambda = 2\pi. \tag{4.14}$$

In general, the phase shift is given by

$$\phi = \frac{2\pi}{\lambda}\Delta x, \tag{4.15}$$

where Δx is the distance by which the two waves are displaced.

In Figs. 4.6*a* and 4.6*b*, we present another example. Let two sources give out the waves in the same phase. Waves meet in phase or out of phase at a point P depending on its distance from two coherent sources. In Fig. 4.6*a*, the path difference at point P is λ and we have constructive interference. In Fig. 4.6*b*, the path difference is 1.5 λ. In general, the condition for constructive interference is that the path difference is an integral multiple of the wavelength λ, i.e.,

$$\text{Path difference} = n\lambda, (n = 0, 1, 2, \cdots) \tag{4.16}$$

and the condition for destructive interference is

$$\text{Path difference} = \left(n + \frac{1}{2}\right)\lambda, (n = 0, 1, 2, \cdots) \tag{4.17}$$

So far we have considered running waves in which a point on the wave moves in the forward direction. Another important class of waves is the standing wave. In a standing wave, certain points on the wave, called 'nodes', remain fixed but the points with non-zero amplitude oscillate at the same location with time period T as shown in Fig. 4.7. In Fig. 4.7, the nodes

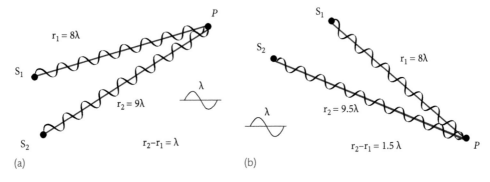

Fig. 4.6 Two waves of wavelength λ originate from points S_1 and S_2. (a) If, at point P the difference in the distance travelled from S_1 and S_2 is equal to λ, there is constructive interference and (b) if the distance is equal to 1.5 λ, there is destructive interference.

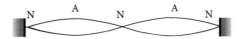

Fig. 4.7 A standing wave is formed between two walls. The nodes N and the antinodes A remain fixed.

and antinodes are labelled as N and A, respectively. The points with maximum amplitudes A are called "anti-nodes." They oscillate from "A" to "$-A$." A standing wave is formed when two waves of equal amplitude propagate in the opposite direction. The wave function for the standing wave is therefore given by

$$\begin{aligned} y(x,t) &= y_1(x,t) + y_2(x,t) \\ &= A\cos(kx - vt) - A\cos(kx + vt) \\ &= 2A\sin(kx)\sin(vt). \end{aligned} \tag{4.18}$$

Here we used the trigonometric identity

$$\cos a - \cos b = -2\,\sin\left(\frac{a+b}{2}\right)\sin\left(\frac{a-b}{2}\right).$$

We can see from Eq. (4.18) that the nodes ($y=0$) are located at

$$x = 0, \frac{\pi}{k}, \frac{2\pi}{k} \cdots = 0, \frac{\lambda}{2}, \frac{2\lambda}{2}, \frac{3\lambda}{2}, \cdots \tag{4.19}$$

and the antinodes are located at

$$x = \frac{\pi}{2k}, \frac{3\pi}{2k} \cdots = 0, \frac{\lambda}{4}, \frac{3\lambda}{4}, \frac{5\lambda}{4}, \cdots \tag{4.20}$$

As an example of a standing wave, we consider a stretched string of length L fixed at both ends. The end points of the string must be nodes. We can set up standing-wave patterns at many frequencies or wavelengths. When the string is displaced at its midpoint and released, the first or fundamental harmonic is excited. The condition for the fundamental harmonic is

$$L = \frac{\lambda}{2}. \tag{4.21}$$

The corresponding frequency is

$$f_1 = \frac{v}{2L},$$

(4.22)

where $v = f_1\lambda$ is the velocity of the wave. The condition for the nth harmonic is

$$L = n\frac{\lambda}{2} \quad (n = 1, 2, \cdots)$$

(4.23)

leading to

$$\lambda_n = \frac{2L}{n} \quad (n = 1, 2, \cdots).$$

(4.24)

The corresponding wave vectors are

$$k_n = \frac{2\pi}{\lambda_n} = \frac{\pi n}{L} \quad (n = 1, 2, \cdots).$$

(4.25)

The frequency for the nth harmonic is

$$f_n = \frac{v}{\lambda_n} = n\frac{v}{2L}.$$

(4.26)

In Fig. 4.8, we plot the standing waves corresponding to $n = 1, 2, 3,$ and 4.

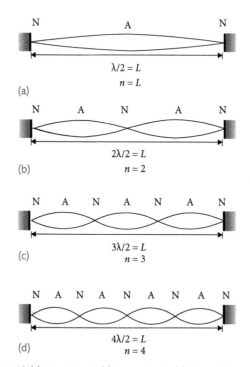

Fig. 4.8 Standing waves with (a) one antinode, (b) two antinodes, (c) three antinodes, and (d) four antinodes.

4.2 Young's Double-slit Experiment

In 1801, Thomas Young carried out an experiment that established the wave nature of light. Its historical significance is discussed in Chapters 6 and 9. Here we discuss the experiment and show how it demonstrates the interference of light waves.

The schematic of Young's double-slit experiment is shown in Fig. 4.9. Light from a source is incident on a screen that contains two narrow slits S_1 and S_2 which are separated from each other by a distance d. As the distances from the source to S_1 and S_2 are equal, the two light waves have the same phase at S_1 and S_2. We also assume that the two slits are identical and the waves emerging from the slits have the same frequency and the same amplitude. The light from these slits forms an interference pattern consisting of bright and dark spots or fringes on a screen a distance L away. The bright fringes are located at those points where the light beams originating from S_1 and S_2 interfere constructively and dark fringes are located at those points where they interfere destructively.

We can derive the conditions whether, at a certain point P on the screen, we get a bright spot or a dark spot with the help of Fig. 4.10. The light intensity at the point P results from light waves arriving from the two slits. The distance from the upper slit S_1 to the observation point P is r_1 and the distance from the lower slit S_2 to the observation point P is r_2. As r_2 is greater than r_1, the light from the lower slit travels a longer distance than the light coming from the upper slit. We can immediately see that, at those locations on the screen where the path difference $r_2 - r_1$ is equal to an integral multiple of the wavelength λ, i.e.,

$$r_2 - r_1 = n\lambda, \tag{4.27}$$

with $n = 0, \pm 1, \pm 2, \cdots$, two waves interfere constructively and we get bright spots. And at those locations where the path difference $r_2 - r_1$ is equal to an odd multiple of $\lambda/2$, i.e.,

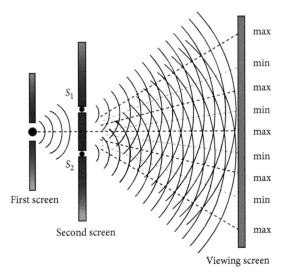

Fig 4.9 In Young's double-slit experiment, light from a source is incident on a screen with two slits. The light from these slits forms an interference pattern on the viewing screen.

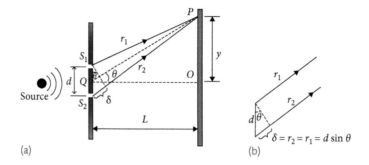

Fig. 4.10 The path difference from light originating from slits S_1 and S_2 is equal to $d \sin \theta$ in the limit when $L \gg d$.

$$r_2 - r_1 = \pm \frac{\lambda}{2}, \tag{4.28}$$

as well as

$$r_2 - r_1 = (2n + 1)\frac{\lambda}{2} \tag{4.29}$$

with $n = \pm 1, \pm 2, \cdots$. The two waves interfere destructively and we get dark spots. The path difference $r_2 - r_1$ can be determined easily in the case when the distance L between the screens is much greater than the distance between the two slits d, i.e., $L \gg d$. In this case, as shown in Fig. 4.10b, the light waves from S_1 to P and from S_2 to P are approximately parallel to each other and the path difference $r_2 - r_1$ is given by

$$r_2 - r_1 = d \sin \theta, \tag{4.30}$$

where θ is the angle that the slits make with the observation point P. The condition for the bright fringes becomes

$$d \sin \theta = n\lambda \tag{4.31}$$

and the condition for the destructive interference becomes

$$d \sin \theta = \pm \frac{\lambda}{2} \tag{4.32}$$

as well as

$$d \sin \theta = (2n + 1)\frac{\lambda}{2}. \tag{4.33}$$

We can see that there is a bright spot at the center point on the screen corresponding to $\theta_{bright} = 0$ where $r_1 = r_2$. The first dark spots are located on either side of the central maximum at

$$\theta_{dark} = \pm \sin^{-1}\left(\frac{\lambda}{2d}\right). \tag{4.34}$$

This is again followed by bright spots located at

$$\theta_{bright} = \pm \sin^{-1}\left(\frac{\lambda}{d}\right) \tag{4.35}$$

and so on.

We can derive the locations of the bright and dark spots explicitly. Again we assume that the distance between the two screens, L, is much larger than the slit separation d. First we note that the vertical displacement y of the observation point P is given by

$$y = L \tan \theta. \tag{4.36}$$

In the limit $L \gg d$, the angle θ is small ($\theta \ll 1$). We discussed in Section 2.2, that, when $\theta \ll 1$, $\tan \theta \approx \sin \theta \approx \theta$. Therefore, we can replace $\tan \theta$ by $\sin \theta$ in Eq. (4.36) to a good approximation. We then obtain

$$y \approx L \sin \theta. \tag{4.37}$$

It follows from condition (4.31) for constructive interference that bright fringes are located at vertical positions

$$y_{bright} = n \frac{\lambda L}{d} \tag{4.38}$$

with $n = 0, \pm 1, \pm 2, \cdots$. The locations for the dark fringes are obtained for the vertical positions at

$$y_{dark} = (2n + 1) \frac{\lambda L}{2d}. \tag{4.39}$$

The separation between the two bright fringes corresponding to $(n + 1)$ and n is thus

$$\left(\Delta y \right)_{bright} = (n + 1) \frac{\lambda L}{d} - n \frac{\lambda L}{d} = \frac{\lambda L}{d}. \tag{4.40}$$

Similarly, the separation between the two dark fringes is also given by

$$\left(\Delta y \right)_{dark} = (2 (n + 1) + 1) \frac{\lambda L}{2d} - (2n + 1) \frac{\lambda L}{2d} = \frac{\lambda L}{d}. \tag{4.41}$$

The fringes are thus equally spaced as shown in Fig. 4.11.

Next we calculate the intensity distribution in the interference pattern. Light consists of waves whose complex amplitudes can be written as

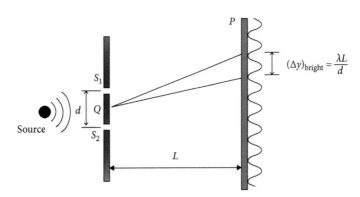

Fig. 4.11 The spacing between the maxima of the adjacent fringes in Young's double-slit experiment is equal and given by $\lambda L/d$.

$$u(x,t) = ue^{i(kx - vt)}.$$

$$(4.42)$$

The amplitude of the light field at point P consists of two contributions, one coming from slit 1 and the other from slit 2 as shown in Fig. 4.10. Since we assume the two slits to be identical, the field amplitudes at P from the two sources are the same. The only difference is the additional phase shift equal to

$$\phi = kd \sin \theta = \frac{2\pi}{\lambda} d \sin \theta$$

$$(4.43)$$

that the light coming from slit 2 acquires due to the additional distance it has to travel. We thus have

$$u_1(t) = ue^{-ivt}$$

$$(4.44)$$

$$u_2(t) = ue^{-ivt + i\phi}.$$

$$(4.45)$$

The intensity at point P is

$$I_p = |u_1(t) + u_2(t)|^2 = I_0 \cos^2 (\phi/2),$$

$$(4.46)$$

where $I_0 = 4u^2$ is the maximum intensity. On substituting for ϕ from Eq. (4.43), we obtain

$$I = I_0 \cos^2 \left(\frac{\pi}{\lambda} d \sin \theta \right).$$

$$(4.47)$$

We can obtain all the above results relating to the location and spacing of the dark and bright spots from this expression for the intensity.

First, we see that the maximum intensity occurs when

$$\frac{\pi}{\lambda} d \sin \theta = n\pi \quad (n = 0, \pm 1, \pm 2, \cdots)$$

$$(4.48)$$

or

$$d \sin \theta = n\lambda.$$

$$(4.49)$$

This is shown in Fig. 4.11. This is the same result as Eq. (4.31). Similarly dark fringes are formed at those angles where

$$\frac{\pi}{\lambda} d \sin \theta = (2n + 1) \frac{\pi}{2} \quad (n = 0, \pm 1, \pm 2, \cdots)$$

$$(4.50)$$

or

$$d \sin \theta = (2n + 1) \frac{\lambda}{2}.$$

$$(4.51)$$

Again this condition is identical to condition (4.33) for dark fringes.

When $y \ll L$, we have $\sin \theta \approx \tan \theta \approx y/L$,

$$I = I_0 \cos^2 \left(\frac{\pi d}{\lambda} \frac{y}{L} \right).$$

$$(4.52)$$

From this expression, we can see that maxima are obtained at vertical distances

$$y_n = n \frac{\lambda L}{d}.$$

$$(4.53)$$

The distance between neighboring maxima (and minima) is

$$y = \frac{\lambda L}{d} \qquad\qquad (4.54)$$

as before. An advantage of this approach is that we have the explicit expression for the intensity on the screen.

4.3 Diffraction

Another important consequence of the wave nature of light is diffraction. When a light beam is incident on an obstacle, it does not travel in a straight line. Instead it bends and light is scattered in those areas where we expect a shadow. This is seen in Fig. 4.12, where incoming light passes through a slit and forms a sort of interference pattern consisting of a broad, intense central band flanked by a series of narrower, less intense secondary bands and a series of dark bands, or minima. This phenomenon cannot be explained if we treat light as consisting of particles, as the particles traveling in straight lines should cast a sharp image of the slit on the screen. Here we consider the slit to have a width a unlike the slits considered in our discussion of Young's double-slit experiment where we assumed the slits to have negligible widths.

Like interference, diffraction can also be understood from Huygens' principle, according to which each point on the slit acts as a source of a wave. Light from one point can interfere with light from another point constructively or destructively depending upon the relative phase, thus leading to the spreading of the wave with an associated interference pattern. The resultant intensity on the screen is the sum of the waves generated by all the points within the slit and thus depends on the direction θ.

In general, it is complicated to analyze the diffraction pattern on the screen as we have to add the contribution of light waves from all the infinite points within the slit of width a. However there is a simple argument to find the locations of the dark points on the screen.

We first divide the slit into halves as shown in Fig. 4.13. We assume that all the points within the slit have the same phase. We also assume that the distance from the slit to the screen, L, is much larger than the width of the slit a, i.e., $L \gg a$. We can then, as in the analysis of Young's

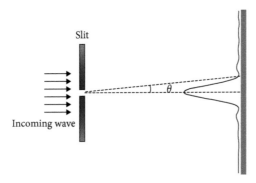

Fig. 4.12 Diffraction of light from a single slit.

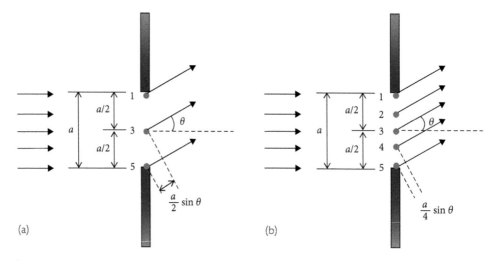

Fig. 4.13 Condition for a dark spot in the diffraction of light from a single slit. (a) The condition for destructive interference is satisfied for light originating from points 1 and 3 as well as 3 and 5 and (b) the condition for destructive interference is satisfied for light originating from points 1 and 2, 2 and 3, 3 and 4, and 4 and 5.

double-slit experiment, treat the waves originating from all the points on the slit to be almost parallel to each other.

We see that, for destructive interference at the screen, the difference in path length between the waves originating from the bottom and the center should be half the wavelength. As the path difference is $(a/2) \sin \theta$ (see Fig. 4.13a) the condition for the destructive interference is

$$\frac{a}{2} \sin \theta = \frac{\lambda}{2}. \tag{4.55}$$

If this condition is satisfied between points 1 and 3, then it must also be automatically satisfied between points 3 and 5. As a matter of fact, for each point in the lower half of the slit, there is a point $a/2$ above in the upper half of the slit, such that the waves originating from the pair of points interfere destructively. This proves that the overall condition for the destructive interference, and consequently the dark spot on the screen, is Eq. (4.55). We can rewrite this condition as

$$\sin \theta = \frac{\lambda}{a}. \tag{4.56}$$

Next we divide the slit in four parts. The path difference between points 1 and 2, 2 and 3, 3 and 4, and 4 and 5 is equal to $(a/4) \sin \theta$ as shown in Fig. 4.13b, and the condition for destructive interference is

$$\frac{a}{4} \sin \theta = \frac{\lambda}{2}, \tag{4.57}$$

which can be rewritten as

$$\sin \theta = \frac{2\lambda}{a}. \tag{4.58}$$

Similarly dividing the slit into six parts, the condition for destructive interference becomes

$$\sin\theta = \frac{3\lambda}{a}. \tag{4.59}$$

In general those points on the screen are dark where the angle θ is given by

$$\sin\theta_{dark} = \frac{n\lambda}{a}, \tag{4.60}$$

where $n = \pm 1, \pm 2, \cdots$.

The locations of the dark spots in the vertical direction can be found by assuming, as before,

$$\sin\theta \approx \tan\theta = \frac{y}{L}. \tag{4.61}$$

It then follows from Eq. (4.60) that

$$y_{dark} = n\frac{\lambda L}{a}, \tag{4.62}$$

where $n = \pm 1, \pm 2, \cdots$. Thus the dark spots are equidistant from each other, the separation being $\lambda L/a$. This is shown in Fig. 4.14.

This simple argument provides the location of the dark spots on the screen. It does not, however, provide information about the intensity distribution. The intensity distribution as a function of the angle θ is given by[1]

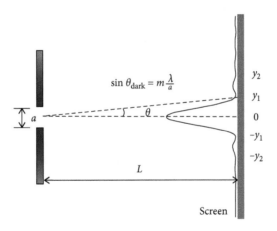

Fig. 4.14 In the diffraction pattern from a slit of width a, the dark spots are equidistant.

---

[1] The derivation for the expression of intensity follows the same method as we used to derive the intensity (4.46) in the Young's double-slit experiment. The difference is that, in the case of Young's double-slit experiment, we added the field contributions at P from only two points. In the case of the diffraction from a single slit of width a, we have to add the contributions from each point within the slit. The result is the following integration:

$$I = \left| u\frac{1}{a}\int_{a/2}^{-a/2} dx\, e^{-i\nu t + ikx\sin\theta} \right|^2 = I_0 \left[\frac{\sin(\pi a \sin\theta/\lambda)}{\pi a \sin\theta/\lambda} \right]^2,$$

where $I_0 = |u|^2$ and we used $k = 2\pi/\lambda$.

$$I = I_0 \left[\frac{\sin\left(\pi a \sin\theta/\lambda\right)}{\pi a \sin\theta/\lambda} \right]^2. \tag{4.63}$$

This distribution is shown in Fig. 4.15. The diffraction from a slit thus produces an intensity pattern with a broad, central, bright fringe flanked by much weaker bright fringes alternating with dark fringes. The location of the dark fringes is given by Eq. (4.62) and the locations of the bright fringes satisfy

$$\frac{\pi a \sin\theta}{\lambda} = (2n+1)\frac{\pi}{2} \tag{4.64}$$

or

$$\sin\theta_{bright} = (2n+1)\frac{\lambda}{2a} \tag{4.65}$$

yielding, with $\sin\theta \approx \tan\theta = y/L$,

$$y_{bright} = (2n+1)\frac{\lambda L}{2a} \tag{4.66}$$

where $n = \pm1, \pm2, \cdots$. These expressions are not valid for $n = 0$. When $\theta = 0$ or $y = 0$, the intensity is maximum and is equal to I_0. Thus the bright and dark fringes alternate on each side of the central point $y = 0$.

So far we have considered a planar slit. For a circular aperture of diameter D, we obtain a similar diffraction pattern as a slit of width a as shown in Fig. 4.16. The difference, of course, is that the pattern consists of bight rings surrounded by dark rings as shown in Fig. 4.17a instead of a pattern consisting of alternate bright and dark fringes. The calculation of the diffraction pattern with a circular aperture is complicated. The resulting intensity distribution, called the Airy pattern, is given by

$$I = I_0 \left[\frac{J_1(w)}{w} \right]^2, w = \frac{\pi D r}{L\lambda}. \tag{4.67}$$

Here $J_1(w)$ is a function and is called the Bessel function of order 1, D is the diameter of the aperture, r is the distance from the center line, and L is the distance from the aperture to the

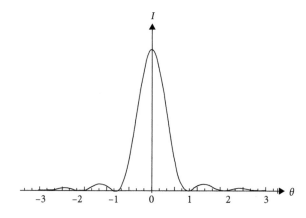

Fig. 4.15 The diffraction pattern as given by Eq. (4.63).

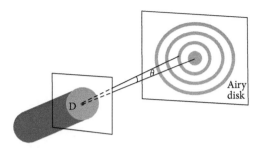

Fig. 4.16 Diffraction from a circular aperture.

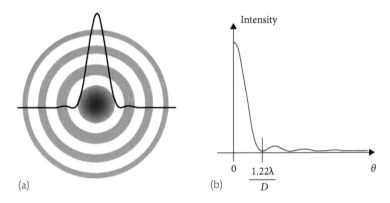

Fig. 4.17 The diffraction pattern from a circular aperture is the Airy pattern (4.67). (a) There is a bright spot at the center followed by alternating dark and bright spots. (b) The first dark spot is located at $\theta = 1.22\lambda/D$.

screen. The first zero of the Bessel function $J_1(w)$ occurs at $w = 3.832$. Therefore the condition for the first dark ring is

$$w = \frac{\pi Dr}{L\lambda} = 3.832. \tag{4.68}$$

The radius of the Airy disc (the distance from the center to the first dark ring) is then given by

$$r_{min} = 1.22\frac{\lambda L}{D}. \tag{4.69}$$

First minimum for the diffraction pattern of a circular aperture of diameter D for small angle $\left(\theta_{min} \ll 1\right)$ occurs for (Fig. 4.17b)

$$\sin\theta_{min} \approx \theta_{min} = 1.22\frac{\lambda}{D}. \tag{4.70}$$

This expression is similar to the corresponding angle for the first dark fringe in the case of diffraction from a single slit, i.e., $\sin\theta = \lambda/a$.

The angular radii of the next two dark rings are given by

$$\sin\theta_2 = 2.23\frac{\lambda}{D}; \sin\theta_3 = 3.24\frac{\lambda}{D}.$$

In between there are bright rings with angular radii given by

$$\sin\theta = 1.63\frac{\lambda}{D}; \sin\theta = 2.68\frac{\lambda}{D}.$$

4.4 Rayleigh Criterion

It is a common experience for all of us that we can distinguish two objects placed at a certain distance from us. However, if these objects come closer, there is a point when we are unable to distinguish them. We just see a blurred object. We experience a similar resolution limit if the two objects keep their separation intact but move farther away from us. How do we understand these limits on resolution? The fundamental limit, as we see below, comes from diffraction.

We have seen that diffraction can cause the spread of a light beam when it passes through an aperture. An image of a point object is no more a point but instead an Airy pattern, as discussed in the last section. Such a circularly symmetric pattern is shown in Fig. 4.18a. If, instead we have two point sources, each one makes an Airy pattern of its own. If the two objects are separated by a sufficiently large distance, the two Airy patterns can be resolved, as seen in Fig. 4.18b. However, when the two objects come very close, the Airy patterns they generate can no longer be resolved, as shown in Fig. 4.18c. We want to find the minimum distance between the two objects when they can be resolved.

For this purpose, we consider light from two point objects passing through a circular aperture and forming an image on a screen as shown in Fig. 4.19. The image formed by each point source is an Airy pattern with a bright spot in the center surrounded by a dark ring. If the aperture diameter is D and the wavelength is λ, the angular radius θ_1 of the first dark ring is given by

$$\sin\theta_1 = 1.22\frac{\lambda}{D}. \tag{4.71}$$

According to the Rayleigh criterion, formulated by Lord Rayleigh in 1879, two point sources are regarded as just resolved when the principal diffraction maximum of one image coincides with the first diffraction minimum of the second image. Thus the condition of minimum angular resolution is

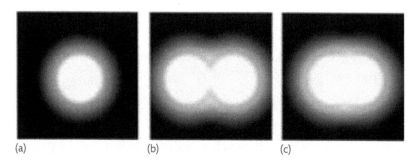

(a)   (b)   (c)

Fig. 4.18 (a) Diffraction pattern from a single point. (b) Diffraction pattern from two points whose separation is large enough that they can be resolved, and (c) when they cannot be resolved.

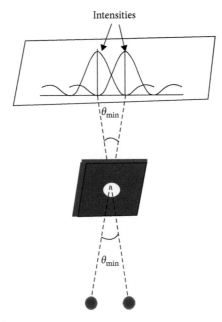

Fig. 4.19 Images formed by two point sources through a circular aperture are Airy patterns. According to the Rayleigh criterion, the two point sources can be resolved if the principal diffraction maximum of one image coincides with the first diffraction minimum of the second image.

$$\sin \theta_{min} \approx \theta_{min} = 1.22 \frac{\lambda}{D}. \tag{4.72}$$

When the objects are separated by a distance s and the aperture is at a distance L away, the angular separation θ is given by

$$\theta = \frac{s}{L}. \tag{4.73}$$

The minimum resolvable distance between two objects is thus

$$s_{min} = L\theta_{min} = 1.22 \frac{\lambda L}{D}. \tag{4.74}$$

As our first example, we calculate the resolution of the eye. The pupil diameter D ranges from 3–4 mm during the day to 5–9 mm at night. An eye is sensitive to optical wavelengths ranging between 0.38 and 0.80 μm. The angular resolution of an eye for light of wavelength $\lambda \sim 0.55$ μm is

$$\theta_{min} = 1.22 \frac{\lambda}{D} = 1.22 \frac{0.55 \times 10^{-6}}{3 \times 10^{-3}} \text{ rad} = 0.224 \times 10^{-3} \text{ rad.} \tag{4.75}$$

and the minimum resolvable separation between two point objects, such as two lines drawn on a paper or two letters on a billboard, at a distance L is given by

$$s_{min} = L\theta_{min} = 1.22 \frac{\lambda L}{D} = 0.224 \times 10^{-3} \times L. \tag{4.76}$$

We can therefore resolve two points at a distance of 1 m if they are separated by a distance of 0.224 mm or more.

The Rayleigh limit also puts restriction on microscopes and telescopes. For example, we can calculate the resolving power of two celestial objects by the orbiting Hubble space telescope. The primary mirror of the Hubble telescope has a diameter $D = 2.5$ m. Thus, for average light wavelength of 0.55 μm, the Rayleigh criterion gives the minimum resolvable angle between two point sources such as far away stars as

$$\theta_{min} = 1.22\frac{\lambda}{D} = 1.22\frac{0.55 \times 10^{-6}}{2.40} = 2.80 \times 10^{-7} \text{rad.} \tag{4.77}$$

The distance s between two objects a distance L away and separated by an angle θ is $s = L\theta$. Thus the Hubble telescope can resolve two stars with minimum separation

$$s_{min} = L\theta_{min} = 2.80 \times 10^{-7} \times L. \tag{4.78}$$

The Andromeda galaxy is about 2 million light years away from us. One light year (ly) is the distance light of velocity 3×10^8 m/s travels during one year ($365 \times 24 \times 60 \times 60$ seconds). The Hubble telescope can therefore resolve stars that are separated by distance

$$s_{min} = 2.80 \times 10^{-7} \times 2 \times 10^6 \text{ ly} = 0.56 \text{ ly.} \tag{4.79}$$

Typical separation between neighboring stars ranges from 1 ly to about 4 ly. Thus the Hubble telescope is able to resolve essentially all the stars in the Andromeda galaxy even when it is about 2 million light years away from those stars.

Problems

4.1 Water waves in a pond travel a distance of 5.5 m in 2.2 s. The time it takes for one complete wave to pass a certain point is 1.5 s. (a) Find the velocity of the wave; (b) find the frequency of the wave; (c) find the wavelength.

4.2 The equation of a traveling wave is $y(x, t) = 0.04 \cos(37.68\, t + 12.56\, x)$, where x and y are in meters, and t in seconds. Determine the following: (a) the direction of the wave; (b) the wave number and the wavelength; (c) the angular frequency; (d) the frequency; (e) the speed of the wave; and (f) the time when $y(0.25, t) = 0.02$.

4.3 A string of length 0.28 m is fixed at both ends. The string is plucked and a standing wave is vibrating at its second harmonic. The speed of the traveling waves that make up the standing wave is 140 m/s. What is the frequency of vibration?

4.4 In a Young's double-slit experiment, the angle that locates the second-order bright fringe is 2.0°. The slit separation is $d = 3.8 \times 10^{-5}$ m. What is the wavelength of the light?

4.5 In a double-slit experiment, the wavelength is $\lambda = 586$ nm, the separation between the two slits is $d = 0.10$ mm, and the distance from the slits to the screen is $L = 20$ cm. On viewing the screen, what is the distance between the fifth maximum and the seventh minimum from the central maximum?

4.6 A slit is illuminated by light of wavelength $\lambda = 586$ nm. What is the distance between the two slits d if the first minimum occurs at $\theta = 15.0°$?

BIBLIOGRAPHY

R. P. Feynman, R. Leighton, and M. Sands, *The Feynman Lectures on Physics, Vol. II* (Addison-Wesley, Reading, MA 1965).

D. C. Giancoli, *Physics: Principles with Applications* (Pearson 2013).

H. D. Young and R. A. Freedman, *University Physics* (Pearson 2015).

R. P. Feynman, *Six Easy Pieces: Essentials of Physics Explained by its Most Brilliant Teacher* (Basic Books 2011).

PART

Fundamentals of Quantum Mechanics

5 Fundamentals of Quantum Mechanics

The laws of quantum mechanics were formulated in the year 1925 through the work of Werner Heisenberg, followed by Max Born, Pascual Jordan, Paul Dirac, and Wolfgang Pauli. A separate but equivalent approach was independently developed by Erwin Schrödinger in early 1926. The laws governing quantum mechanics were highly mathematical and their aim was to explain many unresolved problems within the framework of a formal theory. As mentioned in Chapter 1, a conceptual foundation emerged in the subsequent 2–3 years that indicated (to the consternation of even some of the founding fathers such as Einstein, de Broglie, and Schrödinger) how radically different the new laws were from the classical physics. In Chapter 17, we discuss the Schrödinger equation and its solutions for some simple problems, thus indicating how any problem in physics can be mathematically solved within the framework of quantum mechanics. In this introductory text we do not delve into these details and concentrate on the fundamental and conceptual aspects of quantum mechanics. An understanding of these concepts, as we see in later chapters, is sufficient in understanding many recent exciting applications in areas such as quantum information and quantum computing.

In this chapter we discuss some of these salient features of quantum mechanics. This discussion should indicate how different and counterintuitive its fundamentals are from those of classical physics.

5.1 Quantization of Energy

In classical mechanics, any type of motion (translation, rotation, vibration) can have any value of energy associated with it, i.e., there is a continuum of energy states. One of the most fundamental aspects of quantum mechanics is that, in most situations, the energy comes in discrete units called quanta of energy. The first three major contributions that led to the birth of quantum mechanics all involved quantization of energy. This will be the subject of discussion in the next chapter. Here we summarize these results to indicate how quantization of energy was the common theme of these works.

First we mention the work of Max Planck to explain the spectrum of light emitted from a so-called blackbody. An object that absorbs all *radiation* falling on it, at all wavelengths, is called a *blackbody*. When a *blackbody* is at a uniform temperature, it emits a characteristic frequency distribution that depends on the temperature. The emitted radiation is called *blackbody radiation*. According to classical mechanics, the emitted radiation arises due to the oscillation of electrons of atoms and molecules at any frequency, from zero to infinity. As a result, the amount of energy emitted has no upper limit. Thus an infinite amount of energy should be emitted at large frequencies (or small wavelengths). However this does not happen. As a matter

Quantum Mechanics for Beginners: With Applications to Quantum Communication and Quantum Computing. M. Suhail Zubairy.
© M. Suhail Zubairy 2020. Published in 2020 by Oxford University Press. DOI: 10.1093/oso/9780198854227.001.0001

of fact it was observed that the emitted energy peaks at a certain frequency depending on the temperature and goes down to almost zero at high frequencies. Max Planck explained this result by making a bold ansatz that the energy associated with the oscillations of electrons which give rise to the radiation come in packets or quanta of energy. Each packet has energy proportional to the frequency of oscillation of the electron. The quantization of energy thus explained the blackbody spectrum.

Second we discuss how Einstein used Planck's quantization condition to explain the phenomenon of the photoelectric effect. In metals electrons are free to move and are therefore good conductors of electricity. In the photoelectric effect, it was observed that when light was incident on a metallic surface, electrons were ejected from the surface with some kinetic energy. This is not surprising as light has energy and, when this energy is imparted to electrons inside the metal, they acquire sufficient kinetic energy and are released. There was however a result that could not be explained with this simple common-sense approach. It was observed that, for a given metal, there exists a certain minimum frequency of incident radiation below which no photoelectrons are emitted no matter how intense the radiation field is. This frequency is called the *threshold frequency*. Einstein invoked Planck's hypothesis to explain this effect by arguing that light should come in packets or quanta of energy and this energy should be proportional to the frequency. Electrons are emitted if the energy of the quantum of light, which has come to be known as a photon, is higher than a critical value.

The third problem relates to the atomic structure of hydrogen. It was known through the work of Rutherford that an atom consists of a positively charged nucleus with electrons orbiting around it. What was also known was that hydrogen atoms produced some discrete spectral lines corresponding to light energy radiated at some well specified frequencies. How to reconcile the Rutherford model with the observation of the spectral lines? This was a challenge around 1913 when Bohr applied a quantization condition to explain this curious effect. Bohr's condition implied that the electron can exist only in some well-defined orbits whose radii can be obtained from the quantization condition. The spectral distribution of the radiation emitted can then be explained by assuming that, when an atom in a higher orbit jumps to a lower orbit, a photon of discrete frequency is emitted. Bohr could show that the frequencies predicted based on the quantization condition matched the experimentally observed frequencies.

Thus we see that the theme of quantization of energy, so alien to classical mechanics, is fundamental to the quantum mechanical description. As we discuss in the final chapter, when we solve the Schrödinger equation for the simplest of all problems, a particle such as an electron moving freely inside a box, can lead us to the surprising result that the particle can only have discrete amount of energies. This is in sharp contrast to our everyday observance that, in the macroscopic world, particles can have any amount of energy.

5.2 Wave–Particle Duality

In classical mechanics, particles and waves are distinguishable phenomena, with different, characteristic properties and behaviors. Particles are massive objects that can occupy well-defined positions and can move with well-defined velocities. They can carry momentum and can collide with each other (Chapter 3). We cannot associate any of these properties with waves.

Waves are described in terms of wavelength and frequency and lead to phenomena such as interference and diffraction (Chapter 4).

In quantum mechanics, such distinction is done away with: A truly mind-boggling result is that both light and matter display dual behaviors, i.e., in some experiments they behave like waves and in others they behave like particles. For example, interference and diffraction phenomena can only be explained by treating light as a wave. These phenomena cannot be explained by treating light as consisting of particles. On the other hand, Einstein explained the photoelectric effect in 1905 by treating light as composed of quanta or photons. The wave picture of light fails to explain how light of frequency below a critical level, no matter how intense, cannot eject an electron from the metallic surface. In 1924, the other side of wave–particle duality was argued by de Broglie when he showed that a massive object can behave as a particle with momentum as well as a wave with an associated wavelength, which is usually referred to as the de Broglie wavelength. Therefore a particle can act both as a particle and a wave.

A mysterious consequence of quantum mechanics is that the choice whether light behaves as a wave or as a particle depends on the experimental set-up. The quintessential experiment that demonstrates wave–particle duality is Young's double-slit interference experiment. When a single photon goes through the slits, it registers as a point-like event on the screen (measured say by a CCD array). An accumulation of such events over repeated trials builds up a probabilistic fringe pattern that is characteristic of wave interference. However, if we arrange to measure which slit the photon goes through, the interference always disappears and the pattern on the screen is no different to that if massive particles are incident on the screen. Thus the photons exhibit particle-like behaviour.

The counterintuitive aspect of wave–particle duality is epitomized in the problem of the quantum eraser, as was shown by Marlan Scully and Kai Drühl in 1982. The inability to discern which-path information, or the indistinguishability of interfering pathways, in the double-slit experiment is the key to preserving the wave properties of the photon and the appearance of fringes on the screen. What if, rather than subject the photon to a classical measurement, we can have it interact quantum mechanically with a localized marker particle (such as an atom) and leave behind a record of its path? Whether the interference pattern then survives or not depends on the marker states, which carry the tell-tale information about which path the photon took to the detector. The coherence is destroyed as soon as we have the which-path information. One then wonders whether it might not be possible to retrieve the coherence, and the fringes, by destroying the which-path information contained in the marker—long after the photon is detected on the screen. This is the essence of the quantum eraser idea. We discuss these ideas in detail in Chapter 9 but this brief discussion helps to glean the counterintuitive nature of quantum mechanics.

5.3 End of Certainty–Probabilistic Description

Perhaps the most celebrated and discussed aspect of quantum mechanics is the probabilistic nature of its predictions. As we have seen in Chapter 3, particle trajectory is a central concept in classical physics. If a particle moves under the action of known forces then the position and the velocity of the particle can be simultaneously measured at any time with arbitrary accuracy. The motion of the particle is therefore fully deterministic and follows a known trajectory.

In quantum mechanics, a "particle" (e.g., an electron) does not follow a definite trajectory. The complete description of a particle is contained in the "wave function" $\psi(\mathbf{r})$ which represents the spatial distribution of a "particle". As an example, electrons in an atom are described by wave functions centered on the nucleus. The wave function $\psi(\mathbf{r})$ is a function of the coordinates defining the position of the classical particle. The wave function contains all the information about the system. If we know ψ, we can determine any observable property (e.g., energy, momentum, ...) of the system. Quantum mechanics, through the Schrödinger equation, provides the tools to determine ψ computationally and then to use ψ to determine properties of the system.

Immediately after Schrödinger derived the equation for $\psi(\mathbf{r})$ in 1926 and correctly derived the energy levels of the hydrogen atom, the debate ensued about the meaning of the wave function. How do we interpret the wave function $\psi(\mathbf{r})$? This was one of the most important problems facing physicists immediately after Schrödinger formulated his equation for the wave function. The breakthrough came through the work of Max Born who presented the probabilistic interpretation of $\psi(\mathbf{r})$, according to which the modulus square of the wave function, $|\psi(r)|^2$, at any point in space is proportional to the probability of finding the particle at that point. The wave function ψ itself has no physical meaning. This amounts to the end of the era of certainty that we associate with the predictions in classical mechanics. So whereas, on the basis of Newtonian mechanics, we can predict with certainty where a tennis ball will land on a wall if we know the initial velocity and the acceleration due to gravity, we have no such prediction for an electron. If an electron is accelerated through an electric potential, quantum mechanics does not allow us to describe the motion of the electron in terms of a trajectory and does not tell us where the electron will hit the screen. Quantum mechanics only gives us the *probability* that the electron will hit a certain point on the screen.

This end of certainty and the probabilistic nature of quantum mechanics has been truly exasperating to many, including Einstein who, in a famous quote, said "God does not play dice".

5.4 Heisenberg Uncertainty Relations and Bohr's Principle of Complementarity

An important consequence of quantum mechanics is the Heisenberg uncertainty relations. According to these relations, there are complementary pairs of variables such that if we know one of them very precisely then the other becomes uncertain. One such pair of observables is position and momentum. According to Heisenberg, "it is impossible to specify simultaneously, with precision, both the momentum and the position of a particle" no matter how precise our measurement devices are. This is again in sharp contrast to Newtonian mechanics where we can measure any observable property as precisely as we like—the only restriction comes from the limitation of our measurement devices. Mathematically, the Heisenberg uncertainty relation for position and momentum can be written as

$$\Delta x \Delta p \geq \frac{h}{4\pi}, \tag{5.1}$$

where Δx is the uncertainty in the location of the particle, Δp is the uncertainty in the momentum, and h is a constant called Planck's constant. Thus if we know the position, x, of a particle exactly, we are completely uncertain about its momentum, p_x, and if we know the momentum, p_x, exactly, we are completely uncertain about the position, x. This is consistent with the earlier assertion that there is no concept of a particle trajectory $\{x(t), p_x(t)\}$ in quantum mechanics. Since Planck's constant is very small, the uncertainty relation is valid for small particles. For macroscopic objects, Δx and Δp_x can be very small when compared with x and p_x. We can therefore define a trajectory to a very good approximation.

Similar uncertainty relations exist between energy E and time t, i.e.,

$$\Delta E \Delta t \geq \frac{h}{4\pi} \tag{5.2}$$

and between angular momentum L and angular displacement θ, i.e.,

$$\Delta L \Delta \theta \geq \frac{h}{4\pi}. \tag{5.3}$$

Next we address the question: What is the origin of the Heisenberg uncertainty relation? We answer this question in detail in Section 7.3. Here we only mention that the process of measurement is an inherent source of uncertainty. According to Heisenberg (in his 1927 paper):

> If one wants to be clear about what is meant by "position of an object", for example of an electron..., then one has to specify definite experiments by which the "position of an electron" can be measured; otherwise this term has no meaning at all.

If, for example, we want to measure the position of an electron we have to come up with a device like a microscope to precisely measure the location of the electron. This can be done by first shining light on the electron and measuring the position of the electron by looking at the scattered light. However, as discussed above, a consequence of the wave–particle duality is that a photon has momentum that can be imparted to the electron during the process of observation. This disturbance is the source of uncertainty in both position and momentum of the electron. A careful analysis, as we do in Section 7.3, leads to an uncertainty relation of the form Eq. (5.1). No one has come up with an experimental arrangement where the position and momentum of a particle can be simultaneously determined with precision that violates this inequality.

When Heisenberg formulated his uncertainty relations, Niels Bohr formulated the principle of complementarity. According to this principle, two observables are *complementary* if precise knowledge of one of them implies that all possible outcomes of measuring the other one are equally probable.

We have seen above that light and electrons can behave both as particle or wave in different experiments. For example, light can behave like a wave in interference and diffraction experiments but behaves like a particle (quantum of energy) in experiments such as photoelectric emission. According to the principle of complementarity, light (and electrons) can behave either as a wave or a particle but never both in a given experiment.

The uncertainty relation and complementarity appear to be related concepts. However, they have been shown to be quite different from each other.

5.5 Coherent Superposition and Quantum Entanglement

In classical mechanics, the system is always in a well-defined state—an object can be either here or there. There are, of course, instances when we do not have full knowledge whether the system is in one state or another. For example, a door is open 36% of the time and closed 64% of the time. If we approach the door, we do not know with certainty whether we shall find it open or closed, we only know the probability that we shall find it open or closed. However, when we get to the door we find it either open or closed. More importantly we can infer the past: If we find the door open, we can conclude that the door was open before we got there. This is the essence of classical mechanics.

The quantum mechanical description is however quite different. A quantum mechanical system is, as we discussed above, described by a wave function ψ. The wave function can however be in a coherent superposition of states:

$$\Psi = c_1\psi_1 + c_2\psi_2 + \cdots \tag{5.4}$$

Here $c_1, c_2, \cdots$ are complex numbers and $\psi_1, \psi_2, \cdots$ are possible states of the system. The system therefore exists simultaneously in all the possible states $\psi_1, \psi_2, \cdots$. This description is quite different from the classical description. However, when we make a measurement, the system is found in only one of the states ψ_i with probability $|c_i|^2$. Moreover, if the system is found in a certain state, say ψ_1, we cannot conclude that, even before we made the measurement, it was in the state ψ_1.

We illustrate this strange behavior with an example. Without going into any details, we mention that an electron has a property called "spin". An electron can either be in a state with spin up, designated by $\psi_\uparrow$, or in a state with spin down, $\psi_\downarrow$. Let us assume that, for a given electron, the state of the electron is given by

$$\Psi = c_\uparrow\psi_\uparrow + c_\downarrow\psi_\downarrow \tag{5.5}$$

where $c_\uparrow = 0.6$ and $c_\downarrow = 0.8$. Thus the probability of finding the electron with spin up (in the state $\psi_\uparrow$) is $|c_i|^2 = 0.36$ and with spin down (in state $\psi_\downarrow$) is $|c_i|^2 = 0.64$. We can now do an experiment to find whether the spin is up or it is down by passing it through an inhomogeneous magnetic field pointing in the z-direction. If the spin is up, it will be deflected in the upward $+z$-direction and if the spin is down, it will be deflected in the downward $-z$-direction, as shown in Fig. 5.1. This is an actual experiment that was first done by Otto Stern and Walter Gerlach in 1922. According to quantum mechanics we cannot say anything in advance about whether the electron will be deflected in $+z$-direction or in the $-z$-direction. However, when the experiment is done, the electron is deflected in the $+z$-direction OR in the $-z$-direction. Let us suppose that it is deflected in the $+z$-direction and we conclude that the electron spin is in state $\psi_\uparrow$.

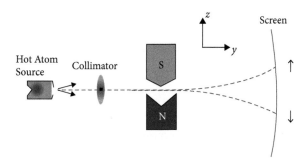

Fig. 5.1 Schematics of Stern–Gerlach experiment. An electron is deflected by an inhomogeneous magnetic field in the "up" direction or in the "down" direction depending on whether it is in the state $\psi_\uparrow$ or in the state $\psi_\downarrow$

The situation so far appears to be completely analogous to our example of "door open" and "door closed". There is, however, a major difference. Let this same electron that is found in state $\psi_\uparrow$ pass through the inhomogeneous magnetic field pointing in the x-direction. Then we get the deflection of the electron in the $+x$-direction or $-x$-direction. If the electron is deflected in the $+x$-direction, we conclude that the electron is in some state which is neither $\psi_\uparrow$ nor $\psi_\downarrow$. This is a bit surprising. However, the amazing result is that, if after this measurement on the electron with magnetic field oriented in the x-direction, we pass the same electron again through the magnetic field in the z-direction, there is a probability that the electron will be deflected in the downward direction, implying that it is in state $\psi_\downarrow$. How can that be? We already knew that the electron is in state $\psi_\uparrow$ and now we find the same electron in state $\psi_\downarrow$. This is very strange, to say the least. Can we assign a definite state $\psi_\uparrow$ or $\psi_\downarrow$ to the electron? The answer is no! Whether the state is $\psi_\uparrow$ or $\psi_\downarrow$ depends on the orientation of the apparatus or, more specifically, the measurement device. Can we say that the electron existed in the state $\psi_\uparrow$ before we made the first measurement? Again the answer is no! This counterintuitive behavior which has no analogue in classical physics has been a matter of great discussion and debate since the earliest days of quantum mechanics. We shall discuss this issue in more detail in later chapters, particularly Chapter 12.

Another highly counterintuitive feature of quantum mechanics is that two or more objects can form an "entangled" state. In our classical world, if we have two objects such as two balls that are placed a large distance apart, they are strictly independent of each other. No matter what we do to the one, it cannot influence the other. Quantum mechanically we can form states for the two objects such that, even when the two objects are very far from each other, what we do to one object can influence the state of the other.

Consider two objects described by wave functions ψ_1 and ψ_2. Classically, the total state of the system is

$$\Psi = \psi_1(r_1)\psi_2(r_2) \text{ or } \psi_1(r_2)\psi_2(r_1)$$

i.e., the two objects are independent of each other. Quantum mechanically, we can have an entangled state:

$$\Psi = \frac{1}{\sqrt{2}}(\psi_1(r_1)\psi_2(r_2) + \psi_1(r_2)\psi_2(r_1)) \tag{5.6}$$

This is a "coherent superposition" for the two objects. The total system can no longer be factorized—and is entangled. Such a state has no analog in classical mechanics. Both coherent superposition and quantum entanglement lie at the heart of quantum computing.

Problems

5.1 A quantum system is found in the superposition of states ψ_1 and ψ_2 as

$$\Psi = c_1\psi_1 + c_2\psi_2$$

If $c_1 = 0.7 + 0.5i$, what are the probabilities of finding the system in states ψ_1 and ψ_2? (Please note that the conservation of probabilities leads to $|c_1|^2 + |c_2|^2 = 1$.)

5.2 The Rydberg formula for the hydrogen atom is,

$$\frac{1}{\lambda} = R_H\left(\frac{1}{m^2} - \frac{1}{n^2}\right)$$

where $R_H = 1.097 \times 10^7$ m^{-1}. For $m = 2$ (Balmer series), find the frequency of the first four wavelengths corresponding to $n = 3, 4, 5, 6$.

5.3 If the position of an electron is measured to an accuracy of $\Delta x = 10^{-10}$ m, what is the minimum uncertainty in its momentum and velocity? (Mass of an electron, $m_e = 9.1 \times 10^{-31}$ kg, Planck's constant $h = 6.626 \times 10^{-34}$ J $\cdot$ s.)

5.4 Consider a tennis ball of mass 100 g. If its position is known to an accuracy of $\Delta x = 0.01$ mm, what is the minimum uncertainty of the speed with which it is moving?

 BIBLIOGRAPHY

J. Polkinghorne, *Quantum Theory: A Very Short Introduction* (Oxford University Press 2002).

K. A. Peacock, *The Quantum Revolution: A Historical Perspective* (Greenwood Press 2008).

H. F. Hameka, *Quantum Mechanics: A Conceptual Approach* (John Wiley 2004).

Birth of Quantum Mechanics—Planck, Einstein, Bohr

As we mentioned in Chapter 1, the basic laws of physics related to mechanics, thermodynamics, electromagnetism, and light were firmly in place at the end of the nineteenth century. There was a feeling among scientific circles that nothing new was expected as far as the foundation of physics is concerned. No one could imagine that physics stood at the threshold of a revolution that would replace the existing laws that had been formulated as far back as Isaac Newton with laws that would be highly counterintuitive and transformational. In particular there were certain experimental observations that could not be explained by the existing laws referred to as the classical laws of physics. These observations related to different areas of physics, but they had one point in common—they all involved light. A resolution of these observations would lead to the birth of quantum mechanics.

In this chapter, we first present a brief history of light. We then describe the unresolved phenomena of blackbody radiation, the photoelectric effect, and the spectrum of light emitted by the hydrogen atom. Next we discuss the novel and revolutionary postulates regarding quantization of energy aimed at resolving those problems by Max Planck, Albert Einstein, and Niels Bohr, respectively.

6.1 Brief History of Light

The nature of light has been a subject of interest going back to antiquity. Until around the seventeenth century, studies of light were mainly concerned with vision. The Greeks were among the first to address the question: How do we perceive objects? Plato, Euclid, Ptolemy, and their followers believed that light consisted of rays emitted by the eyes. The striking of the rays on the object allows the viewer to perceive things such as the color, shape, and size of the object. Our vision is initiated by our eyes reaching out to "touch" or feel something at a distance. This is the essence of the extramission theory of light.

The extramission theory remained influential for almost a thousand years until Alhazen conclusively proved it to be wrong in the beginning of eleventh century. Alhazen, a Persian scientist, proved, that, unlike the conventional theory of vision, light originated, not from the eye but from the illuminated objects. He did this by carrying out a simple experiment in a dark room where light was sent through a hole by two lanterns held at different heights outside the room. He could then see two spots on the wall corresponding to the light rays that originated from each lantern passing through the hole onto the wall. When he covered one lantern, the bright spot corresponding to that lantern disappeared. He thus concluded that light does not emanate from the human eye, but is emitted by objects such as lanterns and travels from these

Quantum Mechanics for Beginners: With Applications to Quantum Communication and Quantum Computing. M. Suhail Zubairy.
© M. Suhail Zubairy 2020. Published in 2020 by Oxford University Press. DOI: 10.1093/oso/9780198854227.001.0001

objects in straight lines. Based on these experiments, he invented the first pinhole camera (that Kepler would use and call a *camera obscura* in the seventeenth century) and explained why the image in a pinhole camera was upside down.

The first studies on the nature of light were carried out in the seventeenth century through the works of Rene Descartes (1596–1650), Isaac Newton (1642–1727), and Christian Huygens (1629–1695). Descartes' main contribution to optics is his book *Dioptrics* that was published in 1637. It deals with many topics relating to the nature of light and the laws of optics.

Newton and Huygens were contemporaries but they came up with very different views on the nature of light. Newton advocated a corpuscular theory of light in his classic and influential book *Opticks* (published in 1704). According to him, light is made up of extremely small particles or corpuscles, whereas ordinary matter was made of larger particles or corpuscles. He speculated that through a kind of alchemical transmutation they change into one another. It is surprising that Newton advocated the corpuscular theory of light when there was evidence that supported wave behavior. For example, Francesco Grimaldi (1618–1663) made the first observation of the phenomenon of diffraction of light. He showed through experimentation that when light passed through a hole, it did not follow a rectilinear path as would be expected if it consisted of particles but took on the shape of a cone. Newton tried to explain the phenomenon of diffraction using questionable assumptions.

While Newton was advocating a corpuscular nature of light, his contemporary, Huygens, suggested a wave picture of light. He considered light waves propagating through the ether just as sound waves propagate through air or water waves propagate through a lake. Light waves, according to Huygens, were thus longitudinal waves (like sound waves), as opposed to the later studies by Fresnel and Maxwell that showed light to consist of transverse waves. Huygens formulated a principle (that now bears his name) which describes wave propagation. Waves consist of wave fronts which are surfaces on a wave with the same phase and amplitude. Huygens' principle provides a geometric construction for determining at some instant the position of a new wave front from the knowledge of the wave front that preceded it. When applied to the propagation of light waves, this principle states that: Every point on a wave front may be considered a source of secondary spherical wavelets which spread out in the forward direction at the speed of light (Fig. 6.1). The new wave front is the tangential surface to all of these secondary wavelets.

Newton's status as a scientist was so great, particularly in the British Isles, that few dared to challenge his corpuscular theory of light. The situation continued for almost 100 years, until, in 1802, Thomas Young conclusively demonstrated the wave nature of light through his double-slit experiment. Young's double-slit experiment was however regarded highly controversial and counterintuitive in his own time.

As discussed in Chapter 4, when the two slits in Young's double-slit experiment are open then there are regions of bright spots where there is constructive interference from light coming from the two point sources as well as the dark spots where, due to destructive interference, no light is present. The dark spots remain dark no matter how intense is the incident light field. Now if one of the slits is covered, then only half the light is incident on the screen. However, in this case, the dark spots are no longer dark. In Young's time, when the principle of interference was not understood, this appearance of brightness when the total light incident is only half was very mysterious. How can a screen uniformly illuminated by a single aperture develop

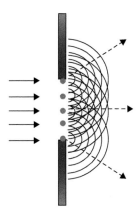

Fig. 6.1 According to Huygens' wave theory, every point on a wave front is a source of a secondary wave in the forward direction.

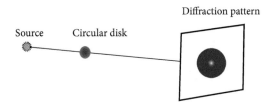

Fig. 6.2 Poisson spot: Light incident on an opaque circular disk forms a bright spot along the axis as a result of Fresnel diffraction.

dark fringes with the introduction of a second aperture? And how could the addition of *more* light result in *less* illumination?

Young's theory would eventually find broad acceptance, particularly through the works of Fresnel in France. Augustin Jean Fresnel (1788–1827), a contemporary of Young, championed the wave nature of light based on his own work on diffraction.

An episode indicates the stunning success of the wave nature of light as formulated by Fresnel. In 1819, Fresnel presented his work on wave theory of diffraction in a competition by the French Academy of Sciences. The committee of judges, headed by Francois Arago, included Jean-Baptiste Biot, Pierre-Simon Laplace, and Simeon-Denis Poisson. They were all prominent advocates of Newton's corpuscular theory and were not well disposed to the wave theory of light. Poisson was however impressed by Fresnel's submission and extended his calculations to come up with an interesting consequence: "*Let parallel light impinge on an opaque disk, the surrounding being perfectly transparent. The disk casts a shadow—of course—but the very center of the shadow will be bright. Succinctly, there is no darkness anywhere along the central perpendicular behind an opaque disk (except immediately behind the disk)*". According to the corpuscular theory, there could be no bright spot behind the disk. As Chair of the Committee, Arago asked Fresnel to verify Poisson's prediction and amazingly Fresnel found the bright spot as predicted. This discovery was an impressive vindication of the wave theory and Fresnel won the competition. This spot is now known as the "Poisson spot" (Fig. 6.2).

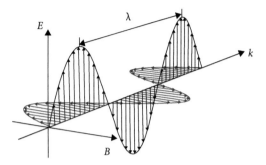

Fig. 6.3 Light is an electromagnetic wave with the mutually perpendicular electric and magnetic fields oscillating in the directions normal to the direction of propagation.

It was left to James Clerk Maxwell (1831–1879) to complete the classical picture of light as consisting of electric and magnetic waves. This was a truly remarkable outcome of his efforts to unify the two known forces of nature: electric force and magnetic force. It was known through the work of Michael Faraday that a change of magnetic field yielded an electric force. The insight due to Maxwell was that if electricity and magnetism were the two sides of the same coin then a change of electric field should similarly result in a magnetic field. This motivated him to add a term in Ampere's law that corresponded to a time rate of change of the electric field. This addition immediately yielded a wave equation for an electromagnetic wave propagating at the same velocity as known for light, 3×10^8 m/sec. The picture of light that emerged was thus that of undulations of mutually perpendicular electric and magnetic fields propagating. The direction of propagation was perpendicular to both the electric and the magnetic fields. Maxwell's results were published in 1865. The electromagnetic waves of Maxwell were shown to be transverse waves in line with Young and Fresnel as opposed to the picture adopted by Huygens where light was seen as a longitudinal wave propagating through the medium *ether*. The description of light as an electromagnetic wave (Fig. 6.3) was experimentally demonstrated by Heinrich Hertz (1857–1894) in 1888.

At the end of the nineteenth century, most phenomena were understood on the basis of the classical theory of Newton and Maxwell. However there were some phenomena involving light that could not be explained with the existing theories. One such phenomenon was the spectrum of light emitted by heated objects, to which we turn next.

6.2 Radiation Emitted by Heated Objects

A puzzle confronting physicists at the turn of the century (1900) was just how do heated bodies radiate? A solid consists of atoms and molecules, and heat causes them to vibrate. However, atoms and molecules are themselves complicated patterns of electrical charges. Oscillating charges emit electromagnetic radiation. This radiation travels at the speed of light and from this we realize that light itself, and the closely related infrared heat radiation, are actually electromagnetic waves. The picture, then, is that when a body is heated, the consequent vibrations on molecular and atomic scales inevitably induce charge oscillations. These oscillating charges radiate, giving off the heat and light that is observed.

In 1859, Kirchhoff addressed this problem. Central to Kirchhoff's studies was the concept of a *blackbody*, an object that absorbs all of the electromagnetic field that falls on it. In practice there is no object that is ideally a blackbody, but many objects in the real world come close to exhibiting blackbody behavior. A perfect blackbody can also emit radiation with a spectral distribution at a given temperature. A spectral distribution means a curve of intensity as a function of frequency or wavelength. Kirchhoff proved by general thermodynamic arguments that the spectrum of the emitted radiation from a blackbody depends only on its temperature and is independent of the material. The spectral distribution function $J(\nu, T)$ therefore depends only on the emitted frequency ν and the temperature T. Kirchhoff posed it as a challenge to find an explicit expression for this function. A search of the function $J(\nu, T)$ would lead to the birth of quantum mechanics, literally at the end of the nineteenth century, in December 1899, by the German physicist Max Planck.

In order to understand the radiation emitted by heated bodies, we notice that a near-blackbody at room temperature emits radiation that peaks at near infrared, which we call heat radiation. As we increase the temperature, the peak shifts to higher frequencies (or lower wavelengths). Around 550 °C an object like an iron rod begins to glow red and at a much higher temperature, around 10 000 °C, the peak of the emitted light moves to higher frequency blue light (see the spectrum in Fig. 6.4). The shifting of the emitted radiation peak to higher frequencies as the temperature increases is described by Wien's displacement law, named after Wilhelm Wien (1864–1928). According to Wien's displacement law, the peak frequency ν_{max} is directly proportional to the temperature T (in Kelvin), i.e.,

$$\nu_{max} = \alpha T, \tag{6.1}$$

where $\alpha = 2\pi \times 5.9 \times 10^{10} \mathrm{K}^{-1} \cdot \mathrm{s}^{-1}$ is the proportionality constant. The peak wavelength is inversely proportional to the temperature, i.e.,

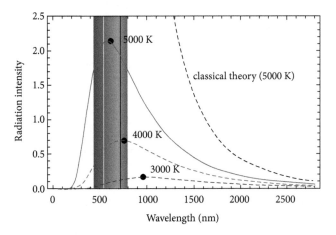

Fig. 6.4 The spectral distribution $J(\nu, T)$ as a function of wavelength. The emitted radiation peaks at shorter wavelength (higher frequency) with increasing temperature. According to classical theory the emitted radiation approaches infinity as the wavelength decreases leading to "ultraviolet catastrophe".

$$\lambda_{max} = \frac{\beta}{T},$$
(6.2)

where $\beta = 2.90 \times 10^{-3}$ m $\cdot$ K is a constant.

Using the laws of classical physics, Rayleigh and Jeans developed a theory of blackbody radiation and derived the following expression for the spectral distribution function $J(\nu, T)$:

$$J(\nu, T) = \frac{8\pi\nu^2}{c^3} k_B T.$$
(6.3)

Here $c = 3 \times 10^8$ m/s is the speed of light in vacuum and $k_B = 1.38 \times 10^{-23}$ m^2 kg s^{-2} K^{-1} is the so-called Boltzmann constant. This result agreed with the observed spectral distribution at low frequencies. However, this result predicted that the emitted intensity should approach infinity at very large frequencies. Experimentally it was observed that the intensity actually drops to zero at large frequencies. There was therefore no agreement between theory and experiment in the ultraviolet region of the blackbody spectrum.

By 1900, this failure, known as the Rayleigh–Jeans ultraviolet catastrophe, had caused people to question the basic concepts of classical physics and thermodynamics. It was, however, Max Planck (1858–1947) who eventually presented the radiation formula that matched the experimentally observed blackbody radiation spectra for the entire range of the frequency spectrum. Planck presented his results that would eventually revolutionize our understanding of the laws of nature.

When Planck addressed the problem of blackbody radiation, he realized that, since the results were independent of the nature of the material in the cavity, one could use a simple model for the cavity. He chose to consider a damped harmonic oscillator as a model for the material in the walls. Central to Planck's derivation of the blackbody formula was the assumption that the total energy of the oscillators was made up of finite energy elements, and each element had an energy E that is equal to $n\hbar\nu$. Here n is an integer and $\hbar$ is a constant that eventually carried Planck's name and is called Planck's constant. Its value is 1.055×10^{-34} J·s. With this quantization condition, Planck derived the spectral distribution function

$$J(\nu, T) = \frac{8\pi\nu^2}{c^3} \frac{\hbar\nu}{e^{\hbar\nu/k_B T} - 1}.$$
(6.4)

Planck's equation gave excellent agreement with the experimental observations for all temperatures.

It is important to realize that the Planck relation

$$E = n\hbar\nu,$$
(6.5)

for integer values of n, is a significant departure from classical thought in two ways. First, it postulates that energy is proportional to frequency, not intensity, as would be expected for a classical oscillator. Second, for a given frequency, ν, the energy is quantized, i.e., it comes in units of $\hbar\nu$. Planck would later describe it as "an act of desperation" to get the correct expression for the Kirchhoff function that agreed with experiments. At the time he proposed his radical hypothesis, Planck could not explain why energies should be quantized. However his hypothesis solved the long-standing problem of explaining the blackbody radiation spectrum with amazing success.

Central to Planck's quantization of energy is Planck's constant. In this book we use two different notations for the Planck's constant h and $\hbar$ (hbar). The relation between the two is

$$\hbar = \frac{h}{2\pi}. \tag{6.6}$$

Planck's constant is one of the most important constants in physics and THE most important constant in quantum mechanics. It is related to the quantization of energy as we saw above. Its small magnitude ($h = 6.626 \times 10^{-34} \text{J} \cdot \text{s}, \hbar = 1.055 \times 10^{-34} \text{J} \cdot \text{s}$) is responsible for the fact that we do not see quantum mechanical effects in our everyday life and quantum effects are mostly confined to microscopic scales.

For example, a harmonic oscillator of frequency $\nu = 540 \times 10^{12}$ Hz (corresponding to green light) has an energy $E = \hbar\nu = 1.055 \times 10^{-34} \times 540 \times 10^{12}$ J $= 5.69 \times 10^{-19}$J. This is a very small amount of energy from our everyday experience.

Planck's constant h or $\hbar$ has dimensions of angular momentum:

$$\hbar = \frac{E}{\nu} \propto \frac{mv^2}{\nu} \propto mvvT \propto px \propto L. \tag{6.7}$$

In order to see how quantum effects can be observable only for very small particles, such as atoms and electrons, we calculate the angular momentum in different situations and compare them with the value of Planck's constant.

As our first example, we consider a small fly flying slowly in tight circles. Suppose its mass is $m = 0.01$ gram $= 0.00001$ kg, and v $= 10$ cm/s $= 0.1$ m/s, and the circle radius is $r = 1$ cm $= 0.01$ m. The angular momentum of the fly is then

$$L_{fly} = mvr = 10^{-5} \times 10^{-1} \times 10^{-2} \text{J} \cdot \text{s} = 10^{-8} \text{ J} \cdot \text{s}$$

Thus the tiny fly has an angular momentum that is about 26 orders of magnitude greater than Planck's constant $h = 6.626 \times 10^{-34}$ J $\cdot$ s.

Next we consider an electron (of mass $m = 10^{-30}$ kg) orbiting around a nucleus in a radius of 10 Angstrom (10 Angstrom $= 10^{-9}$ m). Then the angular momentum is given by

$$L_e = mvr = 10^{-30} \times \text{v} \times 10^{-9} \text{J} \cdot \text{s} = \text{v} \times 10^{-39} \text{ J} \cdot \text{s}$$

This is equal to h for v $\approx 6.6 \times 10^5$ m/s (much less than the velocity of light $c = 3 \times 10^8$ m/s). Planck's constant h is about right for electrons!!!!

Planck's hypothesis to explain the blackbody spectrum was proposed in 1899. It was a revolutionary idea that energy of a harmonic oscillator should be quantized in units of $\hbar\nu$. However it was not perceived as such at the time it was proposed. When Planck proposed his theory of blackbody radiation, there was no dancing on the streets, no major headlines in the newspapers. Even the scientific community at large did not grasp the significance of the quantization condition.

For almost 5 years, Planck's hypothesis could not find any application until Albert Einstein used the quantum condition $E = \hbar\nu$ to explain the photoelectric effect in his well-known Nobel Prize winning paper of 1905. Planck's derivation for the blackbody spectrum was based on the quantization of the harmonic oscillator that modelled the material of the cavity and not the radiation itself. However, it would have far-reaching consequences for the ultimate description of the nature of light through the work of Albert Einstein and others.

6.3 **Einstein and the Photoelectric Effect**

In the 1890s, Heinrich Hertz (and later Philipp Lenard) observed that when light of frequency v is incident on a metallic surface, electrons with kinetic energy T_e are ejected from the surface. The model that was used to understand this phenomenon was that electrons are part of the atom (we present the model of the atom known at the turn of the twentieth century in the next section) and if they are provided with sufficient energy, known as work function Φ for the given metal, these electrons, called photoelectrons, will be released and leave the metal with a kinetic energy T_e that, according to the law of conservation of energy, will be the difference between the incident energy E_i and the work function, i.e.,

$$T_e = E_i - \Phi. \tag{6.8}$$

It was observed that, below a certain critical frequency v_c, no photoelectrons are emitted, no matter how intense the light field (Fig. 6.5). This critical frequency depended on the metal. This frequency v_c is called the *threshold frequency*.

It was also observed that, if the frequency of the incident beam is increased, the maximum kinetic energy of the photoelectrons emitted is also increased if we keep the incident intensity of light fixed.

Another interesting observed effect was that the emission of photoelectrons takes place almost instantaneously after the light shines on the metal, with no detectable time delay even if a very low intensity of light is incident.

Further the kinetic energy of each photoelectron remains constant even if the intensity of the incident radiation of a given frequency is increased.

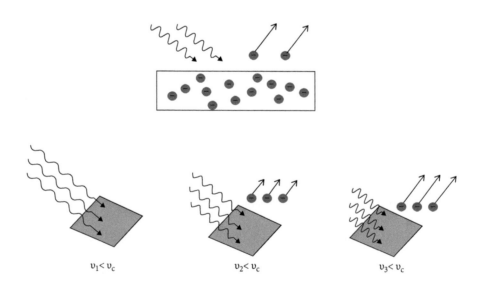

$v_1 < v_c$ $\qquad\qquad v_2 < v_c$ $\qquad\qquad v_3 < v_c$

Fig. 6.5 Photoelectric effect: When light is incident on a metal, photoelectrons are emitted. No electrons are emitted if the frequency of light is below a critical frequency. At the critical frequency, the electron emission starts. As the frequency of light is increased further, the kinetic energy of the emitted electrons is also increased.

These observations were extremely surprising and could not be explained on the basis of classical laws of physics known at the end of the nineteenth century. For example, how could it be that photoelectrons are not emitted below a critical frequency of light even if it has a large intensity but then photoelectrons are emitted even with a feeble light beam above the critical frequency? The idea that the kinetic energy of electrons did not depend on intensity, but rather upon the frequency has no explanation within the classical theory of light and matter. And finally the most mysterious effect was the instantaneous emission of photoelectrons even when a very weak beam of light was incident on the metal.

Classical mechanical laws implied that it would take a substantially long time for the incident field to pile up enough energy to eject a photoelectron. To illustrate this point we consider an example, in which light of intensity I is incident on potassium. The work function of potassium is $\Phi = 2.22 \ eV$ (equal to $2.22 \times 1.6 \times 10^{-19} J = 3.55 \times 10^{-19} J$). The incident energy over a time τ should be at least equal to the work function Φ for the electron to be ejected. Therefore

$$I \times \pi r^2 \times \tau \geq \Phi. \tag{6.9}$$

Here r is the radius of the atom which, as we shall see in Section 6.5, is about the order of 1 Angstrom (10^{-10} m). For a beam of light of intensity, $I = 10^{-2} W/m^2$, the minimum time required is

$$\tau = \frac{\Phi}{I \times \pi r^2} = \frac{3.55 \times 10^{-19}}{10^{-2} \times \pi \times 10^{-20}} s = 1.13 \times 10^3 s \approx 19 \text{ minutes}$$

Thus the observed instantaneous emission of electrons cannot be explained using an argument based on classical physics.

In 1905, Einstein explained the photoelectric effect using Planck's hypothesis of quanta of energy. Einstein assumed that light consists of quanta, called photons, with each photon carrying an energy equal to $\hbar \nu$. When one of these photons penetrates the metal, it gives all its energy to the electron. If the energy $\hbar \nu$ is greater than the work function Φ of the metal, the electron is ejected with the kinetic energy T_e. The condition for the conservation of energy is satisfied by the relation:

$$\hbar \nu = \Phi + T_e. \tag{6.10}$$

A careful reflection on this equation shows that, in one stroke, it explains all the observed characteristics of the photoelectric effect. It explains that, for light of frequency below a critical value

$$\nu_c = \frac{\Phi}{\hbar}, \tag{6.11}$$

there is no photoelectron emission, no matter how intense the light. If the frequency is increased above ν_c, the kinetic energy of electrons increases, and above the critical frequency, the rate of emission is proportional to the intensity of incident light. This equation also explains the instantaneous emission of electrons when light is incident on the metal as only a single photon is required for the emission of an electron if the frequency is above the critical frequency ν_c.

Einstein's explanation of the photoelectric effect was the first vindication of Planck's hypothesis. It was the first time that light quanta were introduced. The idea that light consists of photons had a great impact on subsequent developments in the full formulation of quantum theory. However, the concept of a photon, on one hand, explained the photoelectric effect so beautifully, but, on the other hand, it could not explain the phenomena of interference and diffraction. This was a dilemma whose complete resolution, via a formal theory that would rigorously explain all these phenomena within the structure of a single theory, had to wait almost a quarter century—till the birth of quantum mechanics in the summer of 1925. How light knows when to behave as a wave, as in an interference experiment, and when to behave as a particle, as in the photoelectric effect, has however remained a perplexing question for over a hundred years. We shall discuss some counterintuitive aspects of the wave–particle duality in later chapters.

Soon we turn to the "third coming" of Planck's quantum hypothesis through the work of Niels Bohr on the hydrogen atom. But first a brief review of the history of the atom!

6.4 History of the Atom till the Dawn of the Twentieth Century

The history of the atom, like the history of light, goes back to antiquity. It starts around 450 BC when a Greek philosopher named Democritus wondered what would happen if an object is cut into smaller and smaller pieces. He thought that a point would be reached where the object could not be cut into still smaller pieces. He called these "uncuttable" pieces "*atomos*." This is where the modern term "atom" comes from. Democritus thought that atoms were infinite in number, uncreated, and eternal, and that the qualities of an object result from the kind of atoms that composed it.

Almost a hundred years later, the Greek philosopher, Aristotle, came up with his own idea of matter which was in contradiction with Democritus' concept of atoms. Aristotle believed that four elements—earth, air, fire, and water—made up everything. For example, a heavy substance such as iron and other metals were made up in large part of the element, earth, and in smaller parts, the other three elements. Similarly, lighter objects could be largely made up of lighter elements, air and fire, and a small amount of heavy elements, earth and water.

Aristotle's influence on our scientific thinking dominated for almost 2000 years. His thoughts about the four constituents of matter were accepted till almost the beginning of the scientific revolution in the seventeenth and eighteenth centuries. By that time, Democritus' ideas were more or less forgotten, but were revived around 1800 by a British chemist, John Dalton. On the basis of his studies on the pressure of gases, he concluded that the gases must consist of tiny particles, atoms, in constant motion. His main interest was in studying the properties of compounds. He concluded that a compound consists of the same elements in the same ratio. Another compound would be made up of different elements in different ratios. The main points of Dalton's atomic theory can be summarized as follows:

- All elements are made of extremely tiny particles called atoms. Atoms are the smallest particles of matter. They cannot be divided into smaller particles. They also cannot be subdivided, created, or destroyed.

- All atoms of the same element are identical in size, mass, and other properties; atoms of different elements differ in size, mass, and other properties.
- Atoms of different elements join together to form compounds. A given compound always consists of the same kinds of atoms in the same ratio.

Many aspects of Dalton's theory were correct and it became a widely accepted theory. However he was incorrect in assuming that atoms are the smallest particles and are indivisible. Dalton assumed that atoms are like solid spheres. This model had great difficulties in explaining how atoms can be joined together to make compounds. He thought that atoms could have holes and joined together using hooks. This was too simple a model with no experimental support. Dalton's model was shown to be incorrect when smaller particles like electrons were discovered through the work of J. J. Thomson in 1897, and it was realized that atoms have a much more complicated structure.

Thomson carried out experiments in which he applied a voltage between two metallic plates inside a vacuum tube (Fig. 6.6). He observed that an electric current flows between the two plates, traveling much further than we would expect for a current consisting of atom-size particles. His experiments suggested that the mass of these negatively charged particles should be about 1/1000 times those of a hydrogen atom. He also observed that the mass of these particles was the same regardless of the metal they came from. Thomson had discovered electrons. He also concluded that these particles cannot be the atoms but come from inside the atoms. Electrons are therefore subatomic particles. This was an important discovery.

Next question was how to incorporate the existence of tiny electrons inside the atom. Atoms are electrically neutral, so how could atoms contain negative charges and still be electrically neutral?

Thomson proposed a plum-pudding-type model of an atom in which a spherical atom is like a homogeneously positively charged pudding and electrons are embedded in it like plums. This

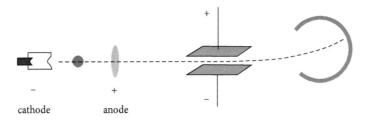

Fig. 6.6 Schematics of Thomson's experiment. A beam of electrons is deflected by an external electric field.

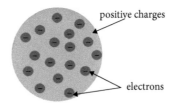

Fig. 6.7 Thomson's plum-pudding model. The atom consists of a sphere of uniform distributed positive charge and negatively charged electrons are embedded like plums in a pudding.

helped to explain the charge neutrality of an atom (Fig. 6.7). Thomson assumed that most of the mass of an atom was due to the positively charged sphere and electrons made only a small contribution.

This was the picture of the atom at the beginning of the twentieth century.

6.5 **The Rutherford Atom**

Ernest Rutherford, a physicist from New Zealand, made the next major discovery about atoms. He discovered the nucleus.

In 1899, Rutherford discovered that certain elements emitted positively charged particles. He called them alpha particles. In 1911, he carried out experiments in which a beam of alpha particles was incident on a very thin sheet of gold (Fig. 6.8). Outside the gold foil he placed an array of detectors of alpha particles. If Thomson's plum-pudding model was correct then most of the alpha particles should pass through the foil with very small deflection, the deflection caused by a repulsion due to the positively charged "pudding." The experimental results were dramatically different. It was observed that most of the alpha particles passed through the gold foil without any significant deflection. However, a few alpha particles were scattered at very large angles and some even scattered in the backward direction. It was as if the atom was mostly empty space through which the alpha particles pass through without any hindrance. But then there were points which sharply repulsed the alpha particles. This clearly showed that Thomson's model of the atom being a sea of positive charge with light electrons embedded in it was incorrect and a new model of atomic structure was required.

Based on the gold foil experiment, Rutherford proposed a new atomic model. His model for an atom was similar to the planetary model (Fig. 6.9). He proposed that most of the mass and the positive charge was concentrated in a small area at the center of the atom. He called this area the "nucleus." Negatively charged electrons revolved around the positively charged nucleus like planets revolve around the sun. Thus most of the atom consisted of almost empty space and most of the mass was concentrated in the small nucleus. This model could explain his experiment—the alpha particles could pass through the empty space without deflection and a few particles were repulsed by the massive, positively charged nucleus and scattered at very large angles, including in the backward direction.

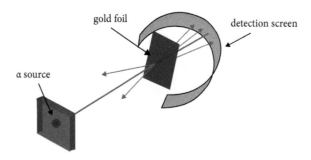

Fig. 6.8 Schematics of Rutherford's experiment. A beam of alpha particles is incident on a thin sheet of gold. An array of detectors detect alpha particles after they scatter from the gold foil.

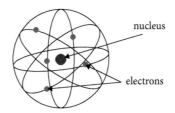

Fig. 6.9 Rutherford model of the atom: It consists of a massive positively charged nucleus surrounded by electrons in random orbits.

Rutherford also showed that the nucleus consisted of protons. These particles are almost 1000 times heavier than electrons but carry the same but opposite charge as electrons. The number of protons in the nucleus is equal to the number of electrons orbiting the nucleus in random orbits. Later it was discovered that, in addition to positively charged protons, there existed electrically neutral particles called neutrons inside the nucleus. The mass of these particles is almost the same as protons. Due to their neutral charge, neutrons were difficult to detect and their discovery had to wait till James Chadwick detected them in 1932.

Rutherford's picture of the atom could explain his gold foil experiments. However, this model was inadequate to explain some other experimental results, most notably the light emitted by various atoms.

6.6 The Hydrogen Spectrum

Atoms are extremely tiny objects with a typical size of 1 Angstrom or 10^{-10} m. It is therefore difficult to study directly the properties of atoms. In the nineteenth century, the tools to study their internal structure were very limited. One important source was the radiation emitted by the atoms. When an electric discharge was sent through a gas, light of different frequencies was emitted, yielding an emission spectrum as shown in Fig. 6.10. The emission spectra for different gases were different. In the late nineteenth century, it was recognized that emission spectra consisted of specific spectral lines.

In 1885, Johann Balmer was aware that the visible spectrum of light from hydrogen displays four wavelengths, 410 nm, 434 nm, 486 nm, and 656 nm. Balmer used this limited information and fitted them into an empirical formula:

$$\lambda = B\left(\frac{n^2}{n^2 - 2^2}\right). \tag{6.12}$$

Here $B = 364.50862$ nm and $n = 3, 4, 5, 6$. Later when he became aware of many other spectral lines (Fig. 6.11), he could see that many, if not all, satisfied this same equation.

In 1888, Johannes Rydberg generalized the Balmer formula for all the emitted spectral lines of hydrogen. The equation commonly used to calculate the Balmer series is a specific example of the Rydberg formula:

$$\frac{1}{\lambda_{nm}} = R_H\left(\frac{1}{m^2} - \frac{1}{n^2}\right), \tag{6.13}$$

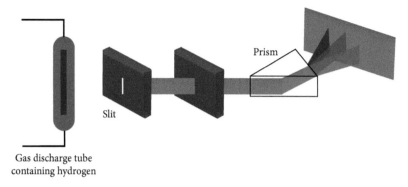

Fig. 6.10 Schematics for observing the emission spectrum of hydrogen. A gas discharge tube containing hydrogen emits radiation which is first collimated by narrow slits and then passed through a prism which deflects light of different frequencies in different directions.

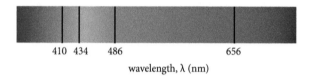

Fig. 6.11 The line spectrum of the hydrogen atom.

where λ_{nm} is the wavelength of the emitted light and depends on integers n and m such that $n > m$. From experimental data the value for the constant R_H was deduced and was given by

$$R_H = 4/B = 10973731.57 \text{ m}^{-1}. \tag{6.14}$$

The constant R_H is called the Rydberg constant. The series of spectral lines for $m = 1$, $m = 2$, and $m = 3$ (with the condition $n > m$) are called the Lyman, Balmer, and Paschen series, respectively.

We recall that the prevailing model of the atom at the time this equation was written was Thomson's model—electrons embedded like plums in a positively charged pudding. There was no way that this model could explain the existence of discrete energy lines for hydrogen and derive the Rydberg formula. Even the Rutherford model with a positively charged nucleus with electrons orbiting around in random orbits could not explain this result. A major breakthrough came through the work of Neils Bohr in 1913, to which we turn next.

6.7 Quantum Theory of the Atom: Bohr's Model

In 1913, Niels Bohr, a Danish scientist, discovered evidence that the orbits of electrons are located at fixed distances from the nucleus. This was in contrast to Rutherford's atomic model in which electrons orbit the nucleus at random. According to Bohr's model, electrons can exist in well-defined energy levels. These energy levels correspond to orbits of fixed radii. Electrons can only exist in these orbits and not in between. The picture is similar to a ladder where one can stand on one rung or another but not in between the rungs. Similarly electrons can exist in one energy level or another but never in between these levels.

Thus Bohr's model of atom had the following features:

- An atom consists of a positively charged nucleus with electrons revolving around it in fixed orbits. Each orbit corresponds to an energy level. Electrons should exist only in these levels and not in between these levels (Fig. 6.12).

- The level closest to the nucleus has the least amount of energy. As the radius increases, the energy of the atomic level also increases.

- Atoms can jump from one energy level to another. When an atom jumps from a higher level to a lower level, it emits a light quantum or a photon whose energy $\hbar\nu$ is equal to the energy difference between the two levels. Similarly an atom absorbs a photon of energy $\hbar\nu$ and an electron jumps to a higher level such that the level energy difference is equal to the photon energy.

The challenging task was to develop a theory that could answer the following questions:

- What are the radii of these energy levels?
- What is the energy of electrons in a given level?
- Can the theory explain the hydrogen spectrum and derive the Rydberg formula?

The simplest atom is the hydrogen atom which consists of a proton and an electron. In Bohr's model, the electron revolves around the proton in fixed orbits. Bohr invoked a quantum postulate to find the radii of electron orbits for this simplest of systems. Using this postulate, he could also find the energy of the energy levels in a hydrogen atom. The most remarkable aspect of this model was that it could explain the spectrum of light emitted by a hydrogen atom.

We now turn to the calculation of a two-particle system of a proton and an electron with the electron circling around the proton. We know that the attractive force between an electron and a proton is described by the Coulomb force:

$$\frac{e^2}{4\pi\varepsilon_0 r^2}$$

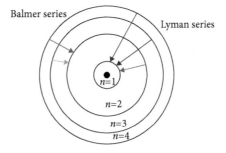

Fig. 6.12 According to the Bohr model of atom, the electrons exists only in prescribed orbits of radii r_n. Each orbit has a definite energy. When an electron jumps from a higher level to a lower level, a photon of energy equal to the energy difference between the two levels is emitted.

where e is the charge of the electron, ε_0 is the permittivity of free space, and r is the radius of the electron orbit. For an electron of mass m_e revolving in a steady orbit around the nucleus, the Coulomb force provides the centripetal force

$$\frac{m_e v^2}{r}$$

where v is the speed of the electron. Equating the two we obtain

$$\frac{m_e v^2}{r} = \frac{e^2}{4\pi\varepsilon_0 r^2}. \tag{6.15}$$

Thus the velocity of the electron in the orbit around the proton with radius r is given by

$$v = \sqrt{\frac{e^2}{4\pi\varepsilon_0 m_e r}}. \tag{6.16}$$

This simple model is, however, very problematic. As the electron moves in a circular orbit around the nucleus it accelerates, as the direction of motion is constantly changing even if the speed is constant. However the electromagnetic theory of Maxwell predicts that an accelerated electron emits radiation. Thus the electron should continuously emit radiation and lose energy. It should therefore spiral around until it falls into the nucleus. Therefore the electron cannot orbit the proton in a stable and stationary (time-independent) orbit.

At this point, Bohr made a very bold move, perhaps one of the most daring in the history of physics. He postulated that only those orbits are stable for which angular momentum $L = m_e v r$ is an integral multiple of the Planck's constant $\hbar$, i.e.,

$$m_e v r = n\hbar, \tag{6.17}$$

where $n = 1, 2, 3, \cdots$. The integer n is called the principal quantum number. Here quantum mechanics makes its first appearance in atomic physics. On substituting the expression of the electronic speed, v, from Eq. (6.16) in this quantum condition, we obtain

$$m_e \sqrt{\frac{e^2}{4\pi\varepsilon_0 m_e r}} r = n\hbar, \tag{6.18}$$

yielding the following expression for the radii of the allowed orbits:

$$r_n = n^2 \frac{\hbar^2}{m_e}\left(\frac{4\pi\varepsilon_0}{e^2}\right). \tag{6.19}$$

This is a departure from anything known about atomic structure until that point. According to Bohr's theory of the hydrogen atom, the electron could not orbit the nucleus in just any orbit, but only in orbits with radii given by Eq. (6.19) for integral values of n. This condition is a direct result of the condition in Eq. (6.17).

The radius of the lowest level ($n = 1$),

$$a_B = r_1 = \frac{\hbar^2}{m_e}\left(\frac{4\pi\varepsilon_0}{e^2}\right), \tag{6.20}$$

is called the Bohr radius and gives an estimate of the size of the atom. With $\hbar = 1.06 \times 10^{-34}$ J · s, $m_e = 9 \times 10^{-31}$ kg, $e = 1.6 \times 10^{-19}$ C, and $1/4\pi\varepsilon_0 = 9 \times 10^9$ N · m^2/C^2,

$$a_B \approx 0.5 \times 10^{-10} m = 0.5 \text{ Angstrom.} \tag{6.21}$$

The next question we address is: What is the energy of the electron in the quantized levels of the hydrogen atom? This question is straightforwardly answered by calculating the total energy of the electron in the nth orbit (orbit with radius r_n).

The energy of the electron consists of two parts: One is the kinetic energy, $KE = m_e v^2/2$, and the other is the potential energy,

$$PE = -e^2/4\pi\varepsilon_0 r. \tag{6.22}$$

The total energy is therefore equal to

$$E = \frac{1}{2}m_e v^2 - \frac{e^2}{4\pi\varepsilon_0 r}. \tag{6.23}$$

On substituting for v from Eq. (6.16), we obtain

$$E = -\frac{e^2}{8\pi\varepsilon_0 r}. \tag{6.24}$$

Finally, on substituting the expression for the radius of the nth orbit from Eq. (6.19), we obtain the resulting expression for the energy of the nth energy level,

$$E_n = -\frac{m_e e^4}{2(4\pi\varepsilon_0)^2 \hbar^2} \frac{1}{n^2}. \tag{6.25}$$

This is the major result Bohr obtained for the hydrogen atom—the energy of the electron is "quantized" as it depends on the quantum number n.

The hydrogen spectrum can now be explained by using Bohr's postulate: When an electron jumps from an excited state described by the quantum number n to a lower level with the quantum number m ($n > m$), a photon of frequency ν_{nm} is emitted which satisfies the relation,

$$E_n - E_m = \hbar\nu_{nm}. \tag{6.26}$$

On the right-hand side we have used Einstein's relation $E = \hbar\nu$. We can substitute for E_n from Eq. (6.25) and obtain

$$E_n - E_m = \hbar\nu_{nm} = \frac{m_e e^4}{2(4\pi\varepsilon_0)^2 \hbar^2}\left(\frac{1}{m^2} - \frac{1}{n^2}\right). \tag{6.27}$$

We can thus find the allowed values of wavelength when jumping from level n to level m by recalling that $\nu = 2\pi c/\lambda$. The result is

$$\frac{1}{\lambda_{nm}} = R_H\left(\frac{1}{m^2} - \frac{1}{n^2}\right). \tag{6.28}$$

Here R_H is the Rydberg constant which is given by

$$R_H = \frac{m_e e^4}{8\varepsilon_0^2 h^3 c} = 10973731.57 \text{ m}^{-1} \tag{6.29}$$

This is identical to the value that was experimentally observed. Here it has been derived from Bohr's model of the hydrogen atom and is related explicitly to the charge and the mass of the electron. This is truly a remarkable achievement—that Bohr's assumption, only those orbits are allowed for which an electron's angular momentum is an integral multiple of Planck's constant $\hbar$, leads to the experimentally observed spectral lines of the hydrogen atom.

Bohr's success in proposing a model that could explain the radiation emitted by the hydrogen atom and derive the Rydberg formula was a major achievement of the quantization postulate of Planck and Einstein. Bohr had additionally postulated that the electrons in the hydrogen atom can exist only in some allowed orbits given by the quantization condition in Eq. (6.17).

The results of Planck, Einstein, and Bohr gradually created a realization that Newtonian mechanics must be an inadequate theory to explain phenomena at the atomic level. The period between 1913 and 1925 was a time when new phenomena were being discovered that could not be explained by classical theory. The foundations of Newtonian mechanics were crumbling and the need for a new theory was being felt very urgently. The breakthrough came in 1925/26 when quantum mechanics was born through the works of Heisenberg, Schrödinger, Born, Dirac, and others.

One big test of the new theory was to solve the problem of the hydrogen atom, not through a postulate as Bohr did, but as a result of a formal theory—a theory that could be applied equally well to essentially all the problems of physics even at the level of our daily experience.

The new theory of quantum mechanics showed that, contrary to Rutherford–Bohr's model of atom, electrons do not travel in fixed orbits. In fact, each electron with energy E_n is described by a wave function, $\psi_n(\mathbf{r})$ such that $|\psi_n(\mathbf{r})|^2$ is the probability density of finding the electron at position $\mathbf{r}$. Thus we have a probabilistic description for the location of each electron. However, the energies E_n matched the expression obtained through Bohr's theory. The full quantum mechanical theory also explained why electrons do not radiate and fall into the nucleus. It is the "wave nature" of electrons that allows them to exist only at certain distances from the nucleus. We discuss these features in Chapter 17 when we consider the Schrödinger equation for the wave function as applied to the hydrogen atom.

Problems

6.1 Using Wien's displacement law, show that the peak radiation is in the infrared region at room temperature and is sizzling red at 5000 °C.

6.2 The sun, with an effective temperature of approximately 5800 K, can be regarded as a blackbody. Show that the emission spectrum peaks in the central, yellow-green part of the visible spectrum.

6.3 Derive the Rayleigh–Jeans law

$$J(\nu, T) = \frac{8\pi\nu^2}{c^3} k_B T$$

from Planck's law

$$J(\nu, T) = \frac{8\pi\nu^2}{c^3} \frac{\hbar\nu}{e^{\hbar\nu/kT} - 1}$$

in the limit

$$\frac{h\nu}{kT} \ll 1.$$

6.4 Sketch the following graphs for the quantities involved in the photoelectric effect. (Explain your sketches using Einstein's theory):

(a) Kinetic energy (max) vs. intensity at constant frequency (assuming frequency ν is greater than the threshold frequency ν_c)

(b) Kinetic energy (max) vs. wavelength at constant intensity

(c) Photoelectric current vs. intensity at constant frequency (assuming ν is greater than the threshold frequency ν_c)

(d) Photoelectric current vs. wavelength at constant intensity.

6.5 What is the longest light wavelength (sometimes referred to as the cutoff wavelength, λ_c) that can result in the production of a photocurrent?

6.6 A light source of wavelength λ illuminates a metal and ejects photoelectrons with a maximum kinetic energy of 1.00 eV. A second light source with half the wavelength of the first ejects photoelectrons with a maximum kinetic energy of 4.00 eV. Determine the work function of the metal.

6.7 A photon is emitted as the electron in a hydrogen atom makes a transition from $n = 4$ to $n = 2$ level. What are the frequency, wavelength, and energy of the emitted photon?

6.8 For the Balmer series, i.e., the atomic transitions where the final state of the electron is $n = 2$, what is the longest and shortest wavelength possible? Are any of the frequencies in the Lyman series, which corresponds to transitions where the electron ends up in $n = 1$ level, in the visible region? (The visible region is characterized by wavelengths in the range 400 nm to 700 nm.)

6.9 Using Bohr's quantization rule, derive a formula for an electron's speed in the quantized Bohr orbits of the hydrogen atom. By putting in the values of constants explicitly, derive the value of an electron's speed in the $n = 1$ orbit. What fraction of the speed of light is it?

 BIBLIOGRAPHY

O. Darrigol, *A History of Light: From Greek Antiquity to the Nineteenth Century* (Oxford University Press, 2012).

A. M. Smith, *From Sight to Light: The Passage from Ancient to Modern Optics* (University of Chicago Press, 2015).

M. S. Zubairy, *A very brief history of light*, in *Optics in Our Time*, Edited by M. D. Alamri, M. M. El-Gomati, and M. S. Zubairy (Springer Nature 2016).

D. M. Greenberger, N. Erez, M. O. Scully, A. A. Svidzinsky, and M. S. Zubairy, *The rich interface between optical and quantum statistical physics: Planck, photon statistics, and Bose-Einstein condensate*, in Progress in Optics, Vol. 50, Edited by E. Wolf (Elsevier, Amsterdam 2007), p. 275.

7 De Broglie Waves: Are Electrons Waves or Particles?

In a speech, accepting the 1929 Nobel Prize for Physics, Louis de Broglie described his discovery of de Broglie waves in these words:

> On the one hand the quantum theory of light cannot be considered satisfactory since it defines the energy of a light particle (photon) by the equation $E = hf$ containing the frequency f. Now a purely particle theory contains nothing that enables us to define a frequency; for this reason alone, therefore, we are compelled, in the case of light, to introduce the idea of a particle and that of frequency simultaneously. On the other hand, determination of the stable motion of electrons in the atom introduces integers, and up to this point the only phenomena involving integers in physics were those of interference and of normal modes of vibration. This fact suggested to me the idea that electrons too could not be considered simply as particles, but that frequency (wave properties) must be assigned to them also.

De Broglie's postulate that particles can behave like waves complemented the observation by Einstein in 1905 that light can behave like particles. This wave–particle duality aspect for both particles and waves had a deep impact on the subsequent development of quantum mechanics. Some highly counterintuitive results, like the Heisenberg uncertainty relation and the Bose–Einstein condensation, that we discuss in this chapter, were motivated by the wave–particle duality. Perhaps the most significant outcome of de Broglie's observation was the search for an equation for the de Broglie waves by Schrödinger resulting in the Schrödinger equation. This important development is discussed in Chapter 17.

7.1 De Broglie waves

Recall that, in 1905, Einstein explained the photoelectric effect by showing that light waves sometime act like particles. The energy of these particles, the photons, is proportional to the frequency of the light field and is given by

$$E = \hbar \nu, \tag{7.1}$$

where E is the energy, $\hbar$ is Planck's constant (1.055×10^{-34} J · s), and ν is the frequency. In 1924, Louis de Broglie argued in his Ph.D. thesis that if light can behave both like waves, as in interference and diffraction, and like particles, as in the photoelectric effect, then particles should also behave like both particles and waves. The de Broglie hypothesis completed the wave–particle duality description of both waves and particles. As has been discussed before, the characteristics of particles are mass and momentum, whereas waves are characterized by

Quantum Mechanics for Beginners: With Applications to Quantum Communication and Quantum Computing. M. Suhail Zubairy.
© M. Suhail Zubairy 2020. Published in 2020 by Oxford University Press. DOI: 10.1093/oso/9780198854227.001.0001

frequency and wavelength. De Broglie showed that a particle of mass m moving with velocity v is characterized by a wave of wavelength

$$\lambda_{dB} = \frac{h}{mv}. \tag{7.2}$$

This wavelength is called the de Broglie wavelength.

De Broglie "derived" this relation by invoking Einstein's theory of relativity. We do not give details of the theory of relativity here but some necessary discussion is given in the section on Compton scattering later in this chapter. A consequence of this theory is that the momentum p of massless objects such as a photon is related to the energy E via the relation

$$p = \frac{E}{c}. \tag{7.3}$$

However the energy of a photon is proportional to its frequency, $E = \hbar \nu$. Thus, for photons,

$$p = \frac{\hbar \nu}{c}. \tag{7.4}$$

If we substitute $2\pi c/\lambda$ for the frequency ν we obtain

$$\lambda = \frac{h}{p}, \tag{7.5}$$

where $h = 2\pi \hbar$. De Broglie's giant leap was to conjecture that this relation, which is true for photons, is also true for any particle with mass, such as an electron, with momentum $p = mv$, i.e., a particle of mass m moving with a velocity v can be characterized by the wavelength

$$\lambda_{dB} = \frac{h}{p} = \frac{h}{mv}. \tag{7.6}$$

Just as a particle moving with a velocity can be described by the de Broglie wavelength, a photon of wavelength λ has a momentum (see Eq. (7.5))

$$p = \frac{h}{\lambda} = \hbar k, \tag{7.7}$$

where we used the relations $k = 2\pi/\lambda$ and $\hbar = h/2\pi$. Therefore a photon can not only have energy, it also has momentum, p, which is inversely proportional to the wavelength, λ, and directly proportional to the wave vector, k, or frequency, $\nu = ck$.

The de Broglie hypothesis that particles behave like waves seems quite mysterious. We do not seem to see the particles around us as waves. The particles, no matter how small, are well defined objects and cannot, for a moment, be perceived as a wave. A baseball, or even a dust particle, cannot be described as waves. Why? The reason, as we see below, is that the corresponding de Broglie wavelength is small, unimaginably small, and therefore the wave nature is completely masked.

First we calculate the de Broglie wavelength of a pitched baseball, a ball of mass $m = 0.15$ kg moving at a speed of v $= 40$ m/s (≈ 90 miles/hour). The corresponding de Broglie wavelength is

$$\lambda_{dB} = \frac{h}{mv} = \frac{6.626 \times 10^{-34}\, \text{J} \cdot \text{s}}{(0.15\, \text{kg})\, (40\, \text{m/s})} = 1.1 \times 10^{-34}\, \text{m}.$$

As discussed in Section 6.7, the atomic diameter is of the order 10^{-10} m and the diameter of a nucleus is of the order 10^{-14} m. The de Broglie wavelength of the baseball is thus a billionth of a billionth of the size of an atom. Such a small wavelength is extremely difficult to observe or measure. This is why we do not see a baseball as a wave.

What about a dust particle of mass 10^{-10} kg moving at 10^{-4} m/ sec (speed at room temperature)? The corresponding de Broglie wavelength is

$$\lambda_{dB} = \frac{h}{mv} = \frac{6.626 \times 10^{-34} \text{ J} \cdot \text{s}}{\left(10^{-10} \text{ kg}\right)\left(10^{-4} \text{ m/s}\right)} = 6.6 \times 10^{-20} \text{ m}.$$

Again the wavelength is too small (a billionth of the size of an atom) and we cannot therefore even see a dust particle as a wave!!!

What about an electron of mass $m_e = 9.11 \times 10^{-31}$ kg, accelerated through a 100V voltage, and moving with a speed of $v = 5.6 \times 10^{6}$ m/s? The de Broglie wavelength of such an electron is

$$\lambda_{dB} = \frac{h}{mv} = \frac{6.626 \times 10^{-34} \text{ J} \cdot \text{s}}{\left(9.11 \times 10^{-31} \text{ kg}\right)\left(5.6 \times 10^{6} \text{ m/s}\right)^3} = 1.2 \times 10^{-10} \text{ m} = 0.12 \text{ nm}.$$

Thus at the nanometer scale, we should be able to see the wave nature of an electron.

A big success of the de Broglie hypothesis is, as de Broglie pointed out in his Nobel lecture mentioned above, that it provides an insight into the Bohr quantization condition. We recall that Bohr postulated that only those orbits of the electron around the nucleus in a hydrogen atom are stable for which the angular momentum, L, is an integral multiple of the reduced Planck's constant $\hbar$. There was no justification for this postulate except that it gave the correct expressions for the emission spectrum of the hydrogen atom. De Broglie waves provided some justification for Bohr's postulate by first arguing that a condition for the stable orbit should be

$$2\pi r = n\lambda_{dB}, \tag{7.8}$$

i.e., the circumference of the allowed orbit of an electron should be an integral multiple of the de Broglie wavelength of the electron as shown in Fig. 7.1. This seems reasonable as the orbits would be stationary with a standing wave. If we substitute the expression of the de Broglie wavelength

$$\lambda_{dB} = \frac{h}{mv} \tag{7.9}$$

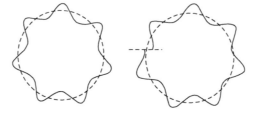

Fig. 7.1 (a) The Bohr condition $mvr = n\hbar$ yields stable orbits such that the circumference of the allowed orbit of an electron is an integral multiple of the de Broglie wavelength of the electron. (b) When the Bohr condition is not satisfied, a stable orbit cannot be obtained.

into the stability condition (7.8), we recover the Bohr condition

$$mvr = n\hbar. \tag{7.10}$$

When de Broglie presented his hypothesis that particles can behave like waves in 1924, he had no experimental evidence to support his conjecture. The situation soon changed. In 1927, Clinton J. Davisson and Lester H. Germer carried out an experiment that could only be explained using de Broglie's conjecture. In their experiment, Davisson and Germer shot electrons at a nickel crystal. What they observed was the diffraction of the electrons similar to wave diffraction against crystals. In the same year, an English physicist, George P. Thomson, carried out a similar experiment in which electrons were fired towards a thin metal foil and obtained the same results as Davisson and Germer.

In the Davisson-Germer experiment, a gun shoots a beam of electrons at a nickel target. These electrons rebound and are detected at various angles of detection as shown in Fig. 7.2. The experiment involves counting the number of electrons at the detector. What Davisson and Germer observed was the following:

- As the angle θ was varied the number of detected electrons exhibited a periodic behavior as long as the energy of the incident electros was fixed. Electron counts oscillated between zero and a maximum number in rapid succession as the angle was varied.

- For a fixed angle, the counts increased and decreased as the energy of the incident electrons changed.

These results were very surprising, almost shocking. They were very different from what we expect if electrons behaved like particles. For example, if electrons behaved like billiard balls scattering from a fixed surface, we would expect a large fraction bouncing back at $\theta = 0$ and the rest scattering almost uniformly with respect to the angle θ. The observed angular distribution of the electrons was the characteristic diffraction fringe pattern. There was absolutely nothing whatsoever to predict such a result if electrons behaved like particles. Only a wave behavior with a characteristic wavelength that agreed with the de Broglie wavelength could explain the observed distribution of counts.

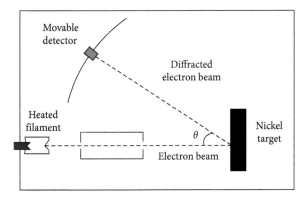

Fig. 7.2 Schematics of the Davisson–Germer experiment.

So how do we explain these results using the wave nature of electrons? First we realize that the nickel crystal consists of layers of atoms in a cubic structure, the side of each cube being equal to 2.15 Angstroms. If electrons behave like particles, they would scatter from each lattice site on the surface, as well as from sites inside the crystal, in all directions giving a uniform angular distribution of electrons. However, if electrons behaved as de Broglie waves of wavelength $\lambda_{dB} = h/mv$, they see all the atoms and are scattered at those angles θ_n where the condition of constructive interference is satisfied. The condition for constructive interference of scattering from the lattice sites is (see Fig. 7.3)

$$d \sin \theta_n = n\lambda_{dB}, \tag{7.11}$$

where $n = 0, 1, 2, \cdots$ Similarly the condition for destructive interference is

$$d \sin \theta_n = (2n + 1)\frac{\lambda_{dB}}{2}. \tag{7.12}$$

Thus the electrons scatter only at those angles for which

$$\sin \theta_n = \frac{n}{d}\lambda_{dB} = \frac{n}{d}\frac{h}{mv}. \tag{7.13}$$

For example, when electrons are accelerated through a voltage $V = 54V$, we obtain

$$\theta_0 = 0^\circ, \theta_1 = 50^\circ, \tag{7.14}$$

in agreement with the experimental results.

The second observation that, for a fixed angle, the counts increased and decreased as the energy of the incident electrons changed, can also be verified from Eq. (7.13). If we choose a fixed θ, then by varying v we can periodically satisfy the condition for constructive interference (7.11) followed by the condition for destructive interference (7.14).

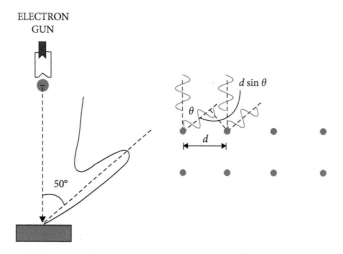

Fig. 7.3 The result of the Davisson–Germer experiment can be explained by treating electrons as de Broglie waves of wavelength $\lambda_{dB} = h/mv$ that are diffracted by the atoms inside the crystal and are scattered at those angles θ_n where the condition of constructive interference is satisfied.

The importance of this remarkable experiment was realized early and Clinton Davisson and George P. Thomson were awarded the Nobel Prize in 1937. An interesting historical note is that J. J. Thomson was awarded a Nobel Prize in 1906 for the discovery of the electron as a particle and his son G. P. Thomson received the same prize 31 years later for proving that the electron behaved like a wave.

Another milestone following the discovery by de Broglie that electrons can behave like waves came in 1932 when Ernst Ruska invented the first electron microscope by replacing light (photons) by electrons. The advantage of the electron microscope is that it operates at much smaller wavelengths, corresponding to the de Broglie wavelength of the electrons, than those employed in conventional microscopes, thus increasing the sensitivity immensely. In 1986 (more than fifty years after the discovery), Ruska received the Noble prize for his invention.

7.2 Wave-Particle Duality–A Wavefunction Approach

De Broglie's description of particles, such as an electron, as waves is very intriguing, to say the least. However this description is validated by the electron diffraction experiment of Davisson and Germer as discussed above. In Chapter 8 we discuss another landmark experiment that shows electrons exhibiting interference in a double-slit experiment. But then what about the experiments discussed in Chapter 3 where we used the classical laws of physics to describe electron deflection in the presence of electric and magnetic fields? There, we treated an electron like a particle with definite mass and charge. A natural question is: How do we reconcile both the wave and particle natures of an electron, as well as for any other massive particle? A related question is: How do we describe a localized particle which exists in a finite space within the context of wave–particle duality? These questions lie at the heart of quantum mechanics and the answer to these questions are rigorously given by the wavefunction approach discussed in Chapter 17. Here we motivate such an approach via describing an electron as a wave packet.

In the above discussion about the de Broglie waves, we assumed that the massive particle can be described by a wave of wavelength

$$\lambda = \frac{h}{mv}.$$

Such a wave with a precise wavelength can be described by a wave function

$$\psi(x, t) = A \sin(kx - \nu t), \tag{7.15}$$

where $k = 2\pi/\lambda$ is the wavevector and $\nu = ck$ is the frequency. We restrict ourselves to a one-dimensional description for the sake of simplicity. In a more complete three-dimensional picture, we should have components of the wavevector $\mathbf{k}$ in all three directions. Equation (7.15) cannot describe a localized object as this wave extends from $x = -\infty$ to $+\infty$. The wavefunction is distributed in the entire space.

A localized particle can, however, be described as a superposition of waves with multiple wavelengths. As the simplest such example, we consider two waves of equal amplitudes but with slightly different wavelengths (or equivalently wavevectors):

$$\psi_1(x) = A \sin((k + \Delta k)x), \tag{7.16}$$

$$\psi_2(x) = A \sin((k - \Delta k)x). \tag{7.17}$$

According to the principle of superposition, the resulting wave is

$$
\begin{aligned}
\psi(x) &= \psi_1(x) + \psi_2(x) \\
&= A \sin((k + \Delta k)x) + A \sin((k - \Delta k)x) \\
&= 2A \sin(kx) \cos((\Delta k)x),
\end{aligned}
\tag{7.18}
$$

where we used the trigonometric identity

$$\sin(\alpha) + \sin(\beta) = 2 \sin\left(\frac{\alpha + \beta}{2}\right) \cos\left(\frac{\alpha - \beta}{2}\right) \tag{7.19}$$

with $\alpha = (k + \Delta k)x$ and $\beta = (k - \Delta k)x$. The wavefunction $\psi(x)$, as shown in Fig. 7.4, consists of two terms. The first term, $\sin(kx)$, oscillates at the average wavevector k

$$k = \frac{1}{2}[(k + \Delta k) + (k - \Delta k)]. \tag{7.20}$$

It is modulated by the slowly varying second term $\cos((\Delta k)x)$, which oscillates at half the difference of the two wavevectors

$$\Delta k = \frac{1}{2}[(k + \Delta k) - (k - \Delta k)]. \tag{7.21}$$

We thus have a wave of wavevector k which is modulated with a wave of a much smaller wavevector Δk. Thus the superposition of the two waves with close wavelengths together breaks up the continuous wave into a series of packets. The width of each packet is of the order $\pi/\Delta k$. As the number of waves increases, the superposition can lead to constructive interference in a small region as shown in Fig. 7.5.

To describe a single electron (or any particle) confined to a highly localized region, we need a single wave packet that is zero or nearly zero everywhere in space except for one localized region. Such a wave packet can be constructed by superposing waves having a continuous distribution of wavelengths, or wavevectors within the order Δk, centered around

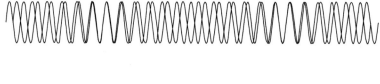

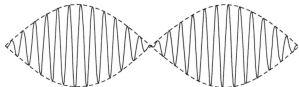

Fig. 7.4 When two waves of the same amplitude but slightly different wavevectors are added, they lead to a wave that is modulated by a wave of much smaller wavevectors.

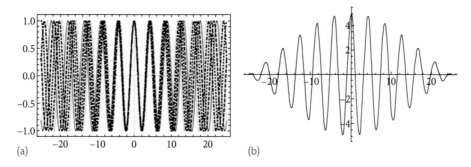

Fig. 7.5 Superposition of 5 waves leads to the formation of a wave packet of width $\pi/\Delta k$.

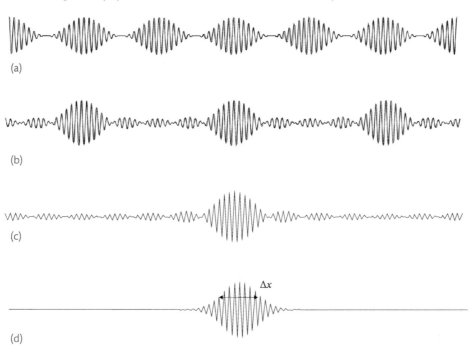

Fig. 7.6 Superposition of (a) 3, (b) 5, (c) 9, and (d) infinite number of waves.

the wavevector k. In this case, the waves are out of phase after a distance of order $\pi/\Delta k$, but since they have many different wavelengths, they never get back in phase again.

In Fig. 7.6, we show a superposition of 3, 5, 9, and an infinite number of waves, centered around k_0 and with k values ranging from $k_0 - (\Delta k/2)$ to $k_0 + (\Delta k/2)$. Thus, for a superposition of 3 and 5 waves,

$$\psi_3(x) = \cos\left(\left(k_0 - \frac{\Delta k}{2}\right)x\right) + 2\cos\left(k_0 x\right) + \cos\left(\left(k_0 + \frac{\Delta k}{2}\right)x\right), \tag{7.22}$$

$$\psi_5(x) = \cos\left(\left(k_0 - \frac{\Delta k}{2}\right)x\right) + \cos\left(\left(k_0 - \frac{\Delta k}{4}\right)x\right) + \cos\left(k_0 x\right)$$
$$+ \cos\left(\left(k_0 + \frac{\Delta k}{4}\right)x\right) + \cos\left(\left(k_0 + \frac{\Delta k}{2}\right)x\right). \tag{7.23}$$

As the number of waves in the superposition increases, the total wavefunction becomes more and more localized. For a continuous distributions of the wavelengths, or wavevectors within Δk, the resulting wave packet is localized in space in the region $\Delta x = \pi/\Delta k$.

As an example, a wave packet can be of the form

$$\psi(x) = \frac{1}{(2\pi\sigma^2)^{1/4}} e^{-\frac{x^2}{4\sigma^2}} e^{ik_0 x}. \tag{7.24}$$

This is called a Gaussian wave packet centered at $x = 0$ with a width $\Delta x = \sigma$. Here k_0 is the carrier wave number. The associated de Broglie wavelength is $\lambda_{dB} \cong 2\pi/k_0$. Such a wave packet can be formed by a continuum of wavevectors with a distribution

$$\phi(k) = \left(\frac{2\sigma^2}{\pi}\right)^{1/4} e^{-\sigma^2(k-k_0)^2}. \tag{7.25}$$

The spread of the wave packet $\sigma = \Delta x$ is inversely proportional to Δk. This wave packet is clearly particle-like in that its region of significant magnitude is confined to a localized region in space. Moreover, this wave packet is constructed out of a group of waves with an average wave number k_0, and so these waves could be associated with a particle of momentum $p_0 = \hbar k_0$. If this were true, then the wave packet would be expected to move with a velocity of $v_0 = p_0/m$. This is in fact found to be the case, as we discuss in Section 17.2.

7.3 Bose–Einstein Condensation

A remarkable effect predicted on the basis of de Broglie waves is the Bose–Einstein condensation which was first predicted in 1925 by Satyendra Nath Bose and Albert Einstein and was experimentally observed 70 years later in 1995 by Eric Cornell and Carl Wieman, by Wolfgang Ketterle, and by Randy Hulet in separate experiments.

So far we have seen the role of small mass on de Broglie waves—the smaller the mass, the larger is the de Broglie wavelength. But the de Broglie wavelength is also inversely proportional to the velocity of the particle—the smaller the velocity, the larger is the de Broglie wavelength. The velocity of particles inside a gas depends on the temperature. As we decrease the temperature, the velocity of the gaseous atoms or molecules decreases and the de Broglie wavelength of the atoms or molecules increases.

The atoms behave like point particles when they are moving sufficiently fast (Fig. 7.7a). As we lower the temperature, the atoms move very slowly with a significantly large de Broglie wavelength and start behaving like a wave (Fig. 7.7b). The size of the atom is of the order of the de Broglie wavelength. As the temperature is further lowered, the atomic size (de Broglie wavelength) becomes so large that different atoms start to overlap (Fig. 7.7c). Finally, at a very low temperature when the atoms can barely move, they all lose their identity and become one— a condensed state (Fig. 7.7d). This phenomenon is called Bose–Einstein condensation. We can calculate the critical temperature at which the Bose–Einstein condensate starts to form in a gas of atoms of mass m.

Consider N atoms inside a box of volume V moving with an average speed v. On the average, an atom occupies a space V/N. The atomic de Broglie wavelength is given by $\lambda_{dB} = h/mv$. If

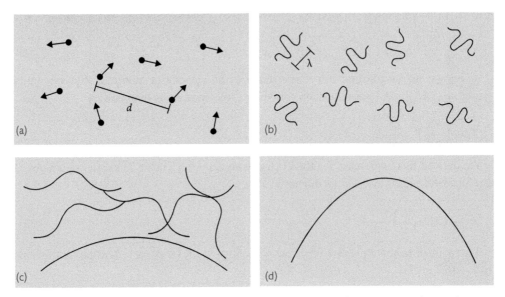

Fig. 7.7 The atomic gas molecules form a Bose–Einstein condensate as the temperature is decreased.

the average space occupied by each molecule starts becoming larger than V/N then they start losing their identity and a condensate is formed. If we assume that the effective radius of the atom, when it is moving with velocity v, is

$$r \approx \frac{\lambda_{dB}}{2} = \frac{h}{2mv} \tag{7.26}$$

then the volume occupied by each atom in the gas is

$$\frac{4\pi}{3}r^3 \approx \frac{4\pi}{3}\left(\frac{h}{2mv}\right)^3. \tag{7.27}$$

The condition for the formation of condensate is

$$\frac{4\pi}{3}\left(\frac{h}{2mv}\right)^3 > \frac{V}{N}. \tag{7.28}$$

But what is the velocity of the atom at a given temperature T? As the atoms move faster with increasing temperature, the kinetic energy of the atoms in the gas is proportional to the temperature T. Thus the higher the temperature, the higher the kinetic energy of the atoms, and vice versa. Therefore the average kinetic energy of each atom inside a gas is

$$\frac{1}{2}mv^2 = \alpha T, \tag{7.29}$$

where α is a constant of proportionality and T is the temperature in Kelvin. The constant α can be found from the kinetic theory of gases and is equal to $(3/2)k_B$ where $k_B = 1.38 \times 10^{-23}$ m$^2 \cdot$ kg $\cdot$ s$^{-2} \cdot$ K^{-1} is called the Boltzmann constant. We obtain

$$\frac{1}{2}mv^2 = \frac{3}{2}k_B T \tag{7.30}$$

or

$$v = \sqrt{\frac{3k_B T}{m}}.$$
(7.31)

It follows, on substituting this expression for the velocity of atoms in the condensate condition (7.28), that a Bose–Einstein condensate is formed at the temperature

$$T_c \cong \left(\frac{\pi N}{6V}\right)^{\frac{2}{3}} \frac{h^2}{3mk_B}.$$
(7.32)

We derived this expression for the critical temperature T_c using heuristic arguments. A careful analysis leads to a slightly different result:

$$T_c \cong 3.3125 \left(\frac{N}{V}\right)^{\frac{2}{3}} \frac{\hbar^2}{mk_B}.$$
(7.33)

The required temperatures are very low, ~ 500 nK $- 2$ μK for particle densities in the range $10^{20} - 10^{21}$ m^{-3}.

7.4 Heisenberg Microscope

Heisenberg's uncertainty principle, formulated in 1927, is one of the cornerstones of quantum mechanics. It is based on the principle that it is impossible to measure anything without disturbing it. For example, if we try to find the location of a moving particle such as an electron, we need to shine light on it. The light scattered after hitting the particle provides information about the location of the particle. However, as we have seen, light consists of photons and photons carry momentum

$$p = \hbar k$$
(7.34)

according to the de Broglie hypothesis. Thus, when photons hit the moving particle, the speed of the particle changes randomly. The consequence is that if we try to measure the position of a particle very precisely, its velocity or momentum changes randomly. Similarly we can argue that if we measure the momentum of a particle precisely, its location becomes uncertain. Heisenberg could show a relationship between the preciseness of measurements of both position and momentum. According to the Heisenberg uncertainty relation, the following inequality should always hold:

$$\Delta x \cdot \Delta p_x \geq \frac{\hbar}{2},$$
(7.35)

no matter how precise our measurement instruments. Here Δx is the uncertainty in determining the position of an object and Δp_x is the corresponding uncertainty in momentum.

It should be mentioned that position and momentum are not the only pair of observable quantities that satisfy an uncertainty relation. There are other observables such as energy and time that obey a Heisenberg uncertainty relation of the type (7.35). However, here we concentrate only on the uncertainty relation between position and momentum.

A formal derivation of the inequality (7.35) may require sophisticated mathematics. However, here is a derivation based on simple physical concepts that we introduced in this and previous chapters. This heuristic derivation of the uncertainty relation is based on an analysis of a microscope to find both the position and momentum of an object, such as an electron, as precisely as we can. This analysis was first presented by Heisenberg himself to elucidate the uncertainty relation named after him and is usually referred to as the Heisenberg microscope.

Let us consider a microscope consisting of a lens of diameter D and focal length F. Suppose an electron is located at a distance d from the lens. The question is how do we look at the electron and, more importantly, how do we determine its location as well as momentum along the x-axis?

In the Heisenberg microscope, light is incident from the side. It is scattered from the electron into the lens as shown in Fig. 7.8. When we see the scattered light through the lens, we can determine the "position of the electron."

Let us first analyze the collision of a photon with an electron, According to de Broglie, the photon of wavelength λ, acting like a particle, has a momentum equal to $\hbar k = h/\lambda$. The electron that we try to locate is assumed to be "at rest." Thus we have a "collision" between two "particles"—a photon and an electron. The photon is assumed to be directed along the x-axis. If, after the collision, the photon is deflected in a direction making an angle θ with the vertical, the conservation of momentum along the x-axis leads to

$$\hbar k + 0 = \hbar k \sin \theta + p_x. \tag{7.36}$$

Here, before the collision, the photon momentum along the x-axis is $\hbar k$ and the electron has zero momentum. After the collision, the x-component of the photon momentum is $\hbar k \sin \theta$ and the electron momentum is $p_x = m v_x$, where v_x is the component of the velocity of the electron in the x-direction.

We assumed that the wavelength of light after the collision is the same as before the collision. As we see, in the next section, this may not be true in general but it is a good approximation. Thus the momentum imparted to the electron is

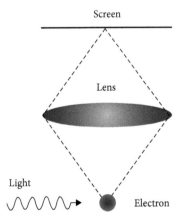

Fig. 7.8 Heisenberg microscope. The position of the electron is determined from the light that is scattered from an electron into the lens.

$$p_x = \hbar k - \hbar k \sin \theta. \tag{7.37}$$

Next we calculate the minimum and maximum momentum imparted to the electron, if we are able to see the electron through the lens.

The photon, after "colliding" with the electron can be deflected in any direction. However, we consider only the range of angles $+\theta$ to $-\theta$ with the vertical as shown in Fig. 7.9. Only the photons that are deflected in this range hit the lens and are detected behind the lens. For the angles of deflection outside this range, the photon is lost and does not contribute to observing the electron.

Thus, for those photons that are deflected in the range $+\theta$ to $-\theta$, the minimum momentum imparted to the electron is for the photon deflection through $+\theta$ and the maximum momentum is imparted for the deflection through $-\theta$. We therefore have

$$\left(p_x\right)_{min} = \hbar k - \hbar k \sin \theta, \tag{7.38}$$

$$\left(p_x\right)_{max} = \hbar k + \hbar k \sin \theta. \tag{7.39}$$

Therefore, in the act of seeing the electron, the electron can acquire a momentum in the range between $\left(p_x\right)_{min}$ and $\left(p_x\right)_{max}$. The uncertainty in the electron's momentum is therefore equal to

$$\Delta p_x = \left(p_x\right)_{max} - \left(p_x\right)_{min} = 2\hbar k \sin \theta. \tag{7.40}$$

For small θ, $\sin \theta \approx \theta$, and we obtain

$$\Delta p_x = \frac{2h\theta}{\lambda}. \tag{7.41}$$

Here, we recall $\hbar = h/2\pi$ and $k = 2\pi/\lambda$. This shows that the uncertainty in the momentum of the electron after colliding with a photon is large for a photon of small wavelength λ. This

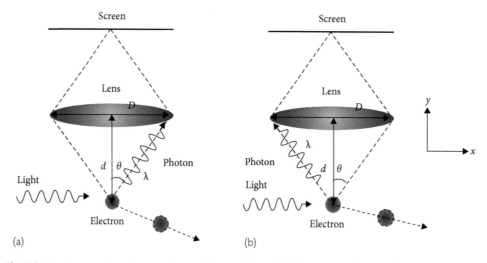

Fig. 7.9 The photon scattered in any angle ranging from $+\theta$ to $-\theta$ with the vertical is detected. (a) The minimum momentum is imparted to the electron by those photons that are scattered at the angle $+\theta$ whereas (b) maximum momentum is imparted by photons scattered at the angle $-\theta$.

happens because a photon with a small λ has a large amount of energy $\left(E = \hbar\nu = hc/\lambda\right)$. Therefore, to minimize the uncertainty in the electron's momentum, we should use light of very large λ.

Next we turn to the position resolution!!

Microscope resolution is the shortest distance between two separate points in a microscope's field of view that can still be distinguished as distinct entities. As we discussed in Chapter 4, according to the Rayleigh criterion, the smallest angle the microscope can resolve is limited by diffraction and is given by

$$\theta_{min} \approx \frac{\lambda}{D}, \tag{7.42}$$

where D is the size of the aperture (or the lens in the present case).

The angle θ_{min} is, however, related to the minimum resolvable position of the electron via (see Fig. 7.10):

$$\tan\left(\theta_{min}/2\right) = \frac{(\Delta x/2)}{d}. \tag{7.43}$$

Therefore the microscope cannot locate the electron in the x-direction any more precisely than

$$\Delta x = 2d\tan\left(\theta_{min}/2\right). \tag{7.44}$$

For small angles, $\tan\left(\theta_{min}/2\right) \approx \theta_{min}/2$, and we obtain

$$\Delta x = d\theta_{min}. \tag{7.45}$$

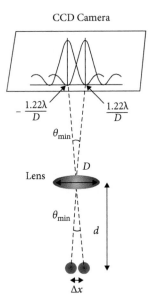

Fig. 7.10 The uncertainty of the position of the electron is determined by using the Rayleigh criterion. The minimum uncertainty Δx subtends an angle θ_{min} with the lens.

However, according to the Rayleigh criterion, $\theta_{min} \approx \lambda/D$. Therefore

$$\Delta x \cong d\frac{\lambda}{D}, \tag{7.46}$$

or since (Fig. 7.9)

$$\frac{d}{D} = \frac{1}{2\tan\theta} \cong \frac{1}{2\theta}, \tag{7.47}$$

the minimum uncertainty in the position is

$$\Delta x = \frac{\lambda}{2\theta}. \tag{7.48}$$

This shows that a photon of long wavelength causes large uncertainty in position. This is in contrast to the measurement of momentum where long wavelength causes small uncertainty in momentum (Eq. (7.41)).

Putting the uncertainties Δx and Δp_x from Eqs. (7.41) and (7.48) together,

$$\Delta x \cdot \Delta p_x \cong \frac{\lambda}{2\theta} \cdot \frac{2h\theta}{\lambda} \cong h. \tag{7.49}$$

We thus obtain the uncertainty relation between the precisions with which we can measure the position and momentum of an electron in a microscope. The product of uncertainties Δx and Δp_x is independent of the wavelength of the light, the size of the lens, and any other geometrical feature of the system—it is equal only to Planck's constant h.

Our derivation is based on heuristic arguments which can be made more rigorous. It turns out that a careful derivation yields

$$\Delta x \cdot \Delta p_x \geq \frac{h}{4\pi} = \frac{\hbar}{2}. \tag{7.50}$$

The uncertainty relation is based on the idea that the very process of measuring one quantity (position) alters a complementary property (momentum). This uncertainty relation is independent of the quality of our measuring instruments and is universally valid.

7.5 Compton Scattering

Although Einstein explained the photoelectric effect by postulating that light consists of quanta of energy, called photons, the first irrefutable proof that photons behave like particles came in 1923 through the experimental work of Arthur Compton, one year before de Broglie's postulate about wave–particle duality. Compton considered the scattering of light by a free electron. What he observed was that the wavelength of the scattered light was different from that of the incident radiation. As we know, a change in wavelength is like changing the color. The amount by which the light's wavelength changes is called the Compton shift. This effect cannot be explained by treating light as a wave. As seen below, the Compton effect can only be explained by treating a photon as a particle with energy and momentum. Compton won the Nobel Prize in Physics in 1927 for the discovery.

What Compton discovered was that, when a photon of wavelength λ_i is incident on an electron of mass m_e, it is scattered at an angle θ with the slightly different wavelength λ_f as shown in Fig. 7.11. The experimental results showed that the difference in the wavelength of the photon is given by

$$\lambda_f - \lambda_i = \frac{h}{m_e c}(1 - \cos\theta). \tag{7.51}$$

Compton derived the wavelength shift of the photon following essentially the same approach as the collision between two particles. Let us therefore first discuss the scattering between particles when a particle of mass m_1 moving with a velocity v_{1i} along the x-axis is incident on another particle m_2 at rest, as shown in Fig. 7.12. After the collision, particle 1 is deflected with a velocity v_{1f} making an angle θ with the x-axis and particle 2 is deflected with a velocity v_{2f} making an angle $-\phi$ with the x-axis. If the collision is perfectly elastic, there should be conservation of momentum and conservation of energy.

The conservation of momentum principle implies that the total momentum of the two particles before the collision should be equal to the total momentum of the two particles after the collision. Since momentum is a vector quantity, momentum is conserved in each direction. We thus obtain

$$m_1 v_{1ix} + m_2 v_{2ix} = m_1 v_{1fx} + m_2 v_{2fx}, \tag{7.52}$$

$$m_1 v_{1iy} + m_2 v_{2iy} = m_1 v_{1fy} + m_2 v_{2fy}, \tag{7.53}$$

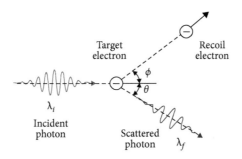

Fig. 7.11 Compton scattering: A photon of wavelength λ_i is scattered by an electron at rest making an angle θ with the horizontal and with a final wavelength λ_f.

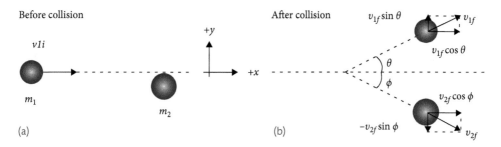

Fig. 7.12 A particle of mass m_1 moving with a velocity v_{1ix} collides with another particle of mass m_2 at rest. The scattered particles move in directions making angles θ and ϕ with the horizontal.

where v_{1ix} and v_{1iy} are the x- and y-components of the initial momentum of particle 1. Similarly v_{1fx} and v_{1iy} are the x- and y-components of the final momentum of particle 1. The same is true for particle 2. The conservation of energy requires that the total energy before the collision should be equal to the total energy after the collision, i.e.,

$$\frac{1}{2}m_1 v_{1i}^2 + \frac{1}{2}m_2 v_{2i}^2 = \frac{1}{2}m_1 v_{1f}^2 + \frac{1}{2}m_2 v_{2f}^2, \tag{7.54}$$

where $v_{1i}^2 = v_{1ix}^2 + v_{1iy}^2$ etc.

In our example above, $v_{1iy} = v_{2ix} = v_{2iy} = 0, v_{1ix} = v_{1i}, v_{1fx} = v_{1f}\cos\theta, v_{1fy} = v_{1f}\sin\theta,$ $v_{2fx} = v_{2f}\cos\phi, v_{2fy} = -v_{2f}\sin\phi$. With these substitutions, we can get simplified equations for the conservation of momentum and energy. These equations can then be solved for any unknown quantities.

We now apply a similar approach to the problem of Compton scattering—scattering of a photon by an electron initially at rest.

Since electrons can move with velocities approaching the speed of light, we cannot use the usual expressions for momentum and energy for the electron. Albert Einstein, in a seminal paper in 1905 (different from his photoelectric effect paper discussed in Chapter 6), had developed a theory of relativity which applied to objects that moved at very high speeds, speeds close to the speed of light, $c = 3 \times 10^8$m/s.

An important consequence of Einstein's theory of relativity is that energy and mass are interconvertible. Typically, there is conservation of energy. Therefore one form of energy can be converted into another form of energy but the total energy should be conserved. Similarly mass is also conserved, i.e., we can convert one form of mass into another but the total mass remains the same. However Einstein showed that energy and mass are interconvertible. For an object of mass m_0 at rest, the equivalent energy is

$$E = m_0 c^2, \tag{7.55}$$

where E is the energy and m_0 is the mass. The amount of energy produced by converting mass into energy is huge. For example, 1 gram of mass is equivalent to 10^{14} J $\equiv$ 25 kilotons of TNT of energy. This conversion of mass into energy is the source of nuclear energy. This is also the source of energy that is released in atomic and hydrogen bombs.

Einstein showed that, when a particle is moving with a velocity v, the energy–mass equivalence relation is modified and is given by

$$E = mc^2 = \frac{m_0 c^2}{\sqrt{1 - \left(\frac{v}{c}\right)^2}}. \tag{7.56}$$

Here m_0 is the mass of the particle at rest (when v = 0) and m is the mass of the moving object. The corresponding momentum of the particle is given by

$$p = mv = \frac{m_0 v}{\sqrt{1 - \left(\frac{v}{c}\right)^2}}. \tag{7.57}$$

We can see that for velocities much smaller than the speed of light, we recover the usual result. For v << c, the expressions of momentum and energy reduce to

$$p \approx m_0 v, \tag{7.58}$$

$$E \approx m_0 c^2 + \frac{1}{2} m_0 v^2. \tag{7.59}$$

In deriving Eq. (7.59), we used

$$\frac{1}{\sqrt{1 - \left(\frac{v}{c}\right)^2}} = \left(1 - \left(\frac{v}{c}\right)^2\right)^{-1/2} \approx 1 + \frac{1}{2}\left(\frac{v}{c}\right)^2. \tag{7.60}$$

when $v \ll c$. This follows from the binomial expansion $(1 + x)^a \approx 1 + ax$ for $x \ll 1$. Thus we see that the usual kinetic energy term is obtained in addition to the rest mass energy, in the small velocity limit.

We can solve Eq. (7.57) for v and find the relativistic factor:

$$1 - \left(\frac{v}{c}\right)^2 = \frac{m_0^2 c^2}{p^2 + m_0^2 c^2}. \tag{7.61}$$

On substituting this factor in Eq. (7.56), we obtain

$$E = \sqrt{\left(m_0 c^2\right)^2 + \left(cp\right)^2}. \tag{7.62}$$

This is the expression that we use for the energy of the electron in the conservation law. We note that, for massless particles ($m_0 = 0$) such as a photon, the energy–momentum relation is

$$E = pc.$$

This relation was used in Section 7.1 to "derive" the de Broglie relation

$$p = \frac{h}{\lambda}.$$

The energy–momentum relations are usually called the dispersion relations. The dispersion relations for the massive particle (Eq. (7.62)) and for the photon (Eq. (7.3)) are plotted in Fig. 7.13.

Next we apply the laws of conservation of energy and momentum in Compton scattering. According to the conservation of energy:

$$E_{pi} + E_{ei} = E_{pf} + E_{ef}, \tag{7.63}$$

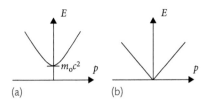

Fig. 7.13 The dispersion relation (a) for a massive particle as given by Eq. (7.62) and (b) for a photon as given by Eq. (7.3).

where

$$E_{pi} = \hbar v_i = \frac{hc}{\lambda_i} \tag{7.64}$$

is the initial energy of the photon,

$$E_{ei} = m_e c^2 \tag{7.65}$$

is the initial energy of the electron,

$$E_{pf} = \hbar v_f = \frac{hc}{\lambda_f} \tag{7.66}$$

is the final energy of the photon, and

$$E_{ef} = \sqrt{(m_e c^2)^2 + (p_{ef} c)^2} \tag{7.67}$$

is the final energy of the electron. It follows, on substituting these expressions in Eq. (7.63), that

$$\frac{hc}{\lambda_i} + m_e c^2 = \frac{hc}{\lambda_f} + \sqrt{(m_e c^2)^2 + (p_{ef} c)^2}. \tag{7.68}$$

A rearrangement of Eq. (7.68) leads to the following equation

$$p_{ef}^2 = \left(\frac{h}{\lambda_i} - \frac{h}{\lambda_f} + m_e c \right)^2 - (m_e c)^2. \tag{7.69}$$

Next we consider the conservation of momentum:

$$\boldsymbol{p}_{pi} + 0 = \boldsymbol{p}_{pf} + \boldsymbol{p}_{ef}, \tag{7.70}$$

where the initial momentum of the electron, p_{ei}, is equal to zero. In Eq. (7.70)

$$p_{pi} = \frac{h}{\lambda_i} \tag{7.71}$$

is the magnitude of the initial momentum of the photon, and

$$p_{pf} = \frac{h}{\lambda_f} \tag{7.72}$$

is the magnitude of the final momentum of the photon. It follows from Eq. (7.70) that the final momentum of the electron, p_{ef} can be calculated as

$$\begin{aligned}
p_{ef}^2 &= \left(\boldsymbol{p}_{pi} - \boldsymbol{p}_{pf} \right) \cdot \left(\boldsymbol{p}_{pi} - \boldsymbol{p}_{pf} \right) \\
&= p_{pi}^2 + p_{pf}^2 - 2 p_{pi} p_{pf} \cos \theta \\
&= \left(\frac{h}{\lambda_i} \right)^2 + \left(\frac{h}{\lambda_f} \right)^2 - 2 \left(\frac{h}{\lambda_i} \right) \left(\frac{h}{\lambda_f} \right) \cos \theta.
\end{aligned} \tag{7.73}$$

On equating the two expressions for p_{ef}^2, one derived on the basis of conservation of energy (Eq. (7.69)) and the other on the basis of conservation of momentum (Eq. (7.73)), we obtain

$$\left(\frac{hc}{\lambda_i} - \frac{hc}{\lambda_f} + m_e c^2\right)^2 - (m_e c^2)^2 = \left(\frac{hc}{\lambda_i}\right)^2 + \left(\frac{hc}{\lambda_f}\right)^2 - 2\left(\frac{hc}{\lambda_i}\right)\left(\frac{hc}{\lambda_f}\right)\cos\theta. \qquad (7.74)$$

On rearranging Eq. (7.74), we obtain the result:

$$\lambda_f - \lambda_i = \frac{h}{m_e c}(1 - \cos\theta)$$

which agrees with the experimental results of Compton. The quantity $h/m_e c = 2.43 \times 10^{-12}$ m is called the Compton wavelength of an electron.

As discussed at the beginning of this section, the significance of Compton's result is that the photon has been treated as a particle of momentum, $p = h/\lambda$.

Problems

7.1 What is the de Broglie wavelength of a 12.0 gram bullet traveling at the speed of sound? The speed of sound is 331 m/s.

7.2 What is the de Broglie wavelength of an electron with 13.6 eV of kinetic energy? (1 eV $= 1.60 \times 10^{-19}$J$)$

7.3 Find the temperature at which a gas of rubidium atoms with a density of 10^{20} m^{-3} forms a Bose–Einstein condensate. Rubidium is a chemical element with symbol Rb, atomic number 37, and a standard atomic weight of 85.47. The mass of a proton is 1.67×10^{-27} kg.

7.4 Calculate the de Broglie wavelength of an electron in the first Bohr orbit in the hydrogen atom.

7.5 If a particle of rest mass m_0 is moving with a velocity v, the relativistic energy and momentum are given by the expressions:

$$E = mc^2 = \frac{m_0 c^2}{\sqrt{1 - \left(\frac{v}{c}\right)^2}},$$

$$p = mv = \frac{m_0 v}{\sqrt{1 - \left(\frac{v}{c}\right)^2}}.$$

Here c is the speed of light in vacuum. Using these expressions, show that

$$1 - \left(\frac{v}{c}\right)^2 = \frac{m_0^2 c^2}{p^2 + m_0^2 c^2}.$$

Finally, show that the energy is related to the momentum via

$$E = \sqrt{(m_0 c^2)^2 + (cp)^2}.$$

7.6 Consider the Compton scattering set-up as shown in Fig. 7. 11. Using the laws of conservation of energy and momentum, show that the angle of scattering ϕ is given by

$$\cot\phi = \left(1 + \frac{h}{m_e c \lambda_i}\right)\tan(\theta/2).$$

BIBLIOGRAPHY

L. de Broglie, *The wave nature of the electron*, Nobel Lecture (1929).

E. A. Cornell and C. E. Wieman, *The Bose–Einstein condensate*, Scientific American 278, 40 (1998).

M. Jammer, *The Philosophy of Quantum Mechanics*, (Wiley 1974).

H. C. Corben, *Another look through the Heisenberg microscope*, American Journal of Physics 47, 1036 (1979).

B. G. Williams, *Compton scattering and Heisenberg's microscope*, American Journal of Physics 52, 425 (1984).

8

Quantum Interference: Wave–Particle Duality

The double-slit experiment, first carried out in 1802 by Thomas Young, played a crucial role in establishing the wave nature of light. This was in contrast to Newton's postulate that light consisted of small corpuscles. In the first quarter of the twentieth century, the concept of wave–particle duality firmly took root, motivating a deeper understating of the double-slit experiment, particularly for incident electrons instead of light beams. The experimental observation that incident electrons yield a similar interference pattern as that formed by light waves was a shocking result. Richard Feynman remarked in his famous Feynman Lectures that Young's double-slit experiment with electrons contains the deepest mystery of quantum mechanics. The only way the experimental results could be explained is via a wavefunction description of electrons. But the mystery does not stop there. If, in the same experiment, one can acquire the information about the path the electrons followed, the interference fringes disappear. This is the essence of the wave–particle duality.

Young's double-slit experiment was also at the center of the first of several debates between Albert Einstein and Niels Bohr on the foundations of quantum mechanics. Einstein came up with arguments that challenged Bohr's principle of complementarity by suggesting a clever scheme in which one can have both the wave and particle aspects exhibited in the same experiment. Bohr successfully defended the principle of complementarity by invoking Heisenberg's uncertainty relation.

The wave–particle aspect as embodied in the double-slit experiment has continued to lead to highly counterintuitive notions of delayed choice and quantum eraser effects showing how the availability or erasure of information generated in the past can affect how we interpret the data in the present. All these topics are discussed in the following sections of this chapter.

8.1 Young's Double-slit Experiment for Electrons

In Chapter 4, we discussed in great detail how, when a light beam passes through two slits, it can generate an interference pattern, a pattern of bright fringes separated by dark fringes, on a screen, as shown in Fig. 8.1a. The bright fringes are located at those points on the screen where the path difference between the light waves from the two slits is zero or an integral multiple of the wavelength λ, thus leading to constructive interference. The dark fringes are, on the other hand, located at those points where the path difference is equal to $(n + 1/2)\lambda$ (with n being an integer), leading to destructive interference.

Quantum Mechanics for Beginners: With Applications to Quantum Communication and Quantum Computing. M. Suhail Zubairy.
© M. Suhail Zubairy 2020. Published in 2020 by Oxford University Press. DOI: 10.1093/oso/9780198854227.001.0001

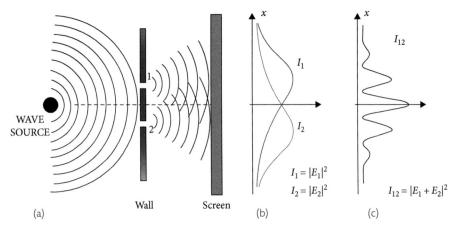

Fig. 8.1 Young's double-slit experiment with waves.

Central to this description is the wave nature of light. Light from a slit incident on the screen is described by an electric field of amplitude E. The complex amplitude of light from slit 1 is given by $E_1 = |E_1| \exp(i\delta_1)$ and from slit 2 is given by $E_2 = |E_2| \exp(i\delta_2)$. The measured intensity is given by $I = |E|^2$.

Thus, the intensity of light at the screen, when slit 2 is blocked and light can only pass through slit 1, is given by

$$I_1 = |E_1|^2 \tag{8.1}$$

and is shown in Fig. 8.1b by the curve I_1. Similarly when slit 1 is covered, light passes through slit 2 only and the light intensity at the screen is given by

$$I_2 = |E_2|^2 \tag{8.2}$$

and is shown by the curve I_2 in Fig. 8.1b.

When both slits are open, the total amplitude of light on the screen is $E_1 + E_2$ and the intensity of light is given by

$$I_{12} = |E_1 + E_2|^2. \tag{8.3}$$

Thus

$$I_{12} \neq I_1 + I_2. \tag{8.4}$$

Instead we have

$$\begin{aligned} I_{12} &= I_1 + I_2 + \left(E_1^* E_2 + E_1 E_2^*\right) \\ &= I_1 + I_2 + 2|E_1||E_2| \cos\delta \\ &= I_1 + I_2 + 2\sqrt{I_1 I_2} \cos\delta, \end{aligned} \tag{8.5}$$

where $\delta = \delta_1 - \delta_2$ is the phase difference between the fields E_1 and E_2. The last term in the bracket is responsible for the interference. The intensity pattern on the screen is depicted by the curve I_{12} in Fig. 8.1c.

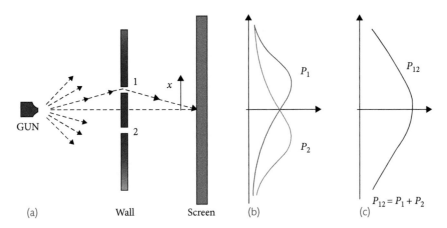

Fig. 8.2 Young's double-slit experiment with bullets.

Next we consider the double-slit experiment with particles like bullets as shown in Fig. 8.2. Here a gun is a source of bullets, sent in the forward direction, which are spread over a wide angle. These bullets can pass through holes 1 and 2 in a wall and hit a screen where they are detected. Unlike the light waves, there is no interference in this case. What we observe is the following.

When hole 2 is covered, bullets pass only through slit 1. The probability of a single bullet hitting the screen at a location x is given by P_1. This is shown by the curve marked P_1. The maximum of P_1 occurs at the value of x which is on a straight line with the gun and slit 1. When a large number of bullets are incident on the screen, their distribution (fraction of the total number of bullets hitting the screen) is given by the curve marked P_1. This curve is identical to the curve I_1 for the waves in Fig. 8.1b. When hole 1 is closed, bullets can only pass through hole 2 and we get the symmetric curve for the distribution P_2. When both holes are open, the bullets can pass through hole 1 or they can pass through hole 2 and the resulting distribution of the bullets on the screen is

$$P_{12} = P_1 + P_2. \tag{8.6}$$

The probabilities just add together. The effect with both holes open is the sum of the effects with each hole open alone. We call this result an observation of "*no interference.*" An important point to note here is that, for each bullet detected on the screen, we know (at least in principle) which hole it came from, i.e., we have the "*which-path*" information for each bullet. Indeed we can determine the full trajectory of each bullet from the point it leaves the gun and hits the screen.

So far, we have considered Young's double-slit experiment with waves and with bullets. In case of waves, the field amplitudes add and we find interference. However, when we repeat the same experiment with bullets, the probabilities add up and we find no interference.

What about Young's double-slit experiment with electrons? Do electrons behave like bullets or do they behave like waves?

We consider electrons being emitted by an electron gun. This beam of electrons passes through a wall with two slits as shown in Fig. 8.3a. The set-up is identical to the set-up for the double-slit experiment for bullets. But do we get the same result as those for the bullets?

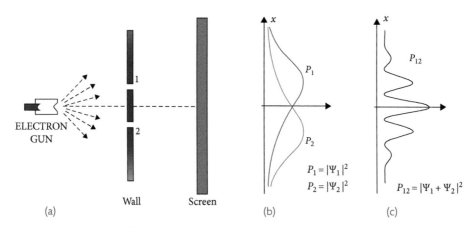

Fig. 8.3 Young's double-slit experiment with electrons.

When slit 2 is closed, electrons can only pass through slit 1. The probability of a single electron to hit the screen at location x is P_1 which is shown in Fig. 8.3b. The similar symmetric curve P_2 is obtained when slit 1 is closed and the electron can pass through slit 2 only. These curves are identical to the corresponding curves when the bullets are incident on the screen and also identical to the intensity distribution when a beam of light is incident.

But what happens, when both slits are open? Do electrons behave like particles as bullets do or they behave like waves as a light beam behaves? The results are shown in Fig. 8.4. Here we see the build-up of electrons on the screen. For 100 electrons, the distribution of the detected electrons on the screen appears to be random. After about 1000 electrons are detected, the distribution on the screen seems to have some regions with a dense distribution compared to other regions. But still it is difficult to conclude anything regarding the particle or wave behavior of the electrons.

After 10 000 electrons are detected on the screen, there is an unmistakable interference pattern with bright fringes, separated by dark fringes. The individual electrons are detected one by one, but instead of giving a pattern that is similar to that corresponding to bullets, we find that the electrons are detected in some regions and not in others. This is a stunning result. How do the electrons know where to hit the screen such that we see an interference pattern emerging after a large number of electrons hit the screen?

This experiment was proposed by Richard Feynman in his famous Feynman Lectures in 1965 in these words:

> We choose to examine a phenomenon which is impossible, *absolutely* impossible, to explain in any classical way, and which has in it the heart of quantum mechanics.

He, however, claimed that the experiment is too difficult to carry out and may never be done. What Feynman apparently did not know was that a double-slit experiment with electrons had already been done by Claus Jönsson in 1961.

The situation becomes more mysterious when a slight variation of this experiment gives us a completely different outcome.

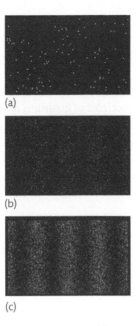

(a)

(b)

(c)

Fig. 8.4 The outcome of the Young's double-slit experiment with (a) 100 electrons, (b) 1000 electrons, and (c) 10 000 electrons.

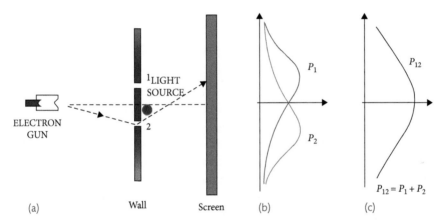

ELECTRON GUN

^{1}LIGHT SOURCE

2

(a)   Wall   Screen   (b)   (c)

P_1

P_2

P_{12}

$P_{12} = P_1 + P_2$

Fig. 8.5 Young's double-slit experiment with which-path information.

Let us place a source of light between the two slits as shown in Fig. 8.5. When an electron passes the slits, light scatters from the electron and provides us the which-path information. In this case, the interference disappears and the result is depicted in Fig. 8.5c, which is identical to the result obtained for the double-slit experiment with bullets. This is in contrast to the experiment depicted in Fig. 8.3, where we had a lack of knowledge about the path each individual electron took. This lack of which-path knowledge seems to be responsible for interference.

Thus if we "look" at *which path* each electron followed, the interference disappears and we get the same distribution on the screen as for particles. So either we get interference when we have *no which-path* information or we lose interference when we have the *which-path* information.

No classical explanation can describe these observations. We can reconcile these observations only with the fundamental principles of quantum mechanics discussed in Chapter 5.

We describe the electron, not as a particle traveling in a well-defined trajectory, but by a wavefunction $\psi(\mathbf{r})$ which is a complex function of position. At any point R on the screen, there are two contributions for the same electron coming from the two slits, $\psi_1(R)$ and $\psi_2(R)$.

When slit 2 is closed, the total wavefunction at the position R is $\psi_1(R)$ and the probability of finding the electron is

$$P_1 = |\psi_1|^2 \tag{8.7}$$

Similarly, when slit 1 is closed, the total wavefunction at the position R is $\psi_2(R)$ and the probability of finding the electron is

$$P_2 = |\psi_2|^2 \tag{8.8}$$

When both slits are open, the total wavefunction of the electron at the position R is

$$\psi(R) = \psi_1(R) + \psi_2(R) \tag{8.9}$$

and the probability of finding the electron is

$$\begin{aligned} P_{12} &= |\psi_1 + \psi_2|^2 \\ &= |\psi_1|^2 + |\psi_2|^2 + (\psi_1^*\psi_2 + \psi_1\psi_2^*) = |\psi_1|^2 + |\psi_2|^2 + 2|\psi_1||\psi_2|\cos\theta. \end{aligned} \tag{8.10}$$

Here $\psi_1 = |\psi_1|\, exp(i\theta_1), \psi_2 = |\psi_2|\, exp(i\theta_2)$, and $\theta = \theta_1 - \theta_2$. The angle θ depends on the location on the screen. The last term is the interference term which, depending on θ, can become equal and opposite to $|\psi_1|^2 + |\psi_2|^2$ at certain locations, giving us a zero probability of finding the electron at those locations and is responsible for the interference. The wavefunctions ψ_1 and ψ_2 seem to play the same role as the complex fields E_1 and E_2 in the case of Young's double-slit experiment with waves. However there is one crucial difference: The quantities $I_1 = |E_1(R)|^2$ and $I_2 = |E_2(R)|^2$ are the intensities of the light coming to the point R from slits 1 and 2 whereas the quantities $|\psi_1(R)|^2$ and $|\psi_2(R)|^2$ are the probabilities that the electron coming from slits 1 and 2 hits the screen at the point R, respectively.

If an experiment is performed which is capable of determining whether the electrons passed through slit 1 or slit 2, the probability of finding the electron at a point R on the screen is the sum of the probabilities for each alternative,

$$P_{12} = P_1 + P_2, \tag{8.11}$$

and the interference is lost.

This concept of wave–particle duality has been a source of intense discussion since the earliest days of quantum mechanics. How the same electron can behave like a wave in one situation and a particle in another is quite mysterious. Wave–particle duality was the subject of a fierce debate between Albert Einstein and Niels Bohr, as we discuss in the next section.

Richard Feynman expresses his amazement at these incredible results in these words:

> One might still like to ask: "How does it work? What is the machinery behind the law?" No one has found any machinery behind the law. No one can "explain" any more than we have just "explained." No one will give you any deeper representation of the situation. We have no ideas about a more basic mechanism from which these results can be deduced.

Here we discussed the experiment with electrons. The same can be said about a similar experiment with light. If we treat the light beam in a Young's double-slit experiment as consisting of a large number of photons, the situation is similar to the interference experiment with electrons. The reason we get interference of a light beam in a double-slit experiment is due to the lack of which-path information for each photon. If somehow we are able to get which-path information for each photon, the interference disappears.

This is the essence of the wave–particle duality or Bohr's principle of complementarity: Electrons and photons can behave like waves when we have no which-path information and they behave like particles when we have the which-path information.

8.2 Einstein–Bohr Debate on Complementarity

In 1927 Niels Bohr introduced the principle of complementarity, which can be stated as follows: In any quantum mechanical experiment, certain physical concepts are complementary. If the experiment clearly illustrates one concept the other concept will be completely obscured. As an example, if the particle nature of an object is exhibited in an experiment then the wave nature will be completely obscured. Thus, in the double-slit experiment, we can either have the which-path information or the existence of an interference pattern. According to Bohr's principle of complementarity, they can never be observed at the same time, in the same experiment.

Einstein, however, came up with a clever scheme such that we can have both which-path information (particle nature) and interference (wave nature) in the same experiment—violating Bohr's principle of complementarity.

In Einstein's proposed experiment a wall with two slits is placed on rollers so that it can move freely in the vertical direction as shown in Fig. 8.6. This is an example of what we call a *thought experiment* or a *gedanken experiment*. In other words, we do not *actually* conduct the experiment, we use only our imagination and reasoning instead. An electron gun shoots electrons towards the wall where they can pass through the two slits and then onto the back screen to create the interference pattern. The electrons have momentum in the forward direction. However the electron beam is spread and can have small momentum components in both $+x$ and $-x$ directions. For example, the electrons passing through slit 1 should have a momentum component along the x-axis equal to p_1 and those passing through slit 2 should have the x-momentum component equal to p_2. After passing through the slits, there is a momentum change in the electrons. For those passing through slit 1, if the final momentum in the x-direction is p_1', the momentum change of that electron is $\delta p_1 = p_1' - p_1$. Similarly for the electrons passing through slit 2, the momentum change is $\delta p_2 = p_2' - p_2$.

Einstein argued that if the wall is on rollers then, by the law of conservation of momentum, it should recoil with a momentum equal to the change in momentum of the electron and in

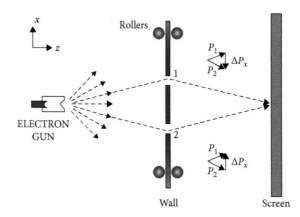

Fig. 8.6 Einstein's *gedanken* experiment. Electrons pass through a narrow slit through a wall that can freely move on a roller. The momentum transfer to the wall provides the which-path information. At the same time electrons from the two slits can form an interference pattern on the screen.

the opposite direction. Thus electrons that pass through the upper slit (slit 1) should impart a momentum equal to $-\delta p_1$ to the wall. If the electron is deflected in the downward direction as shown in Fig. 8.6, then the momentum of the wall should be in the upward direction. Similarly, the electrons passing through the lower slit (slit 2) and deflected in the upper direction will give a downward kick to the wall. Therefore, for every position of the detector on the screen, the momentum received by the wall will have a different value for a traversal via slit 1 than for a traversal via slit 2. Since an electron has a very small mass, the momentum change is very tiny and it may be difficult to measure the momentum change of the wall. However, no matter how small this momentum is, it should in principle be detectable. So without disturbing the electrons *at all*, but just by watching the *wall*, we can tell which path the electron used.

Einstein then argued that, after passage through the slits, the undisturbed electrons can proceed to the screen and give the interference pattern as before. However, we can get the information about which slit the electrons passed through by measuring the momentum of the wall after each electron has passed through. Thus we have both the 'which-path' information and the interference. This is in contradiction to Bohr's principle of complementarity.

This was a forceful argument against the foundational principle of quantum mechanics and Bohr had to respond to it immediately. Leon Rosenfeld records the encounter in his book *Fundamental Problems in Elementary Particle Physics* (Proceedings of the Fourteenth Solvay Conference, Interscience, New York, p. 232) in the following words:

> ... Einstein thought he had found a counterexample to the uncertainty principle. It was quite a shock for Bohr ... he did not see the solution at once. During the whole evening he was extremely unhappy, going from one to the other and trying to persuade them that it couldn't be true, that it would be the end of physics if Einstein were right; but he couldn't produce any refutation. I shall never forget the vision of the two antagonists leaving the club [of the Fondation Universitaire]: Einstein a tall majestic figure, walking quietly, with a somewhat ironical smile, and Bohr trotting near him, very excited ... The next morning came Bohr's triumph.

Bohr invoked the Heisenberg uncertainty relation to refute Einstein's argument and saved the principle of complementarity.

According to the Heisenberg uncertainty relation, if we determine the x-component of its momentum with an uncertainty Δp, we cannot, at the same time, know its x-position more accurately than $\Delta x = \hbar/2\Delta p$ (Section 7.4). In Einstein's argument, it is necessary to know the momentum of the wall before the electron passes through it sufficiently precisely. This is required as we need to know the change in the momentum of the wall in the x-direction after the electron has passed in order to obtain the which-path information. However, according to the Heisenberg uncertainty principle, we cannot know the position of the wall in the x-direction with arbitrary accuracy. Therefore a precise measurement of momentum means that the locations of the slits become indeterminate. The uncertainty in the location of the slits means that the electrons effectively see a blurred pair of slits. The locations where electrons hit the screen consequently become random and the center of the interference pattern has a different location for each electron, thus wiping out the interference pattern. This shows that the which-path information in the Young's double-slit experiment smears the interference pattern.

In order to quantitatively see this result, we consider a slightly different set-up as shown in Fig. 8.7. Here a beam of electrons is first sent along the z-axis through a wall with a narrow opening that selects only those electrons moving along the z-axis. Before hitting the wall the x-component of the momentum of these electrons is zero. After passing through the slit they diffract in the x-direction. Electrons can pass through another wall at a distance L with double slits. The separation between the two slits is d. The electrons are detected on the screen another distance L away. The first wall is placed on a roller such that it can freely move in the x-direction.

The incident electrons move with a momentum

$$p_0 = \frac{h}{\lambda} \tag{8.12}$$

along the z-axis. Here λ is the de Broglie wavelength of the electrons. After passing the first wall, they acquire momentum in the x-direction. Since these electrons pass through the two

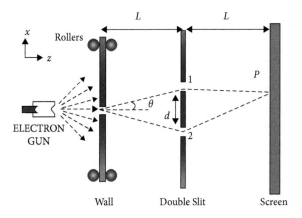

Fig. 8.7 An analysis of Einstein's *gedanken* experiment. Electrons pass through a wall giving it a push in the upward or the downward direction depending on whether the incoming electron is scattered in the downward or upward direction. This provides the which-path information. These electrons then pass the double slit and form a pattern on the screen.

slits at the distance L from the first wall and located at $x = \pm(d/2)$ the x-component of the momentum can range from $-p_0\sin\theta$ for electrons going below the z-axis to $+p_0\sin\theta$ for electrons going above the axis. By the conservation of momentum, the corresponding recoil momentum on the first wall therefore ranges from $+p_0\sin\theta$ for electrons going below the z-axis to $-p_0\sin\theta$ for electrons going above the axis. Thus the limit on the accuracy of measuring the recoil momentum is

$$\Delta p = +p_0 \sin\theta - \left(-p_0 \sin\theta\right) = 2p_0 \sin\theta \approx 2p_0\theta = 2\frac{h}{\lambda}\frac{d}{2L} = \frac{hd}{\lambda L}, \tag{8.13}$$

where we assume $\theta \ll 1$ and $\sin\theta \approx \theta$. According to the Heisenberg uncertainty relation, the minimum uncertainty in the position of the source slit is

$$\Delta x \approx \frac{h}{\Delta p} = \frac{\lambda L}{d}, \tag{8.14}$$

where we substituted for Δp from Eq. (8.13). Thus, if the electron momentum in the x-direction is known with sufficient accuracy to find the *which-slit* information, the location of the slit in the first wall is uncertain by an amount given by Eq. (8.14). This leads to a corresponding uncertainty in the location where the electron hits the screen. We recall that the fringe spacing in the double slit experiment is $\lambda L/d$ (Eq. (4.54). The resulting pattern on the screen becomes blurred to the extent that the interference pattern is lost. This clearly shows that the which-path information leads to the disappearance of the interference pattern—Bohr's principle of complementarity is saved, thanks to the Heisenberg uncertainty relation.

Richard Feynman, in his Lectures, describes the role of the uncertainty relation in keeping the foundations of quantum mechanics secure in the following words:

> The uncertainty principle "protects" quantum mechanics. Heisenberg recognized that if it were possible to measure the momentum and the position simultaneously with a greater accuracy, then quantum mechanics would collapse. So he proposed that it must be impossible. Then people sat down and tried to figure out ways of doing it, and nobody could figure out a way to measure the position and the momentum of anything—a screen, an electron, a billiard ball, anything—with any greater accuracy. Quantum mechanics maintains its perilous but still correct existence.

8.3 Delayed Choice

In the Young's double-slit experiment, whether we get the interference fringes or not depends on whether we have no which-path information or we have the which-path information. Thus a photon behaves like a wave or a particle depending upon what kind of an experiment we decide to do. If we decide not to look at the photon when it is passing through the slits, it behaves like a wave. However, if we decide to find which slit the photon goes through, it behaves like a particle. This wave–particle duality is very mysterious and it becomes even more so when we try to address the question whether the photon knew in advance what behavior it should exhibit. This question was addressed by John Wheeler in his "delayed choice" *gedanken* experiment.

In Wheeler's *gedanken* experiment, photons are generated by cosmic objects like quasars. They are split into two paths with the galaxy acting as a gravitational lens as shown in Fig. 8.8. Photons can follow either path, left of the galaxy or right of the galaxy. After having traveled a distance of billions of miles the photons arrive at earth where we detect these photons in one of the two different experimental set-ups.

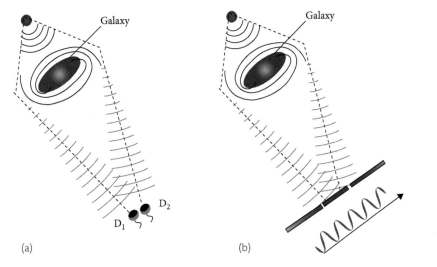

Fig. 8.8 Wheeler's delayed choice experiment. Light that left a quasar millions of years ago can be made to act like a particle or a wave depending on our choice of the experimental set-up.

In the first set-up, we place two detectors D_1 and D_2. The detector D_1 clicks if the photon followed the left path and the detector D_2 clicks if the photon followed the right path. Thus a click at either D_1 or D_2 provides the which-path information. For example, for a click at D_1, we can conclude that the photon was in the left path all along for all those billions of years. Similarly, for a click at D_2, we can conclude that the photon was in the right path all along.

The other possibility is to pass them through the two slits in a Young's double-slit experiment and get an interference pattern. We can then conclude that the photons behaved like waves and they went through both ways around the galaxy.

Therefore, in the first case the photons appear to pass through only one side of the galaxy and behave like particles and in the latter case they behave like waves and go through both ways around the galaxy. The paradoxical situation is that it depends on the experimenter's "delayed choice" whether the photon generated billions of years ago behaves like a particle or a wave. Until the experiment is done, we cannot say whether the photon will behave as a particle or as a wave.

8.4 Quantum Eraser

An even more counterintuitive aspect of wave–particle duality is the notion of the "quantum eraser" introduced by Marlan Scully and Kai Drühl in 1982. In the Young's double-slit experiment, we get an interference pattern if we have no knowledge about which slit the photon went through. However if we somehow obtain the which-path information then the interference is lost. Scully posed the question: Is it possible to "erase" the which-path information and recover the interference pattern *after* the photon has passed through the slits and is detected on the screen? The quantum eraser brings out the counterintuitive aspects related to time in the quantum mechanical domain.

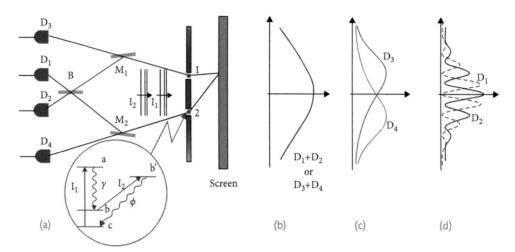

Fig. 8.9 Schematics of the quantum eraser experiment. (*a*) The single-photon pulses l_1 and l_2 incident on two atoms help to generate two photons γ and ϕ by either atoms at site 1 or the atom at site 2. The γ photons proceed to the screen and the ϕ photon to the left. (*b*) The distribution of γ photons when no detection is made at detectors D_1, D_2, D_3, or D_4. This distribution is also obtained when either both mirrors M_1 and M_2 are removed and the clicks are at detectors D_3 and D_4 or the mirrors M_1 and M_2 are in place and the clicks are at detectors D_1 and D_2. (*c*) The distributions of the γ photons for clicks at D_3 and D_4. The which-path information destroys the interference. (*d*) The distributions of the γ photons for clicks at D_1 and D_2. In this case we do not have the which-path information and interference is obtained.

We present a simple description of the quantum eraser as depicted in Fig. 8.9. Instead of two slits, we consider the scattering of light from two atoms on the screen.

The two atoms are placed at sites 1 and 2. Each atom is of the type shown in the inset of Fig. 8.9. There are four atomic levels a, b, b', and c and the atoms are initially in level c. These atomic levels are of the type we discussed for the hydrogen atom in Section 6.5. The atom can absorb a photon and make a jump from a lower level to a higher level if the energy difference between the two levels is the same as the energy of the incident photon. Similarly, an atom in the excited state can jump to the lower level and emit a photon whose energy (and frequency) matches the level spacing. These atoms are excited by pulses l_1 and l_2 which carry just enough energy to excite only one atom from level c to a and from level b to b', respectively.

The photon pulse l_1 tuned to *c-a* transition excites one atom (we do not know which one) to level a. The other atom remains in level c. The excited atom emits a photon by making a jump from level a to level b. We call such a photon a γ photon. The photon pulse l_2 excites the atom from level b to b'. The atom finally makes a transition from level b' to level c emitting a photon we call ϕ photon. Thus, after the passage of the pulses l_1 and l_2, one of the atoms (we do not know which one) has generated two photons, γ and ϕ and both atoms are found in the ground state c after the scattering process is complete.

We repeat this scattering process a large number of times. We consider only those instances where the γ photon proceeds to the right to the screen and the ϕ photon proceeds to the left to the mirrors M_1 and M_2. The γ photons are collected on the screen as in the usual double-slit experiment. The ϕ photon is detected by one of the detectors D_1, D_2, D_3, or D_4 after passing through the optical set-up consisting of the mirrors M_1, M_2, and the beam splitter B. The role of the beam splitter B is to let the photon get transmitted or get reflected with equal probability.

For example, a photon reflected from the mirror M_1 can either get reflected through B and be detected at D_1 or be transmitted and detected at D_2, with equal probability. A detailed analysis of a beam splitter for a single photon is given in Section 9.4. The γ photons from atom 1 or atom 2 play the same role as the light passing through the two slits in the Young's double-slit experiment. The ϕ photons can be employed to manipulate the which-path information as described below.

This experiment yields a distribution of γ photons on the screen as shown in Fig. 8.9. But what about the appearance and disappearance of interference fringes discussed above? For this purpose we look at the ϕ photons that proceed to the left.

The ϕ photon, if emitted by atom 1, proceeds to mirror M_1 and, if emitted by atom 2, to mirror M_2. The distance between the screen (with two atoms) and the mirrors M_1 and M_2 is assumed to be much larger than the distance between the atoms and the screen where the γ photons are detected.

For each ϕ photon, a choice is made: either both mirrors M_1 and M_2 are removed OR they are kept in place. In the case where the mirrors M_1 and M_2 are removed the photon proceeds unhindered and there is a click at either detector D_3 or D_4. On the other hand, if the mirrors M_1 and M_2 are in place, there is a click either at detector D_1 or D_2.

For each detection of a γ photon on the screen, we thus have four possibilities for the detection of the corresponding ϕ photon: It can be detected at detectors D_1 or D_2 or at detectors D_3 or D_4 depending on whether the mirrors M_1 and M_2 are in place or they are removed. Let us examine these cases.

First we consider the case when a decision is made to remove both mirrors M_1 and M_2. In this case there is a click either at D_3 or D_4.

If the ϕ photon is detected at D_3, there is only one path possible, namely $1D_3$. The ϕ photon must have come from atom 1. We thus have the information about the atom that generated the ϕ photon. The corresponding γ photon must have been generated by atom 1 as well and we acquire the which-path information for the γ photon on the screen.

Following the same reasoning, we conclude that, if the ϕ photon is detected at D_4, it must have come from atom 2. The corresponding γ photon must have been generated by atom 1 as well and, again, we acquire the which-path information for the γ photon on the screen.

Next we consider the case when a decision is made to keep both mirrors M_1 and M_2. In this case there is a click either at D_1 or D_2.

If the ϕ photon is detected at D_1, there is an equal probability that it may have come from the atom located at 1 following the path $1M_1BD_1$ or it may have come from the atom located at 2 following the path $2M_2BD_1$. Thus we have erased the information about which atom scattered the ϕ photon and there is no which-path information available for the corresponding γ photon.

The same can be said about the ϕ photon detected at D_2. There is an equal probability that it may have come from the atom located at 1 following the path $1M_1BD_2$ or it may have come from the atom located at 2 following the path $2M_2BD_2$. There is, however, a phase shift of π, as we get two reflections in case 1 and one reflection and one transmission in case 2.[1]

---

[1] The π phase shift is understood using the properties of the beam splitter that we formally derive in Sections 9.3 and 9.4.

After this experiment is done a large number of times, we shall have ϕ photons detected each at detectors D_1, D_2, D_3, or D_4. The spatial distribution for all the collected γ photons in the absence of any sorting is given in Fig. 8.9b. Next we do a sorting process. We separate out all the events where the ϕ photons are detected at detectors D_1, D_2, D_3, and D_4. For these four groups of events, we locate the positions of the detected γ photons on the screen.

And now comes the key result! For the events corresponding to the detection of ϕ photons at detectors D_3 and D_4, the pattern obtained by the γ photons on the screen is the same as we would expect if these photons had scattered from atoms at sites 1 and 2, respectively. This is shown in Fig. 8.9c. That is, there are no interference fringes as would be expected when we have which-path information available. On the contrary, we obtain phase-shifted interference fringes for those events where the ϕ photons are detected at D_1 and D_2. This is shown in Fig, 8.9d. For this set of data there is no which-path information available for the corresponding γ photons.

Mathematically we can understand the essential results of the Scully–Drühl quantum eraser by first realizing that the photon state emitted by the atoms located at sites 1 and 2 is given by

$$\Psi = \frac{1}{\sqrt{2}} \left(\psi_{\gamma_1} \psi_{\phi_1} + \psi_{\gamma_2} \psi_{\phi_2} \right), \tag{8.15}$$

i.e., either the photon pair γ_1, ϕ_1 is emitted by the atom located at site 1 or pair γ_2, ϕ_2 is emitted by the atom located at site 2. Thus if the ϕ photon is detected by D_3, the quantum state reduces to ψ_{γ_1}. A similar result is obtained for the ϕ photon detection by D_4. This is the situation when the which-path information is available and the sorted data yields no interference fringes.

The physics behind the retrieval of the fringes is made clear by rewriting the state Ψ as[2]

$$\Psi = \frac{1}{2} \left(\psi_{\gamma_1} + \psi_{\gamma_2} \right) \psi_{\phi_+} + \frac{1}{2} \left(\psi_{\gamma_1} - \psi_{\gamma_2} \right) \psi_{\phi_-}, \tag{8.16}$$

where

$$\psi_{\phi_+} = \frac{1}{\sqrt{2}} \left(\psi_{\phi_1} + \psi_{\phi_2} \right) \tag{8.17}$$

is the symmetric state of the ϕ photon at the detector D_1 after passage through the beam splitter B, and

$$\psi_{\phi_-} = \frac{1}{\sqrt{2}} \left(\psi_{\phi_1} - \psi_{\phi_2} \right), \tag{8.18}$$

is the antisymmetric state of the ϕ photon at the detector D_2 after passage through the beam splitter B. Thus a click at detectors D_1 or D_2 reduces the state of a γ photon to

$$\psi_{\gamma_+} = \frac{1}{\sqrt{2}} \left(\psi_{\gamma_1} + \psi_{\gamma_2} \right) \tag{8.19}$$

---

[2] Here again the symmetric and antisymmetric states ψ_{ϕ_+} and ψ_{ϕ_-} are obtained at the detectors D_1 and D_2, respectively, by using the properties of the beam splitter that we derive in Sections 9.3 and 9.4.

or

$$\psi_{\gamma_-} = \frac{1}{\sqrt{2}}\left(\psi_{\gamma_1} - \psi_{\gamma_2}\right),$$ (8.20)

respectively, leading to a retrieval of the interference fringes according to Eqs. (8.9) and (8.10).

Thus, in summary, a detection of the ϕ photon at the detectors D_3 or D_4 corresponds to the probabilities $|\psi_{\gamma_1}|^2$ or $|\psi_{\gamma_2}|^2$, respectively, of the γ photon being detected on the screen, leading to no interference. This situation is similar to the double-slit experiment with which-path information as in Fig. 8.4 and the results are depicted in Fig. 8.9c. However, a detection of the ϕ photon at detectors D_1 or D_2 corresponds to the probabilities $|\psi_{\gamma_1} + \psi_{\gamma_2}|^2$ or $|\psi_{\gamma_1} - \psi_{\gamma_2}|^2$, respectively, of the γ photon being detected on the screen, leading to interference in both cases due to a lack of which-path information as in Fig. 8.3, and the present result is shown in Fig. 8.9d.

The remarkable result is that we can place the ϕ photon detectors, D_1, D_2, D_3, and D_4, far away—very far away, such that we can make the decision whether to remove the mirrors M_1 and M_2, thus acquiring the which-path information, or to place the mirrors to lose the which-path information long after the γ photon is detected on the screen. Thus the future measurements on the ϕ photons influence the way we think about the γ photons measured today (or yesterday!). For example, we can conclude that γ photons, whose ϕ partners were successfully used to ascertain which-path information by removing the mirrors M_1 and M_2 resulting in clicks at D_3 or D_4, can be described as having (in the past) originated from site 1 or site 2. We can also conclude that γ photons, whose ϕ partners had their which-path information erased by placing the mirrors M_1 and M_2 resulting in clicks at D_1 and D_2, cannot be described as having (in the past) originated from site 1 or site 2, but must be described, in the same sense, as having come from both sites. The future helps shape the story we tell of the past. This is a highly counterintuitive and startling result. The scheme for the quantum eraser discussed above has been realized experimentally.

In his book, *The Fabric of the Cosmos*, Brian Greene sums up beautifully the counterintuitive outcome of the experimental realizations of the quantum eraser:

> These experiments are a magnificent affront to our conventional notions of space and time. Something that takes place long after and far away from something else nevertheless is vital to our description of that something else. By any classical-common sense-reckoning, that's, well, crazy. Of course, that's the point: classical reckoning is the wrong kind of reckoning to use in a quantum universe. For a few days after I learned of these experiments, I remember feeling elated. I felt I'd been given a glimpse into a veiled side of reality.

Problems

8.1 Electrons of momentum p fall normally on a pair of slits separated by a distance d. What is the distance between adjacent maxima of the interference fringe pattern formed on a screen a distance L beyond the slits? *Note: You may assume that the width of the slits is much less than the electron de Broglie wavelength.*

8.2 In an experiment performed by Jönsson in 1961, electrons were accelerated through a 50 kV potential towards two slits separated by a distance $d = 2 \times 10^{-4}$ cm, then detected on a screen $L = 35$ cm beyond the slits. Calculate the electron's de Broglie wavelength, λ, and the fringe spacing Δy. *Note: kinetic energy of electrons is equal to eV.*

8.3 In an interference experiment with electrons, we find the most intense fringe is at $y = 7.0$ cm. There are slightly weaker fringes at $y = 6.0$ cm and 8.0 cm, still weaker fringes at $y = 4.0$ cm and 10.0 cm. No electron are detected at $y < 0$ cm or $y > 14$ cm. Sketch a graph of $|\psi|^2$.

BIBLIOGRAPHY

R. P. Feynman, R. Leighton, and M. Sands, *The Feynman Lectures on Physics, Vol. IIIA* (Addison-Wesley, Reading, MA 1965).

C. Jönsson, Zeitschrift für Physik, 161, 454 (1961); translated in C. Jönsson, *Electron Diffraction at Multiple Slits*, American Journal of Physics 42, 4 (1974).

J. A. Wheeler, *The "Past" and the "Delayed-Choice Double-Slit Experiment"*, in A.R. Marlow, editor, *Mathematical Foundations of Quantum Theory*, Academic Press (1978).

M. O. Scully and K. Drühl, *Quantum eraser: A proposed photon correlation experiment concerning observation and "delayed choice" in quantum mechanics*, Physical Review A 25, 2208 (1982).

M. O. Scully, B.-G. Englert, and H. Walther, *Quantum optical tests of complementarity*, Nature 351,111 (1991).

Y. Aharonov and M. S. Zubairy, *Time and the quantum: Erasing the past and impacting the future*, Science 307, 875 (2005).

S. P. Walborn, M. O. Terra Cunha, S. Pádua, and C. H. Monken, *Double-slit quantum eraser*, Physical Review A 65, 033818 (2002).

Y.-H. Kim, R. Yu, S. P. Kulik, Y. Shih, and M. O. Scully, *Delayed "choice" quantum eraser*, Physical Review Letters 84, 1 (2000).

R. J. Scully and M. O. Scully, *The Demon and the Quantum* (John-Wiley-VCH, 2010).

M. O. Scully and M. S. Zubairy, *Quantum Optics* (Cambridge University Press, 1997).

B. Greene, *The Fabric of the Cosmos* (Alfred A. Knopf, New York 2004).

9

Simplest Quantum Devices: Polarizers and Beam Splitters

As we discussed in Section 6.1, James Clark Maxwell showed that light consists of electric and magnetic fields that oscillate in directions perpendicular to the direction of propagation. The electric and magnetic fields also oscillate in directions mutually perpendicular to each other. Thus the directions of the electric field and the magnetic field and the direction of propagation are all perpendicular to each other. Light is therefore a transverse electromagnetic wave. Associated with this picture of light as an electromagnetic wave is an important property— the polarization of light. The polarization of light is related to the direction of oscillation of the electric field in an electromagnetic wave. The classical properties of polarization have been studied for more than two centuries. An important question relates to the propagation of polarized light through certain materials called polarizers and the changes in the characteristics of a light beam when it passes through a polarizer.

The polarization property of a single photon is a major resource in our understanding of the basic laws of quantum mechanics. It is interesting to note that one of the founding fathers of quantum mechanics, Paul Dirac, described the passage of a single photon through a polarizer to discuss the mysteries of quantum mechanics in the very first pages of perhaps the very first book on quantum mechanics written by him in 1930. In this chapter, we follow this approach and show how some of the basic laws of quantum mechanics can be understood by an analysis of the simplest of systems—a polarizer. We also discuss the transformation properties of beam splitters. These properties when applied to a single photon can lead to many novel phenomena that are the subjects of later chapters.

9.1 Polarization of Light

Electromagnetic waves are transverse waves as the direction of oscillation of the electric and magnetic fields, E and B, are perpendicular or transverse to the direction of motion k as shown in Fig. 9.1. This is different from longitudinal waves such as sound waves where the wave is oscillating in the direction of propagation.

Polarization of light: An important property of the electromagnetic field is polarization. The polarization of an electromagnetic wave is described as the direction of the electric field oscillation. As the field oscillates in both positive and negative directions, it is equivalent if we describe the direction of polarization to be along (say) +y-axis or along –y-axis. The possible directions of polarization therefore range over π angles (from 0 to π).

Natural light sources such as sunlight as well as many artificial sources such as a light bulb emit light with random directions of polarization. For a light beam from such sources, the electric field vector oscillates in all possible π directions in the plane perpendicular

Quantum Mechanics for Beginners: With Applications to Quantum Communication and Quantum Computing. M. Suhail Zubairy.
© M. Suhail Zubairy 2020. Published in 2020 by Oxford University Press. DOI: 10.1093/oso/9780198854227.001.0001

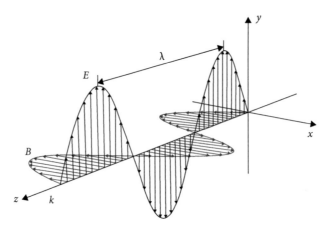

Fig. 9.1 An electromagnetic wave. Electric and magnetic fields, **E** and **B**, oscillate in directions mutually perpendicular to each other as well as perpendicular to the direction of propagation of the wave vector **k**.

to the direction of propagation with equal probability. Such a light beam is described as unpolarized as it lacks a certain well defined direction of polarization. When such a light passes through a material called a polarizer, the electric field of the filtered light oscillates in a well-defined direction depending on the orientation of the polarizer and the light becomes polarized.

Polarizer: A polarizer is typically made of a crystalline structure with a preferred axis called the polarizing axis such that light whose electric field, **E**, is oscillating in the direction perpendicular to this axis is completely absorbed and none of that light is transmitted, whereas the light whose electric field vector is oscillating in the direction parallel to this axis passes through. For a field oscillating in an arbitrary direction, we can decompose the electric field vector in the directions parallel and perpendicular to the polarizing axis. The component of the field along the direction of the polarizing axis passes through and the component perpendicular to the polarizing axis is blocked. The electric field vector of the transmitted field, and hence the polarization, is in the direction of the polarizing axis and its magnitude is reduced by an amount that depends on the angle between the direction of initial polarization and the polarizing axis of the polarizer.

As an example, let us assume that a beam of light, propagating along the z-axis, is incident on a polarizer with its polarizing axis along x-axis. Let the polarization of the incident beam be along a direction **E** making an angle θ with the x-axis as shown in Fig. 9.2. The electric field can then be decomposed along x- and y-directions as follows:

$$\mathbf{E} = E_x \hat{\mathbf{x}} + E_y \hat{\mathbf{y}} = E \cos\theta \, \hat{\mathbf{x}} + E \sin\theta \, \hat{\mathbf{y}}, \tag{9.1}$$

where $E_x = E\cos\theta$ is the x-component of the electric field and $E_y = E\sin\theta$ is the y-component of the field. The component of the field that passes through the polarizer is $E_x = E\cos\theta$ and the field component $E_y = E\sin\theta$ is blocked. The transmitted field is polarized along x-axis and its amplitude, E_T, is

$$E_T = E\cos\theta. \tag{9.2}$$

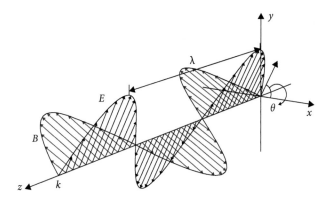

Fig. 9.2 Polarization is along the direction **E** making an angle θ with the x-axis.

The intensity of the transmitted field is proportional to $|E_T|^2$ and is given by

$$I_T = I_0\cos^2\theta, \tag{9.3}$$

where $I_0 = |E|^2$ is the intensity of the incident field. This is called Malus' law.

Next we consider various examples that illustrate the properties of polarizers.

First we consider an unpolarized light beam of intensity I_0 in which all polarization directions in the xy-plane, the plane perpendicular to the z-axis which is the direction of propagation, are equally probable (Fig. 9.3). When such a beam passes through a polarizer with the polarizing axis along the x-axis, only half of the intensity $I_0/2$ is transmitted. This can be shown with the help of Malus' law as follows.

As discussed above, the polarization direction θ is equivalent to the polarization direction $\theta + \pi$. Therefore, for unpolarized light, the polarization angle θ is uniformly distributed between 0 and π. According to Malus' law, the transmitted intensity for the polarization angle θ is equal to $I_0 \cos^2\theta$. Thus for each angle ϕ in the range 0 and $\pi/2$, we have a corresponding angle $\phi + (\pi/2)$ in the range $\pi/2$ and π. The average contribution to the transmission intensity from the two angles is

$$\frac{\cos^2\phi + \cos^2(\phi + (\pi/2))}{2} = \frac{1}{2},$$

where we used $\cos(\phi + (\pi/2)) = -\sin\phi$ and $\cos^2\phi + \sin^2\phi = 1$. The transmitted intensity is thus equal to $I_0/2$ and the light beam is polarized along the x-axis.

We proved this statement for a polarizer oriented along the x-axis. However this is a general result. If an unpolarized beam of intensity I_0 passes through a polarizer oriented along any arbitrary direction in the xy plane, the transmitted beam has an intensity equal to $I_0/2$ and is polarized along the direction of the polarizing axis of the polarizer.

Next we consider a set of three polarizers, A, B, and C as shown in Fig. 9.4. Let an un-polarized light beam be incident on polarizer A whose polarizing axis is along the x-axis. As discussed above, only half the incident intensity is transmitted, and is polarized along the x-axis. In the following we consider the output intensities for various orientations of polarizers B and C.

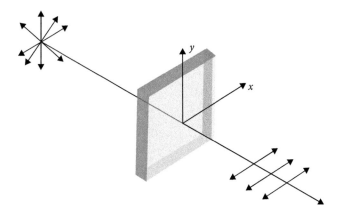

Fig. 9.3 Unpolarized light passing through a polarizer with its polarizing axis along the x-axis.

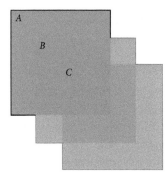

Fig. 9.4 The three polarizers A, B, and C.

First, we consider the case when polarizer B is also oriented along the x-axis (Fig. 9.5a). Since the light is polarized along the x-axis after passage through A, the polarization of the light incident on B is entirely along the x-axis. The result is that all the light incident on B is transmitted which is equal to half the intensity of the incident light on A. This can be verified from Malus' law. If the incident light is unpolarized and has intensity I_0, then after passing through the polarizer A, the light intensity is

$$I_A = \frac{I_0}{2}. \tag{9.4}$$

The intensity of light after passing through the polarizer B, according to Malus' law, is given by

$$I_B = I_A \cos^2 0^\circ = \frac{I_0}{2}\cos^2 0^\circ = \frac{I_0}{2}. \tag{9.5}$$

The AB system is thus transparent to the horizontally polarized light.

Next we consider the situation when the polarizer B is oriented such that its polarization axis is along the y-axis (Fig. 9.5b). Then, like before, the intensity of light after passing through the polarizer A is

$$I_A = \frac{I_0}{2} \tag{9.6}$$

Fig. 9.5 The polarizers A and B are placed next to each other such that (a) the polarization axes of both polarizers are parallel to each other; (b) the polarization axes of both polarizers are perpendicular to each other. (c) The polarizer C is inserted between the polarizers A and B with an orientation of 45° with respect to both A and B in the configuration (b).

and the polarization of the light is along the x-axis. The intensity of light after passing through polarizer B, according to Malus' law, is

$$I_B = I_A \cos^2 90° = \frac{I_0}{2}\cos^2 90° = 0. \qquad (9.7)$$

This AB system is thus opaque to the horizontally polarized light.

Next comes the interesting question: What happens if we insert a third polarizer C between the two polarizers A and B whose polarization axes are oriented along x- and y-axes, respectively (Fig. 9.5c). At first sight, the guess is that, regardless of the orientation of the polarization axis of C, the ABC system should be opaque to the incident unpolarized light. However this happens only if the orientation of the polarization axis of C is either along the x- or y-axis. For any other orientation, the ABC system is no longer opaque.

To see this, let us consider the situation when C is oriented at 45° to the x-axis. Then, as before, the intensity of light after A is

$$I_A = \frac{I_0}{2}. \qquad (9.8)$$

When this x-polarized light passes through C, the transmitted light is

$$I_C = I_A \cos^2 45° = \frac{I_0}{2}\cos^2 45° = \frac{I_0}{4}. \qquad (9.9)$$

Importantly, the emerging light is polarized along the polarization axis of C, namely along the direction making an angle of 45° to the x-axis. In vector notation, the direction of the polarization is along

$$\frac{\hat{x}+\hat{y}}{\sqrt{2}}.$$

When this light passes through the polarizer B whose polarization axis is along the y-axis, the angle between the polarization axes of C and B is 45°. Thus the intensity of the transmitted field after passage through the ABC system is

$$I_B = I_C \cos^2 45° = \frac{I_0}{4}\cos^2 45° = \frac{I_0}{8}. \qquad (9.10)$$

The polarization of the transmitted light is along the y-axis. Thus the insertion of a polarizer at an angle makes the set-up partially transparent.

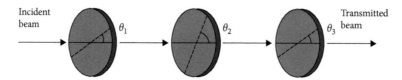

Fig. 9.6 A beam of unpolarized light is incident on a series of three polarizers oriented by angles θ_1, θ_2, and θ_3 with respect to the x-axis.

As another example, we consider a system of three polarizers as shown in Fig. 9.6. For the polarizers, we consider the orientation of the polarization axis to be along axes making angles θ_1, θ_2, and θ_3 with the x-axis as shown in the figure. The incident beam of light of intensity I_0 is unpolarized. We want to find the intensity of the beam transmitted through the three polarizers.

The intensity of light after the first polarizer oriented at θ_1 is $I_0/2$ and the light is polarized along θ_1. According to Malus' law, the intensity of light after passage through the second polarizer with orientation along θ_2 is

$$\frac{I_0}{2}\cos^2(\theta_1 - \theta_2) \tag{9.11}$$

and the light is polarized along θ_2. The intensity of light after passage through the third polarizer with orientation along θ_3 is

$$\frac{I_0}{2}\cos^2(\theta_1 - \theta_2)\cos^2(\theta_2 - \theta_3) \tag{9.12}$$

and the light is polarized along θ_3.

We note that the system depicted in Fig. 9.5c corresponds to $\theta_1 = 0, \theta_2 = 45°$, and $\theta_3 = 90°$.

9.2 Malus' Law for a Single Photon–Dirac's ket-bra Notation

So far we have analyzed the polarization of a beam of light. Malus' law relates the incident intensity to the transmitted intensity when light of certain polarization passes through a polarizer. The fraction of light that is transmitted depends on the orientation of the polarizer.

In this section, we consider the quantum mechanical picture of a light beam consisting of "photons". In this picture, light consists of individual particles. Malus' law can now be interpreted not in terms of intensities but in terms of number of photons passing through the polarizer. For example, the law

$$I_T = I_0\cos^2\theta \tag{9.13}$$

can be rewritten as

$$n_T = n_0\cos^2\theta, \tag{9.14}$$

where n_0 is the number of incident photons, n_T is the number of transmitted photons, and θ is the angle between the polarization of individual photons and the direction of the polarizing axis of the polarizer. The two equations are not different from each other. Here we have implicitly assumed that the intensity is directly proportional to the number of photons, $I_0 \propto n$. This relation can be obtained by realizing that, for a light beam of frequency ν, the energy of an individual photon is $\hbar\nu$, and therefore

$$I_0 = n\hbar\nu. \tag{9.15}$$

On substituting this expression for I into Eq. (9.13), we obtain Eq. (9.14) with $n_T = I_T/\hbar\nu$ and $n_0 = I_0/\hbar\nu$. Thus n_T and n_0 can be interpreted as dimensionless intensities.

The question we ask now is: What happens when a single photon passes through a polarizer? An answer to this question bears on the novel features of quantum mechanics that allow us to use these concepts in several applications, such as in quantum communication.

The most important difference between a single photon and a beam of light is that we do not have a relation of the type in Eqs. (9.13) or (9.14). A single photon cannot be described in terms of intensity. The passage of a light beam through the polarizer is described in terms of the fraction of intensity, a fraction is absorbed and a fraction is transmitted. However a single photon cannot be split such that part of the photon can be absorbed and part of it is transmitted. The photon either goes through the polarizer as a whole or it does not go through at all.

As we show below, we should talk in the language of probability as opposed to the language of intensity. The relevant question for the case of single photon is: What is the *probability* of a photon of certain polarization passing through an appropriately oriented polarizer? The probabilistic description is the hallmark of quantum mechanics that we address in many contexts in the coming chapters.

However, the first question we address is how to describe a single photon of certain polarization. Here we note that a photon in a certain polarization state is described by a vector, the so-called state vector. The properties of this state vector are very similar to a vector in two-dimensional space. We next discuss this close analogy.

As we recall from Chapter 2, two-dimensional space is described by two mutually perpendicular unit vectors. Let us assume that these vectors, $\hat{x}$ and $\hat{y}$ are along x- and y-axes, respectively. The normalization condition states, that $\hat{x}$ and $\hat{y}$ are of unit length, i.e.,

$$\hat{x} \cdot \hat{x} = 1, \hat{y} \cdot \hat{y} = 1 \tag{9.16}$$

and the orthogonality condition states

$$\hat{x} \cdot \hat{y} = \hat{y} \cdot \hat{x} = 0. \tag{9.17}$$

An arbitrary vector $\boldsymbol{A}$ (as shown in Fig. 9.7) can then be decomposed in terms of x- and y-components as follows:

$$\boldsymbol{A} = A_x\hat{x} + A_y\hat{y}, \tag{9.18}$$

where

$$A_x = A\cos\theta, A_y = A\sin\theta. \tag{9.19}$$

We also note that that the choice of the (x, y) coordinate system with unit vectors $\hat{x}$ and $\hat{y}$ is very subjective. We could have chosen another coordinate system (X, Y) with unit vectors $\hat{X}$ and $\hat{Y}$. The (X, Y) coordinate system is rotated by an angle ϕ with respect to the

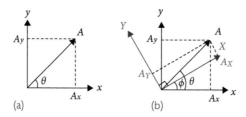

Fig. 9.7 (a) The vector **A** with components A_x and A_y along the x- and y-axes, respectively. (b) The same vector **A** in the two coordinates systems (x,y) and (X,Y) mutually rotated by an angle ϕ.

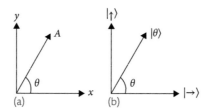

Fig. 9.8 An analogy between a vector **A** and a state vector $|\theta\rangle$ for a single photon. Just as the vector can be decomposed in terms of its x- and y-components, the state vector $|\theta\rangle$ can be decomposed in terms of the state vectors $|\rightarrow\rangle$ and $|\uparrow\rangle$.

(x, y) coordinate system as shown in Fig. 9.7b. The vector A can then be decomposed in terms of X- and Y-components as follows:

$$A = A_X\hat{X} + A_Y\hat{Y}, \tag{9.20}$$

where

$$A_X = A \cos(\theta - \phi), \quad A_Y = A \sin(\theta - \phi). \tag{9.21}$$

The relation between the (x, y) and (X, Y) coordinate systems can be established via the rotation angle between the two coordinate systems. For example, if the (X, Y) coordinate system is rotated by $45°$ with respect to the (x, y) coordinate system, then

$$\hat{X} = \frac{1}{\sqrt{2}}(\hat{x} + \hat{y}), \hat{Y} = \frac{1}{\sqrt{2}}(-\hat{x} + \hat{y}), \tag{9.22}$$

and it follows from Eqs. (9.18) and (9.20) that

$$A_X = \frac{1}{\sqrt{2}}(A_x + A_y), A_Y = \frac{1}{\sqrt{2}}(A_y - A_x). \tag{9.23}$$

Based on this description of an arbitrary vector in a coordinate system, we can first describe Malus' law for a beam of polarized light in vector notation and then answer the question regarding the polarization of a single photon.

First we choose a coordinate system in the xy-plane as shown in Fig. 9.8a. A beam polarized in a direction A making an angle θ with the x-axis can be described by the field

$$E = E_x\hat{x} + E_y\hat{y} = E\cos\theta\,\hat{x} + E\sin\theta\,\hat{y}. \tag{9.24}$$

If this beam passes through a polarizer whose polarization axis $\hat{n}$ is oriented at an angle ϕ with respect to the x-axis,

$$\hat{n} = \cos\phi\,\hat{x} + \sin\phi\,\hat{y}, \tag{9.25}$$

then the field component that passes through the polarizer is given by the dot product

$$\begin{aligned}
\hat{n} \cdot \boldsymbol{E} &= (\cos\phi\,\hat{x} + \sin\phi\,\hat{y}) \cdot (E\cos\theta\,\hat{x} + E\sin\theta\,\hat{y}) \\
&= E\cos\phi\cos\theta\,\hat{x}\cdot\hat{x} + E\cos\phi\sin\theta\,\hat{x}\cdot\hat{y} + E\sin\phi\cos\theta\,\hat{y}\cdot\hat{x} \\
&\quad + E\sin\phi\sin\theta\,\hat{y}\cdot\hat{y} \\
&= E\cos(\theta - \phi).
\end{aligned} \tag{9.26}$$

Here we used $\hat{x}\cdot\hat{x} = 1, \hat{y}\cdot\hat{y} = 1, \hat{x}\cdot\hat{y} = \hat{y}\cdot\hat{x} = 0$ and $\sin\phi\sin\theta + \cos\phi\cos\theta = \cos(\theta - \phi)$. The intensity of the transmitted light is proportional to $|\hat{n}\cdot\boldsymbol{E}|^2$ and is given by

$$I_T = I_0\cos^2(\theta - \phi), \tag{9.27}$$

where I_0 is the intensity of the incident field. This is a generalized form of Malus' law discussed in the previous section.

Next we consider the polarization of a single photon. The description follows the same line as above but we first need a notation to describe the quantum state of the photon. This was done by Paul Dirac and the notation is therefore named after him as the "Dirac notation."

In the Dirac notation, the polarization state of the single photon along an axis making an angle of θ with the x-axis is denoted by $|\theta\rangle$ (Fig. 9.8b). We designate the polarization state along the x-axis as $|\rightarrow\rangle$ and along the y-axis by $|\uparrow\rangle$. An alternate representation for $|\rightarrow\rangle$ and $|\uparrow\rangle$ can be given by $|H\rangle$ and $|V\rangle$, respectively. Here H represents the Horizontal polarization and V represents the Vertical polarization. The polarization state $|\theta\rangle$ can then be written as a linear combination or superposition of the polarization states $|\rightarrow\rangle$ and $|\uparrow\rangle$ as

$$|\theta\rangle = \cos\theta\,|\rightarrow\rangle + \sin\theta\,|\uparrow\rangle. \tag{9.28}$$

This equation can be compared with Eq. (9.24) where the polarization vector is decomposed in terms of the x- and y-components. The notation $|\theta\rangle$ is called Dirac's "ket" notation.

An important point! As discussed above, as the field oscillates in both positive and negative directions, the polarization direction along (say) the +x-axis is equivalent to polarization along the –x-axis. In general a polarization direction along an angle θ is equivalent to polarization directions $\theta + \pi$ or $\theta - \pi$. Thus, for a single photon, $|\rightarrow\rangle \equiv |\leftarrow\rangle$ and $|\uparrow\rangle \equiv |\downarrow\rangle$. In general $|\theta\rangle \equiv |\theta \pm \pi\rangle$.

Next, we define another notation, Dirac's "bra" notation, $\langle\phi|$. The origin of "bra" and "ket" notations is the word "bracket", "bra" for the left side and "ket" for the right side of a "bra c ket", $\langle||\rangle$. Traditionally $\langle\phi||\theta\rangle$ is written as $\langle\phi|\theta\rangle$ and has a clear analogy with the dot product. The states $|\rightarrow\rangle$ and $|\uparrow\rangle$ are normalized like the unit vectors $\hat{x}$ and $\hat{y}$. So in analogy with $\hat{x}\cdot\hat{x} = 1$ and $\hat{y}\cdot\hat{y} = 1$, we have

$$\langle\rightarrow|\rightarrow\rangle = \langle\uparrow|\uparrow\rangle = 1 \tag{9.29}$$

and, similar to the orthogonality relation $\hat{x}\cdot\hat{y} = \hat{y}\cdot\hat{x} = 0$,

$$\langle\rightarrow|\uparrow\rangle = \langle\uparrow|\rightarrow\rangle = 0. \tag{9.30}$$

So what is the physical meaning of the ket and bra notations? The ket notation $|\theta\rangle$ represents the quantum state of the photons and the bra notation $\langle\phi|$ is defined such that $\langle\phi|\theta\rangle$ represents the projection of the $|\theta\rangle$ state onto $\langle\phi|$. The quantity $\langle\phi|\theta\rangle$ is called the "probability amplitude" of the photon in state $|\theta\rangle$ to be found in state $|\phi\rangle$ and its modulus square, $|\langle\phi|\theta\rangle|^2$, is the corresponding probability. This is an important definition that we use extensively in this and later chapters.

In analogy with Eq. (9.25), a polarizer whose polarization axis is oriented at an angle ϕ with respect to the x-axis, is described by

$$|\phi\rangle = \cos\phi\,|\rightarrow\rangle + \sin\phi\,|\uparrow\rangle. \tag{9.31}$$

The probability amplitude for the photon in state $|\theta\rangle$ to pass through a polarizer in a state $|\phi\rangle$ is therefore given by

$$\langle\phi|\theta\rangle = (\cos\phi\,\langle\rightarrow| + \sin\phi\,\langle\uparrow|)(\cos\theta\,|\rightarrow\rangle + \sin\theta\,|\uparrow\rangle) \tag{9.32}$$

Using the orthogonality and normalization properties of the states $|\rightarrow\rangle$ and $|\uparrow\rangle$, as given in Eqs. (9.29) and (9.30), we obtain

$$\langle\phi|\theta\rangle = \cos\phi\cos\theta + \sin\phi\sin\theta = \cos(\theta-\phi). \tag{9.33}$$

The quantum mechanical probability of the photon passing through the polarizer is obtained by taking the modulus square of $\langle\phi|\theta\rangle$, i.e.,

$$P_\theta = |\langle\phi|\theta\rangle|^2 = \cos^2(\theta-\phi). \tag{9.34}$$

This is Malus' law for a single photon.

This equation is very similar to Eq. (9.26), the only difference being that the intensity I_T is replaced by the probability P_θ. It is important to emphasize this difference as this corresponds to the difference between the classical picture and the quantum picture. In the classical picture, the light beam is described in terms of intensity. At the polarizer, part of the beam passes through and the rest is absorbed. We can say with certainty about what fraction will be transmitted and what fraction will be absorbed. This is not true in quantum mechanics. In the quantum description of a single photon, either the whole photon is transmitted or it is absorbed, but we do not have a definite prediction about what will happen. The only quantity we know is the probability whether the photon will be transmitted or it will be absorbed.

As a special case, the probability that a photon polarized at an angle θ with respect to the x-axis passes through a polarizer oriented in the horizontal direction (the polarization axis along the x-axis) is given by

$$|\langle\rightarrow|\theta\rangle|^2 = \cos^2\theta. \tag{9.35}$$

Similarly the probability of the same photon passing through a polarizer oriented in the vertical direction (the polarization axis along the y-axis) is

$$|\langle\uparrow|\theta\rangle|^2 = \sin^2\theta. \tag{9.36}$$

The probability that a vertically polarized photon passes through a polarizer with its polarization axis oriented along the x-axis is zero, i.e.,

$$|\langle\uparrow|0^\circ\rangle|^2 = |\langle\uparrow|\rightarrow\rangle|^2 = \cos^2 90^\circ = 0, \tag{9.37}$$

and the probability that a vertically polarized photon passes through a polarizer with its polarization axis oriented along the y-axis is unity, i.e.,

$$|\langle\uparrow|90°\rangle|^2 = |\langle\uparrow|\uparrow\rangle|^2 = \cos^2 0° = 1. \tag{9.38}$$

We have defined a polarization state $|\theta\rangle$ in terms of the horizontally and vertically polarized states $|\rightarrow\rangle$ and $|\uparrow\rangle$ as

$$|\theta\rangle = \cos\theta|\rightarrow\rangle + \sin\theta|\uparrow\rangle. \tag{9.39}$$

This is just like we defined an arbitrary vector A in terms of unit vectors $\hat{x}$ and $\hat{y}$ as

$$A = A\cos\theta\,\hat{x} + A\sin\theta\,\hat{y}. \tag{9.40}$$

In the vector analysis, $\hat{x}$ and $\hat{y}$ form the basis in which any vector A can be expanded. In the same way we can define $|\rightarrow\rangle$ and $|\uparrow\rangle$ to be the basis for the polarization state such that any arbitrary state of polarization $|\theta\rangle$ can be expanded in terms of these states. As discussed above, the probability of detecting the photon in the state $|\rightarrow\rangle$ is $|\langle\rightarrow|\theta\rangle|^2 = \cos^2\theta$ and the probability of detecting the photon in the state $|\uparrow\rangle$ is $|\langle\uparrow|\theta\rangle|^2 = \sin^2\theta$. An important point to note is that these outcomes depend on the orientation of the polarizer θ.

Recall the amazing result of how, by an insertion of a 45° polarizer (C) between the horizontal (A) and the vertical polarizer (B), light gets through the vertical polarizer that stopped it previously (Fig. 9.5c). What is the corresponding result if, instead of a beam of light, only a single photon is incident? We consider a single photon initially polarized along the x-axis which is incident on a sequence of three polarizers as shown in Fig. 9.9. The state of the incident photon is $|\rightarrow\rangle$.

After passing the first polarizer, whose polarization axis is along the x-axis, the photon is transmitted in state $|\rightarrow\rangle$. The probability that a horizontally polarized photon $|\rightarrow\rangle$ gets through the polarizer oriented at an angle of 45° is

$$|\langle\theta = 45°|\rightarrow\rangle|^2 = |\langle\nearrow|\rightarrow\rangle|^2 = \cos^2 45° = \frac{1}{2}. \tag{9.41}$$

If the photon passes through the second polarizer, its polarization is oriented at an angle 45° with the horizontal. We designate such a polarization state by $|\nearrow\rangle$. Just as a unit vector pointing in a direction making an angle 45° can be written as a superposition of equal vectors in the x- and y-directions, a photon in the state $|\nearrow\rangle$ can be written as a linear combination of $|\uparrow\rangle$ and $|\rightarrow\rangle$, i.e.,

$$|\nearrow\rangle = \frac{1}{\sqrt{2}}(|\rightarrow\rangle + |\uparrow\rangle). \tag{9.42}$$

Therefore, the probability that the photon passes the final vertically oriented polarizer is

$$|\langle\uparrow|\nearrow\rangle|^2 = \frac{1}{2}. \tag{9.43}$$

Here we used $\langle\uparrow|\rightarrow\rangle = 0$ and $\langle\uparrow|\uparrow\rangle = 1$. The probability that a photon emerging from the vertical polarizer passes through the final horizontal polarizer in the presence of an intermediate 45° polarizer can be calculated as

$$|\langle\uparrow|\nearrow\rangle|^2|\langle\nearrow|\rightarrow\rangle|^2 = \frac{1}{2}\cdot\frac{1}{2} = \frac{1}{4}. \tag{9.44}$$

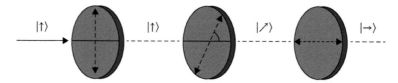

Fig. 9.9 A single photon with initial polarization along the y-axis passes through a sequence of three polarizers whose polarization axes are oriented along the y-axis, making an angle $45°$ with respect to the x-axis, and along the x-axis.

This result is similar to the classical result where we considered a beam of light with incident intensity I_0 and showed that only $I_0/4$ passes through the ACB polarization set-up of Fig. 9.5. For a single photon, the corresponding result is that the probability of transmission is 1/4.

9.3 Input–Output Relation for a Classical Beam Splitter

In this section, we discuss the classical description of the beam splitter. In the next section, we discuss the quantum beam splitters that are important for our analysis of some quantum devices that we discuss in the following chapters.

A beam splitter is a piece of glass on which two light beams of equal frequencies and amplitudes a_1 and a_2 are incident as shown in Fig. 9.10a. We assume that the glass plate is lossless, i.e., it does not absorb any energy. When these two beams pass through the beam splitter, they are reflected and transmitted, resulting in the two beams of amplitudes b_1 and b_2 in the output. We want to find the relation between the output amplitudes b_1 and b_2 and the input amplitudes a_1 and a_2.

From Fig. 9.10a, we see that the amplitude b_1 consists of the reflected part of a_1 and the transmitted part of a_2. Similarly the amplitude b_2 consists of the transmitted part of a_1 and the reflected part of a_2. We can thus write the following input–output relations:

$$b_1 = r_1 a_1 + t_2 a_2, \tag{9.45}$$

$$b_2 = t_1 a_1 + r_2 a_2, \tag{9.46}$$

where r_1 and r_2 are the reflection coefficients and t_1 and t_2 are the transmission coefficients. In general they can be complex numbers. These coefficients are not arbitrary.

For a lossless beam splitter, the input energy is equal to output energy. We recall that the energy of a light field is proportional to the modulus square of field amplitude. The law of conservation of energy therefore requires

$$|b_1|^2 + |b_2|^2 = |a_1|^2 + |a_2|^2. \tag{9.47}$$

On substituting for b_1 and b_2 from Eqs. (9.45) and (9.46)

$$
\begin{aligned}
|b_1|^2 + |b_2|^2 &= (t_2 a_2 + r_1 a_1)\left(t_2^* a_2^* + r_1^* a_1^*\right) + (t_1 a_1 + r_2 a_2)\left(t_1^* a_1^* + r_2^* a_2^*\right) \\
&= \left(|t_1|^2 + |r_1|^2\right)|a_1|^2 + \left(|t_2|^2 + |r_2|^2\right)|a_2|^2 \\
&\quad + \left(t_1 r_2^* + t_2^* r_1\right) a_1 a_2^* + \left(t_1^* r_2 + t_2 r_1^*\right) a_1^* a_2.
\end{aligned}
\tag{9.48}
$$

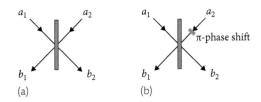

Fig. 9.10 A beam splitter with input amplitudes being a_1 and a_2 and output amplitudes being b_1 and b_2. The relationship between the input and output amplitudes is given by Eqs. (9.45) and (9.46). The two systems are equivalent if we choose (a) $-t_1 = t_2 = \sin\theta$ and $r_1 = r_2 = \cos\theta$ and (b) $t_1 = t_2 = \sin\theta$ and $r_1 = -r_2 = \cos\theta$ along with the π phase shifter at the input a_2.

It follows that the two sides in Eq. (9.47) are equal if

$$|t_1|^2 + |r_1|^2 = |t_2|^2 + |r_2|^2 = 1, \tag{9.49}$$

$$t_1 r_2^* + t_2^* r_1 = t_1^* r_2 + t_2 r_1^* = 0 \tag{9.50}$$

Note that these relations between the reflection coefficients r_1 and r_2 and the transmission coefficients t_1 and t_2 are obtained only under the condition that the beam splitter is lossless. Any lossless beam splitter has to obey these relations. The specific choice of these coefficients within these constraints depends on the thickness and the material of the glass plate.

A choice $-t_1 = t_2 = \sin\theta$ and $r_1 = r_2 = \cos\theta$, satisfying the above conditions, gives

$$b_1 = \cos\theta\, a_1 + \sin\theta\, a_2 \tag{9.51}$$

$$b_2 = \cos\theta\, a_2 - \sin\theta\, a_1. \tag{9.52}$$

Here θ is a parameter such that $\sin^2\theta$ and $\cos^2\theta$ are the transmissivity and reflectivity coefficients of the beam splitter, respectively, i.e., if I is the intensity of light incident on the beam splitter, a fraction $\sin^2\theta$ is transmitted and a fraction $\cos^2\theta$ is reflected.

This is the classical picture.

For a usual beam splitter, $t_1 = t_2 = \sin\theta$ and $r_1 = -r_2 = \cos\theta$, which gives slightly different relations than Eqs. (9.51) and (9.52). We however follow the input–output relations in Eqs. (9.51) and (9.52) in later chapters as they give simple transformation properties when discussing more complicated optical set-ups. The transformations in Eqs. (9.51) and (9.52) can be realized in the laboratory by including a π phase shifter at the input a_2 of a beam splitter with $t_1 = t_2 = \sin\theta$ and $r_1 = -r_2 = \cos\theta$, as shown in Fig. 9.10b.

9.4 Beam Splitter for a Single-photon State

Next we address the question: What happens when there is only single photon in the input of a beam splitter? More precisely, what are the output states for the photon, when the input state at one of the ports is a single photon state $|1\rangle$ and the other port has zero photon state $|0\rangle$. Here we have to be careful as the output states will be different in the two situations whether port 1 or port 2 has the photon state $|1\rangle$. We represent port 1 as the port on the left-hand side and port 2 is the port on the right-hand side as in Fig. 9.10. We represent the total input state $|10\rangle$ as the state when the photon is incident from the left side (Fig. 9.11a) and $|01\rangle$ when the photon is

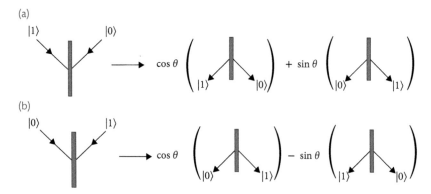

Fig. 9.11 The input-output relation when the input photon is incident from the left-hand side and the right-hand side.

incident from the right side (Fig. 9.11b). As there is only one photon in the input and it cannot be split, there are two possibilities in the output: either it is reflected or it is transmitted. As before, the reflection coefficient is $\cos\theta$ and the transmission coefficient is $\sin\theta$. In analogy with the classical input–output relations (9.51) and (9.52), the following are the input–output relations for a single photon input:

$$|10\rangle \rightarrow \cos\theta\,|10\rangle + \sin\theta\,|01\rangle, \tag{9.53}$$
$$|01\rangle \rightarrow \cos\theta\,|01\rangle - \sin\theta\,|10\rangle. \tag{9.54}$$

The probability that the photon is reflected is equal to $R = \cos^2\theta$ and the probability that it is transmitted is equal to $T = \sin^2\theta$.

9.5 Polarization Beam Splitter and Pockel Cell

In this section, we discuss how we can measure the polarization state. A polarizer is an inconvenient device as the photon is either transmitted or it is absorbed. What is more desirable is a device that is able to send one polarization state (say $|\rightarrow\rangle$) along one way and the other $|\uparrow\rangle$ along a different path. This is done in a polarization beam splitter.

Let us consider a photon that is prepared in the polarization state

$$|\theta\rangle = \cos\theta\,|\rightarrow\rangle + \sin\theta\,|\uparrow\rangle. \tag{9.55}$$

When such a photon is incident on a polarization beam splitter, as shown in Fig. 9.12, it can either go in the forward direction in the horizontally polarized state $|\rightarrow\rangle$ or in the downward direction in the vertically polarized state $|\uparrow\rangle$. We can then find the state of the photon depending on whether we get a click at the detector D_1 or at the detector D_2. A click at D_1 means that the photon is in the state $|\rightarrow\rangle$ and a click at D_2 means that the photon is in the state $|\uparrow\rangle$. The probability of the click at D_1 is $|\langle\rightarrow|\theta\rangle|^2 = \cos^2\theta$ and the probability of click at D_2 is $|\langle\uparrow|\theta\rangle|^2 = \sin^2\theta$.

Unlike a polarizer, the polarizing beam splitter cannot be easily rotated to measure the polarization state along some other axis, say along an axis rotated by an angle α with the

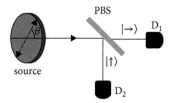

Fig. 9.12 If a photon polarized in a direction making an angle θ with the polarization axis is incident on a polarization beam splitter (PBS), it can pass through as a photon in state $|\rightarrow\rangle$ in the forward direction or get reflected in state $|\uparrow\rangle$ in the downward direction.

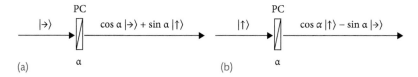

(a)   (b)

Fig. 9.13 A Pockel cell (PC) rotates the polarization of an incoming photon by an angle α. (a) The state $|\rightarrow\rangle$ is transformed to state $|\alpha\rangle$ and (b) the state $|\uparrow\rangle$ is transformed to state $|\alpha + \pi/2\rangle$.

horizontal. For that, we need to insert a polarization rotator before the polarizing beam splitter. This device should be able to rotate the state of the polarization of an incoming photon by an angle α before passing it through the polarization beam splitter. One such device is the Pockel cell. It is an electro-optic device which rotates the polarization of the incident light passing through it in proportion to the applied voltage. As an example, by applying an appropriate voltage, the Pockel cell can rotate the polarization of a photon by an angle α with the horizontal as shown in Fig. 9.13. As a result the horizontally and vertically polarized photons in states $|\rightarrow\rangle$ and $|\uparrow\rangle$, respectively, undergo the following transformations:

$$|\rightarrow\rangle \rightarrow |+\alpha\rangle \equiv |\alpha\rangle = \cos\alpha\,|\rightarrow\rangle + \sin\alpha\,|\uparrow\rangle, \tag{9.56}$$

$$|\uparrow\rangle \rightarrow |-\alpha\rangle \equiv |\alpha + \pi/2\rangle = \cos\alpha\,|\uparrow\rangle - \sin\alpha\,|\rightarrow\rangle. \tag{9.57}$$

We note that, just like the pair of states $\{|\rightarrow\rangle, |\uparrow\rangle\}$, the states $\{|+\alpha\rangle, |-\alpha\rangle\}$ are normalized and are mutually orthogonal, i.e.,

$$\langle +\alpha | +\alpha \rangle = \langle -\alpha | -\alpha \rangle = 1, \tag{9.58}$$

$$\langle +\alpha | -\alpha \rangle = \langle -\alpha | +\alpha \rangle = 0. \tag{9.59}$$

A polarization beam splitter can determine whether the polarization state of the incoming photon is $|\rightarrow\rangle$ or $|\uparrow\rangle$. A question of interest is: How can we determine whether the polarization of the incoming photon is along an angle α or along $\alpha + \pi/2$ with the horizontal? The corresponding states are $|+\alpha\rangle \equiv |\alpha\rangle$ and $|-\alpha\rangle \equiv |\alpha + \pi/2\rangle$. A way of doing this is to first rotate the polarization angle of the incoming photon by an angle $-\alpha$. This should transform the state $|\alpha\rangle$ to $|\rightarrow\rangle$ and the state $|\alpha + \pi/2\rangle$ to $|\uparrow\rangle$. This can be done by passing the photon through a Pockel cell that rotates the polarization by an angle $-\alpha$ with the horizontal. Next the photon passes through a polarization beam splitter as shown in Fig 9.14. If the detector D_1 clicks, the polarization of the incoming photon is along an angle α with the horizontal (in state

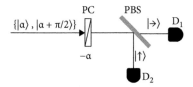

Fig. 9.14 A Pockel cell that rotates the polarization by angle $-\alpha$ followed by a beam splitter can determine the polarization of the incoming photon. A click at D_1 implies that the polarization of the incoming photon is along an angle α with the horizontal and a click at D_2 means that the state of the incoming photon is along an angle $\alpha + \pi/2$ with the horizontal.

$|\alpha\rangle$) and a click at the detector D_2 means that the state of the incoming photon is along an angle $\alpha + \pi/2$ with the horizontal (in state $|\alpha + \pi/2\rangle$).

As an example, if we want to find whether the photon is in the state $|\theta = 45°\rangle \equiv |\nearrow\rangle$ or $|\theta = 135°\rangle \equiv |\nwarrow\rangle$, we consider the set-up in Fig. 9.14 with $\alpha = 45°$. A rotation of the polarization by an angle $-45°$ transforms the state $|\theta = 45°\rangle \equiv |\nearrow\rangle$ to the horizontally polarized state $|\rightarrow\rangle$ and the state $|\theta = 135°\rangle \equiv |\nwarrow\rangle$ to the vertically polarized state $|\uparrow\rangle$. Therefore a click at D_1 implies that the incoming photon is in the state $|\nearrow\rangle$ and a click at D_2 implies that the incoming photon is in the state $|\nwarrow\rangle$.

Problems

9.1 A beam of horizontally polarized light of intensity I impinges on a set-up consisting of N polarizers, where N is very large. The first polarizer is oriented at an angle $\epsilon = \pi/2N$ to the horizontal; the next one is at an angle ϵ from the previous, and so on until the final one, which is exactly vertical. What is the intensity of light at the output of the last polarizer? What is its polarization?

9.2 Consider an unpolarized photon passes through a series of five polarizers oriented at $0°$, $22.5°$, $45°$, $67.5°$, and $90°$ with respect to the horizontal axis. Find the probability that the photon will pass through. What is the probability of transmission if we remove the second and the fourth polarizers oriented at $22.5°$ and $67.5°$, respectively? What happens if we remove the third polarizer oriented at $45°$ as well?

9.3 Consider a photon in the state

$$|\theta\rangle = \sin\theta \, |\uparrow\rangle + \cos\theta \, |\rightarrow\rangle,$$

which, after passing through a Pockel cell that rotates the polarization by an angle α with respect to the horizontal, passes through a polarization beam splitter (PBS) as shown in Fig. 9.15. Find the probability of clicks at detectors D_1 and D_2.

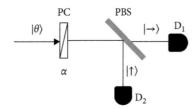

Fig. 9.15 Optical set-up corresponding to problem 9.3.

9.4 Consider the experimental set-up in Fig. 9.16. For an initial state

$$|\theta\rangle = \sin\theta\,|\uparrow\rangle + \cos\theta\,|\rightarrow\rangle,$$

find the probability of clicks at detectors D_1, D_2, and D_3 if the Pockel cells PC_1 and PC_2 rotate the polarizations by angles α and β, respectively.

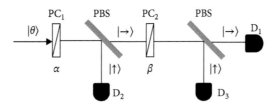

Fig. 9.16 Optical set-up corresponding to problem 9.4.

 BIBLIOGRAPHY

S. Prasad, M. O. Scully, and W. Martienssen, *A quantum description of the beam-splitter*, Optics Communications 62, 139 (1987).

R. A. Campos and B. E. A. Saleh, *Quantum mechanical lossless beam splitter: SU(2) symmetry and photon statistics*, Physical Review A 40, 1371 (1989).

C. H. Holbrow, E. Galvez, and M. E. Parks, *Photon quantum mechanics and beam splitters*, American Journal of Physics 70, 260 (2002).

V. Scarani, L. Chua, and S. Y. Liu, *Six Quantum Pieces: A First Course in Quantum Mechanics* (World Scientific 2010).

W. P. Schleich, *Quantum Optics in Phase Space*, (Wiley VCH 2001), p 350.

10 Quantum Superposition and Entanglement

In Chapter 8, we discussed the novel features of quantum interference. An electron can exhibit an interference pattern in a Young's double-slit type experimental set-up. An explanation of this experiment brought out the need to describe an electron in terms of a wavefunction which can be in a superposition of states of being at both slit 1 and slit 2.

In this chapter, we introduce the notion of quantum superposition of states in a formal way through the example of a polarized photon. This, like the double-slit experiment with electrons, brings out the novel feature that the state of the system depends on how we set up the experiment. The orientation of the polarizer determines the state of the polarization. We also discuss some paradoxical consequences of quantum superposition, namely, a microscopic superposition can be transformed into a superposition of macroscopic objects: A cat can be simultaneously dead and alive. This is the essence of the famous Schrödinger's cat paradox.

This description motivates another important consequence of quantum mechanical description of the multiple objects, namely, their ability to exist in an entangled state. The properties of two objects, like two photons, can remain entangled no matter how far away they are from each other, and thus have the ability to influence each other. Quantum entanglement was first introduced by Schrödinger when he discussed some paradoxical consequences of quantum mechanics. Quantum entanglement saw its revival in recent years when it was shown to be a wonderful resource for quantum communication and quantum computing, topics that we discuss later in this book.

10.1 Coherent Superposition of States

Conventionally, if we have a system that can exist in two possible states then it is found either in one state or another but never simultaneously in both states. For example, a door can be either open or it can be shut but cannot be in a state where it is both open and shut. Similarly a ball can be inside a box or outside the box but never be simultaneously in "inside" and "outside" states. A quantum system, on the other hand, can be in a "coherent superposition of states."

We have seen that a photon polarized at an arbitrary angle θ with respect to the horizontal can be written as a coherent superposition of states with vertical and horizontal polarizations (Fig. 10.1):

$$|\theta\rangle = \cos\theta\,|\!\rightarrow\rangle + \sin\theta\,|\!\uparrow\rangle, \tag{10.1}$$

i.e., the photon simultaneously exists in both horizontal and vertical states of polarization. This continues to be true as long as we do not disturb the system or try to make a measurement.

Quantum Mechanics for Beginners: With Applications to Quantum Communication and Quantum Computing. M. Suhail Zubairy.
© M. Suhail Zubairy 2020. Published in 2020 by Oxford University Press. DOI: 10.1093/oso/9780198854227.001.0001

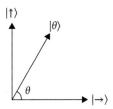

Fig. 10.1 A photon in the state $|\theta\rangle$ is in a coherent superposition of $|\rightarrow\rangle$ and $|\uparrow\rangle$ states.

The situation changes when we make a measurement and try to "see" the state of polarization of the photon. What happens if we try to measure the state of polarization?

First we note that the measurement process involves passing the photon through a Pockel cell and a polarizing beam splitter as discussed in Section 9.5. The set-up, as depicted in Fig. 9.14, can be used to find the polarization state in an arbitrary direction. We can, for example, determine whether the polarization state of the photon is $|\alpha\rangle$ or $|\alpha + \pi/2\rangle$ by rotating the polarization of an incoming photon by an angle $-\alpha$ with the horizontal and then passing it through a polarizing beam splitter. If the photon is found in the state $|\rightarrow\rangle$, we can conclude that the state of the incoming photon was $|\alpha\rangle$. However, if the photon is found in the state $|\uparrow\rangle$, we conclude that the state of the incoming photon was $|\alpha + \pi/2\rangle$. Thus the set-up in Fig. 9.14 corresponds to effectively rotating the polarization beam splitter by an angle α. In the following, when we talk about a polarization beam splitter rotated by an angle α, we have this set-up in mind.

It turns out that the outcome of measuring the photon in state (10.1) depends on the orientation of the polarization beam splitter. If the polarization measurement apparatus is set up with $\alpha = 0$, the state $|\theta\rangle$ "collapses" and the outcome is that we get either a horizontally polarized photon $|\rightarrow\rangle$ or a vertically polarized photon $|\uparrow\rangle$. The superposition is destroyed in the process of measurement.

The measurement process is mathematically represented by applying the "bra" operator $\langle\rightarrow|$ onto the state $|\theta\rangle$ and the result is

$$\langle\rightarrow|\theta\rangle = \cos\theta. \tag{10.2}$$

The probability that the outcome of the measurement is going to be $|\rightarrow\rangle$ is given by

$$P_{|\rightarrow\rangle} = |\langle\rightarrow|\theta\rangle|^2 = \cos^2\theta. \tag{10.3}$$

In a similar manner, the probability of finding the photon in the state with vertical polarization $|\uparrow\rangle$ is

$$P_{|\uparrow\rangle} = |\langle\uparrow|\theta\rangle|^2 = \sin^2\theta. \tag{10.4}$$

Therefore, what we see is that a photon in the polarization state $|\theta\rangle$, when measured, can be found either in state $|\rightarrow\rangle$ with probability $P_{|\rightarrow\rangle} = \cos^2\theta$ or in state $|\uparrow\rangle$ with probability $P_{|\uparrow\rangle} = \sin^2\theta$. The coherent superposition (10.1) persists only if we do not make any measurement. As soon as we make a measurement, the coherent superposition (ability to be simultaneously in both $|\rightarrow\rangle$ and $|\uparrow\rangle$ states) disappears.

At first sight this result does not appear surprising or mysterious. After all we seem to encounter similar situations all the time. If a window remains open 40% of the time and remains closed for the other 60% of the time then before looking at the window we can only say that there is a 40% probability that it is closed and 60% probability that it is open. However when we look at the window, it is found either open or closed—never simultaneously open and closed. So what is so special about the quantum "coherent superposition of states"?

Just as we can represent a vector A in any mutually perpendicular basis as discussed in Section 2.2, we can express the state of polarization in any other basis set of orthogonal states. For example, a vertically polarized photon may be written as a linear superposition of any other orthogonal basis states, for example $|45°\rangle$ and $|135°\rangle$ which we designate $|\nearrow\rangle$ and $|\nwarrow\rangle$, respectively. The polarization states $|\nearrow\rangle$ and $|\nwarrow\rangle$, like $|\rightarrow\rangle$ and $|\uparrow\rangle$, form another set of basis states.

The two basis states $\{|\rightarrow\rangle, |\uparrow\rangle\}$ and $\{|\nearrow\rangle, |\nwarrow\rangle\}$ are related to each other via rotation relations similar to those between the vector coordinate systems (x, y) and (X, Y) discussed in Section 8.2.

$$|\nearrow\rangle \equiv |45°\rangle = \frac{1}{\sqrt{2}}(|\uparrow\rangle + |\rightarrow\rangle), \tag{10.5}$$

$$|\nwarrow\rangle \equiv |135°\rangle = \frac{1}{\sqrt{2}}(|\uparrow\rangle - |\rightarrow\rangle). \tag{10.6}$$

These relations can be verified from Fig. 10.2. We can invert these polarization states and obtain

$$|\uparrow\rangle = \frac{1}{\sqrt{2}}(|\nearrow\rangle + |\nwarrow\rangle), \tag{10.7}$$

$$|\rightarrow\rangle = \frac{1}{\sqrt{2}}(|\nearrow\rangle - |\nwarrow\rangle). \tag{10.8}$$

We can verify

$$\langle \nearrow | \nearrow \rangle = \langle \nwarrow | \nwarrow \rangle = 1, \tag{10.9}$$
$$\langle \nearrow | \nwarrow \rangle = \langle \nwarrow | \nearrow \rangle = 0. \tag{10.10}$$

On substituting for $|\rightarrow\rangle$ and $|\uparrow\rangle$ from Eqs, (10.7) and (10.8), the state

$$|\theta\rangle = \cos\theta \, |\rightarrow\rangle + \sin\theta \, |\uparrow\rangle \tag{10.11}$$

can be rewritten in the new basis as

$$|\theta\rangle = \cos\phi \, |\nearrow\rangle + \sin\phi \, |\nwarrow\rangle \tag{10.12}$$

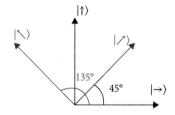

Fig. 10.2 The two set of basis states: $\{|\rightarrow\rangle, |\uparrow\rangle\}$ and $\{|\nearrow\rangle, |\nwarrow\rangle\}$.

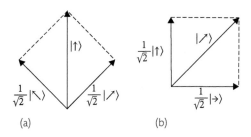

Fig. 10.3 (a) A photon in state $|\uparrow\rangle$ can be looked upon as a superposition of $|\nearrow\rangle$ and $|\searrow\rangle$ states; (b) A photon in state $|\nearrow\rangle$ can be looked upon as a superposition of $|\rightarrow\rangle$ and $|\uparrow\rangle$ states.

with

$$\cos\phi = \frac{1}{\sqrt{2}}(\sin\theta + \cos\theta), \sin\phi = \frac{1}{\sqrt{2}}(\sin\theta - \cos\theta). \tag{10.13}$$

Thus we obtain the same polarization state $|\theta\rangle$ with different amplitudes (cos ϕ and sin ϕ instead of cos θ and sin θ) in the new basis. If we measure the polarization of the photon in the $\{|\rightarrow\rangle, |\uparrow\rangle\}$ basis, the outcome will be $|\rightarrow\rangle$ or $|\uparrow\rangle$ with probabilities $\cos^2\theta$ and $\sin^2\theta$, respectively. However if we measure the polarization of the same photon in the $\{|\nearrow\rangle, |\searrow\rangle\}$ basis, the outcome will be $|\nearrow\rangle$ or $|\searrow\rangle$ with different probabilities $\cos^2\phi$ and $\sin^2\phi$, respectively. Thus the outcome depends on the setting of the measuring apparatus, the orientation of the polarizing beam splitter in this example. The dependence of the experimental outcome of the same object on the orientation of the apparatus is a very important difference between the usual classical description and the quantum description.

This raises an important question: Can we objectively define the state of the polarization of a single photon? This question lies at the heart of the conceptual foundations of quantum mechanics as we see later. In the light of the above discussion, a question such as "What is the state of polarization of the given photon?" is ambiguous and incomplete. A well-defined question would include the description of the experimental apparatus (in this case, the orientation of the polarization beam splitter as the measurement device) as well. So a complete question should be: "What is the state of polarization of the given photon if it passes through a polarization beam splitter oriented in the θ direction?" We can further illustrate the role of measurement in defining the physical property such as the polarization by the following discussion.

Consider a single photon with polarization $|\uparrow\rangle$. What happens when this photon passes through a polarization beam splitter oriented at 45°? The state of the photon $|\uparrow\rangle$ can be decomposed in the $\{|\nearrow\rangle, |\searrow\rangle\}$ basis of the polarization beam splitter which is the measuring device via (see Fig. 10.3a)

$$|\uparrow\rangle = \frac{1}{\sqrt{2}}(|\nearrow\rangle + |\searrow\rangle).$$

Experimentally, the system will be found in $|\nearrow\rangle$ with the probability $|\langle\nearrow|\uparrow\rangle|^2 = 1/2$ or $|\searrow\rangle$ with the probability $|\langle\searrow|\uparrow\rangle|^2 = 1/2$. Let us assume that the photon is found in the polarization state $|\nearrow\rangle$, If, after this measurement, we measure the polarization again in the original $\{|\rightarrow\rangle, |\uparrow\rangle\}$ basis, our classical intuition tells us that the photon should be found in the state $|\uparrow\rangle$

with unit probability, that is with certainty. However this is not what happens. To see this we consider the passage of the photon again through a polarization beam splitter that can measure the polarization state in the $\{|\rightarrow\rangle, |\uparrow\rangle\}$ basis. The polarization state $|\nearrow\rangle$ is decomposed in the $\{|\rightarrow\rangle, |\uparrow\rangle\}$ basis of the polarizer via (see Fig. 10.3b)

$$|\nearrow\rangle = \frac{1}{\sqrt{2}}\left(|\uparrow\rangle + |\rightarrow\rangle\right)$$

The polarization components $|\uparrow\rangle$ and $|\rightarrow\rangle$ are now equally probable, i.e., if we measure the polarization again in the original $\{|\rightarrow\rangle, |\uparrow\rangle\}$ basis then the outcome will be $|\uparrow\rangle$ or $|\rightarrow\rangle$ with equal probability. Thus there is a 50% chance that the photon will be found in the horizontally polarized state $|\rightarrow\rangle$. This is a counterintuitive result that could not be expected in classical mechanics.

This result motivates us to ask a related question: Can we associate two polarization components $|\uparrow\rangle$ and $|\nearrow\rangle$ for the same photon? The answer is an emphatic No!

The above analysis provides a beautiful and simple example of Bohr's principle of complementarity, namely, two observables are *complementary* if precise knowledge of one of them implies that all possible outcomes of measuring the other one are equally probable. In the above example, if we measure the polarization in the $\{|\rightarrow\rangle, |\uparrow\rangle\}$ basis and the outcome is $|\uparrow\rangle$, then subsequently the outcome in the $\{|\nearrow\rangle, |\nwarrow\rangle\}$ basis becomes completely uncertain, with 50% probability each for the outcomes $|\nearrow\rangle$ and $|\nwarrow\rangle$. Measurement disturbs the system. This result provides the foundation for some of the most dramatic successes in the field of secure communication as we discuss in later chapters.

10.2 Quantum Entanglement and the Bell Basis

As we have seen, the ability of a quantum system to exist in a coherent superposition of states is a novel feature of quantum mechanics. Another interesting aspect of quantum systems is that they can exist in an entangled state. Quantum entanglement is not only a counterintuitive effect but, as we see in our discussions on quantum computing in Chapters 15 and 16, it is a remarkable resource.

Let us consider a system of two independent objects. In principle, they may be so far apart that they cannot interact with each other in any way. If this happens then, according to classical mechanics, both objects are independent of each other and their properties are not influenced by what we do to the other object. For example, say we have two balls, one of them red and the other blue. Let them be very far away from each other. No matter what we do to the ball in our possession (paint it, crush it, throw it ...) the properties of the ball far away from us will not be affected in any way. This behavior can be seen in all classical systems.

Quantum mechanical systems can remarkably demonstrate a dramatically different behavior. We illustrate it with a simple example. Let us consider two photons A and B. Suppose we prepare them in the quantum state

$$|\psi_{AB}\rangle = \frac{1}{\sqrt{2}}\left(|\rightarrow_A\rangle|\uparrow_B\rangle + |\uparrow_A\rangle|\rightarrow_B\rangle\right). \tag{10.14}$$

Let these photons propagate long distances from each other such that they cannot interact with each other. Let the photon A stay with Alice and the photon B be far away with Bob. It is clear that if Alice measures her photon, there are two possibilities: If Alice's photon is found in state $|\rightarrow_A\rangle$ then Bob's photon is definitely found in state $|\uparrow_B\rangle$ and if Alice's photon is found in state $|\uparrow_A\rangle$ then Bob's photon is found in state $|\rightarrow_B\rangle$ with certainty.

There is nothing remarkable about this result. We could say the same thing about our two-ball example: If Alice finds the ball in her possession to be red then the ball in Bob's possession is blue and vice versa.

However the situation becomes different if Alice decides to measure her photon in the $\{|\nearrow\rangle, |\nwarrow\rangle\}$ basis by passing the photon through a polarizer oriented at 45° with respect to the horizontal (Fig. 10.1). According to the transformation Eqs. (10.7) and (10.8),

$$|\uparrow\rangle = \frac{1}{\sqrt{2}}\left(|\nearrow\rangle + |\nwarrow\rangle\right),$$

$$|\rightarrow\rangle = \frac{1}{\sqrt{2}}\left(|\nearrow\rangle - |\nwarrow\rangle\right),$$

and the two-photon state becomes

$$|\psi_{AB}\rangle = \frac{1}{2}\left[|\nearrow_A\rangle\left(|\uparrow_B\rangle + |\rightarrow_B\rangle\right) + |\nwarrow_A\rangle\left(|\rightarrow_B\rangle - |\uparrow_B\rangle\right)\right]. \tag{10.15}$$

Thus a measurement outcome of Alice's photon in the state $|\nearrow_A\rangle$ "collapses" Bob's photon in the state

$$\langle\nearrow_A|\psi_{AB}\rangle = \frac{1}{\sqrt{2}}\left(|\uparrow_B\rangle + |\rightarrow_B\rangle\right) = |\nearrow_B\rangle. \tag{10.16}$$

However if Alice's photon is found in the state $|\nwarrow_A\rangle$, then the state of Bob's photon reduces to

$$\langle\nwarrow_A|\psi_{AB}\rangle = \frac{1}{\sqrt{2}}\left(|\rightarrow_B\rangle - |\uparrow_B\rangle\right) = -|\nwarrow_B\rangle. \tag{10.17}$$

Thus the quantum state of Bob's photon depends on what Alice decides to do even when there is no way that Alice's and Bob's photons can interact with each other. The same is also true for Alice's photon—the state of the photon in Alice's possession is influenced by what Bob does to his photon. The two photons are "*entangled*" even if they are far apart. There is no corresponding result for classical objects.

This remarkable result is due to the fact that the two photons were initially created in a quantum state (10.14) which is not separable—instead it is entangled. We can now formally define the "separable" and the "quantum entangled" states.

The state of the AB system is separable if the total state vector of the AB system $|\psi_{AB}\rangle$ can be factorized into separate states for the A system and B system, i.e.,

$$|\psi_{AB}\rangle = |\psi_A\rangle\,|\psi_B\rangle. \tag{10.18}$$

The AB system is entangled if such a separation is not possible. i.e., $|\psi_{AB}\rangle$ cannot be written as a product of $|\psi_A\rangle$ and $|\psi_B\rangle$. We thus define a quantum entangled state as the state that satisfies the condition:

$$|\psi_{AB}\rangle \neq |\psi_A\rangle\,|\psi_B\rangle. \tag{10.19}$$

We illustrate these concepts with some examples. As a first example, we note that the state

$$|\psi_{AB}\rangle = \frac{1}{\sqrt{2}} \left(|\uparrow_A\rangle|\uparrow_B\rangle + |\uparrow_A\rangle|\rightarrow_B\rangle \right) \tag{10.20}$$

can be written as a product of individual states for photons A and B. We can identify the states of the individual photons to be

$$|\psi_A\rangle = |\uparrow_A\rangle \text{ and } |\psi_B\rangle = \frac{1}{\sqrt{2}} \left(|\uparrow_B\rangle + |\rightarrow_B\rangle \right).$$

We say the states of the two photons are separable.

Next we consider the following state of the two photons:

$$|\psi_{AB}\rangle = \frac{1}{\sqrt{2}} \left(|\rightarrow_A\rangle|\uparrow_B\rangle + |\uparrow_A\rangle|\rightarrow_B\rangle \right). \tag{10.21}$$

This state of the AB system cannot be written as a product of two independent states. This state is therefore an example of an entangled state. We can prove the entangled behavior of the state (10.21) as follows:

Let us suppose that $|\psi_{AB}\rangle$ is separable so that

$$|\psi_{AB}\rangle = |\psi_A\rangle \, |\psi_B\rangle. \tag{10.22}$$

The most general superposition states of A and B photons are

$$|\psi_A\rangle = c_1 \, |\uparrow_A\rangle + c_2 \, |\rightarrow_A\rangle \text{ and } |\psi_B\rangle = d_1 \, |\uparrow_B\rangle + d_2 \, |\rightarrow_B\rangle, \tag{10.23}$$

where c_1, c_2, d_1, and d_2 are complex numbers that satisfy the conditions

$$|c_1|^2 + |c_2|^2 = 1, \tag{10.24}$$

$$|d_1|^2 + |d_2|^2 = 1. \tag{10.25}$$

Then if the state $|\psi_{AB}\rangle$ is separable,

$$\frac{1}{\sqrt{2}} \left(|\rightarrow_A\rangle|\uparrow_B\rangle + |\uparrow_A\rangle|\rightarrow_B\rangle \right) = \left(c_1|\uparrow_A\rangle + c_2|\rightarrow_A\rangle \right) \left(d_1|\uparrow_B\rangle + d_2|\rightarrow_B\rangle \right). \tag{10.26}$$

This equality is satisfied if

$$c_1 d_2 = c_2 d_1 = \frac{1}{\sqrt{2}} \text{ and } c_1 d_1 = c_2 d_2 = 0. \tag{10.27}$$

But this is impossible. No four complex numbers c_1, c_2, d_1, and d_2 can satisfy the conditions (10.27) simultaneously. Therefore our assumption that $|\psi_{AB}\rangle = |\psi_A\rangle \, |\psi_B\rangle$ is wrong. The state $|\psi_{AB}\rangle$ is therefore an entangled state.

In general, for a system consisting of two objects 1 and 2, that can have two possible states $|0\rangle$ and $|1\rangle$, there are four possible states: both are in $|0\rangle$ state, first is in $|0\rangle$ state and the second in $|1\rangle$ state, first in $|1\rangle$ state and the second in $|0\rangle$ state, and both are in $|1\rangle$ states. These states

$$|0_1, 0_2\rangle, |0_1, 1_2\rangle, |1_1, 0_2\rangle, |1_1, 1_2\rangle \tag{10.28}$$

form the basis set of states for two-particle systems just as $|0\rangle$ and $|1\rangle$ formed the basis set of states for a single particle system. For two particles, the most general state can be written as

$$|\psi_{12}\rangle = c_{00}|0_1, 0_2\rangle + c_{01}|0_1, 1_2\rangle + c_{10}|1_1, 0_2\rangle + c_{11}|1_1, 1_2\rangle, \tag{10.29}$$

where c_{00}, c_{01}, c_{10}, and c_{11} are complex numbers that satisfy the condition

$$|c_{00}|^2 + |c_{01}|^2 + |c_{10}|^2 + |c_{11}|^2 = 1. \tag{10.30}$$

The physical meaning of the coefficients c_{00}, c_{01}, c_{10}, and c_{11} can be understood as follows. The joint probability $P(1, 2)$ of the two-photon system being in (say) state $|0_1, 0_2\rangle$ is

$$P(0_1, 0_2) = |\langle 0_1, 0_2|\psi_{12}\rangle|^2 = |c_{00}|^2. \tag{10.31}$$

Similarly,

$$P(0_1, 1_2) = |\langle 0_1, 1_2|\psi_{12}\rangle|^2 = |c_{01}|^2, \tag{10.32}$$

$$P(1_1, 0_2) = |\langle 1_1, 0_2|\psi_{12}\rangle|^2 = |c_{10}|^2, \tag{10.33}$$

$$P(1_1, 1_2) = |\langle 1_1, 1|\psi_{12}\rangle|^2 = |c_{11}|^2. \tag{10.34}$$

We can also determine the probability of finding the state of a single particle by using the procedure discussed in Section 2.4. As an example, the probability of finding the first photon in state $|0\rangle$ is obtained by summing the probabilities that the first photon is in the state $|0_1\rangle$ and the second photon is in state $|0_2\rangle$ as well as the first photon is in the state $|0_1\rangle$ but the second photon is in state $|1_2\rangle$, i.e.,

$$P(0_1) = P(0_1, 0_2) + P(0_1, 1_2) = |c_{00}|^2 + |c_{01}|^2. \tag{10.35}$$

Similarly,

$$P(1_1) = P(1_1, 0_2) + P(1_1, 1_2) = |c_{10}|^2 + |c_{11}|^2, \tag{10.36}$$

$$P(0_2) = P(0_1, 0_2) + P(1_1, 0_2) = |c_{00}|^2 + |c_{10}|^2, \tag{10.37}$$

$$P(1_2) = P(0_1, 1_2) + P(1_1, 1_2) = |c_{01}|^2 + |c_{11}|^2. \tag{10.38}$$

Next we ask the question: What is the condition under which the general state (10.29) is an entangled state? For this purpose we define the quantity

$$C = 2|c_{00}c_{11} - c_{01}c_{10}|. \tag{10.39}$$

It turns out that $0 \leq C \leq 1$. The state (10.29) is entangled if and only if $C > 0$. The quantity C is called concurrence and is a measure of entanglement. A two-particle state is unentangled or separable if $C = 0$ and is maximally entangled if $C = 1$.

The following states, called Bell basis states (or simply Bell states), are among the most entangled states of the two particles with $C = 1$:

$$|B_{00}(1,2)\rangle = \frac{1}{\sqrt{2}}\left(|0_1,0_2\rangle + |1_1,1_2\rangle\right), \tag{10.40}$$

$$|B_{01}(1,2)\rangle = \frac{1}{\sqrt{2}}\left(|0_1,1_2\rangle + |1_1,0_2\rangle\right), \tag{10.41}$$

$$|B_{10}(1,2)\rangle = \frac{1}{\sqrt{2}}\left(|0_1,0_2\rangle - |1_1,1_2\rangle\right), \tag{10.42}$$

$$|B_{11}(1,2)\rangle = \frac{1}{\sqrt{2}}\left(|0_1,1_2\rangle - |1_1,0_2\rangle\right). \tag{10.43}$$

These states are all mutually orthogonal. For example, $\langle B_{01}|B_{00}\rangle = \langle B_{01}|B_{10}\rangle = \langle B_{01}|B_{11}\rangle = \langle B_{11}|B_{00}\rangle = 0$. A significance of the Bell basis states is that we can write a general entangled state (10.29) as a linear superposition of Bell states.

10.3 Schrödinger's Cat Paradox

Erwin Schrödinger was one of the inventors of quantum mechanics. However, he always felt uncomfortable with quantum mechanics and came up with paradoxical consequences. Schrödinger's cat paradox (1935) is one such example. Schrödinger's cat experiment is an example of a *gedanken* experiment, i.e., we do not *actually* conduct the experiment, we use only our imagination and reasoning instead.

What Schrödinger was concerned with was the notion of the coherent superposition of states where, as we have seen above, the system can be simultaneously in two quantum states. However when we measure, we find the system in one state or another. Schrödinger argued that if we extend the ideas that are valid for microscopic objects such as an electron, a photon, or an atom to the macroscopic world, it may lead to absolutely amazing consequences. One such consequence is a cat that can simultaneously be in a state of being alive and a state of being dead. Since we do not encounter a cat both alive and dead, there must be something wrong about quantum mechanical interpretation.

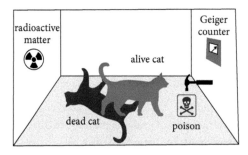

Fig. 10.4 Schrödinger's *gedanken* experiment: A radioactive atom decays and the radiated particle is detected by a Geiger counter that releases a hammer hitting a vial of poison, causing the cat to die. So the state of the cat (dead or alive) depends on the state of the radioactive atom (decayed or not decayed).

In Schrödinger's *gedanken* experiment, a cat is placed inside a steel box (Fig. 10.4). The box also contains a radioactive atom, a Geiger counter, a hammer, and a vial of poison. The radioactive atom can decay in the process by which its unstable atomic nucleus loses energy by emitting massive alpha and beta particles and massless gamma ray photons. Radioactive decay is a random process at the level of single atoms. According to quantum theory it is impossible to predict when a particular radioactive atom will decay. A Geiger counter can detect the emitted particles. In Schrödinger's set-up, when the radioactive substance decays, the Geiger counter detects it and triggers the hammer to release the poison, which subsequently kills the cat. Thus we have two possibilities: the radioactive decay does not take place and the cat is alive OR the radioactive decay does take place and the cat is dead. We thus have the microscopic state of the radioactive atom and the cat:

$$|\psi_{AC}\rangle = c_a |no\ decay\rangle |Alive\ cat\rangle + c_b |decay\rangle |Dead\ cat\rangle. \qquad (10.44)$$

Here the coefficients c_a and c_b are the amplitudes such that $|c_a|^2$ and $|c_b|^2$ are the probabilities that the atom has not decayed and cat is alive or that the atom has decayed and the cat is dead, respectively. Equation (10.44) corresponds to an entangled state between the state of the atom and the state of the cat.

A coherent "superposition" of the cat states of the form

$$|\psi_C\rangle = c_a |Alive\ cat\rangle + c_b |Dead\ cat\rangle \qquad (10.45)$$

can be obtained first by an operation that takes the atomic state $|decay\rangle$ to the state $(|no\ decay\rangle + |decay\rangle)/\sqrt{2}$ and the state $|no\ decay\rangle$ to the state $(|no\ decay\rangle - |decay\rangle)/\sqrt{2}$ and then making a measurement on the radioactive atom. The atom, if found in the state $|no\ decay\rangle$, would lead to the cat in the superposition state (10.45). This is a paradoxical situation, as in real life, we do not see a cat that is simultaneously alive and dead.

Until the box is opened, an observer does not know whether the cat is alive or dead. The cat's fate is intimately tied to the atomic state. As soon as the box is opened we find that the cat is dead or alive but not in a "superposition" state—the cat is alive or is dead but not both.

Here we present the cat paradox in Schrödinger's own words:

One can even set up quite ridiculous cases. A cat is penned up in a steel chamber, along with the following device (which must be secured against direct interference by the cat): in a Geiger counter, there is a tiny bit of radioactive substance, so small, that perhaps in the course of the hour one of the atoms decays, but also, with equal probability, perhaps none; if it happens, the counter tube discharges and through a relay releases a hammer that shatters a small flask of hydrocyanic acid. If one has left this entire system to itself for an hour, one would say that the cat still lives if meanwhile no atom has decayed. The first atomic decay would have poisoned it. The psi-function of the entire system would express this by having in it the living and dead cat (pardon the expression) mixed or smeared out in equal parts... ...It is typical of these cases that an indeterminacy originally restricted to the atomic domain becomes transformed into macroscopic indeterminacy, which can then be resolved by direct observation. That prevents us from so naively accepting as valid a "blurred model" for representing reality. In itself, it would not embody anything unclear or contradictory. There is a difference between a shaky or out-of-focus photograph and a snapshot of clouds and fog banks.

An interesting and revealing comment is by Einstein in his letter to Schrödinger, essentially patting him on the back, for coming up with an elegant argument regarding "reality"[1]:

> You are the only contemporary physicist, besides Laue, who sees that one cannot get around the assumption of reality, if only one is honest. Most of them simply do not see what sort of risky game they are playing with reality—reality as something independent of what is experimentally established. Their interpretation is, however, refuted most elegantly by your system of radioactive atom + amplifier + charge of gun powder + cat in a box, in which the ψ-function of the system contains both the cat alive and blown to bits. Nobody really doubts that the presence or absence of the cat is something independent of the act of observation.

We discuss Einstein's concept of reality and his own criticism of quantum mechanics in detail in Chapter 12.

10.4 Quantum Teleportation

Teleportation is a science fiction concept that was popularized by the 1960s TV program "Star Trek." The phrase "Beam me up, Scotty!!" used by Captain Kirk to direct his Chief Engineer Scott when he wanted to be transported from a faraway planet to the Starship Enterprise, remains one of the most famous phrases of television programming. Teleportation was used as a means of transport—Captain Kirk would disappear on the planet and reappear on a platform inside the Starship.

Can we do teleportation in real life? If teleportation becomes possible, there will be no need for cars or air planes—we could be teleported from one place to another. So far, teleportation of human beings is only science fiction—it would take an infinite amount of energy to teleport even a single atom. However, teleporting a two-state quantum system has become a reality, thanks to the concept of quantum entanglement. Quantum teleportation of a two-state system from one location to another, first proposed by C. H. Bennett, G. Brassard, C. Crépeau, R. Jozsa, A. Peres, and W. K. Wootters, is one of the most beautiful examples of quantum entanglement.

The problem of teleportation can be described as follows. Suppose Alice has an object with an unknown quantum state

$$| \psi(A) \rangle = c_0 | 0_A \rangle + c_1 | 1_A \rangle \tag{10.46}$$

at location A. By unknown we mean that we do not know the coefficients c_0 and c_1. The binary states $|0\rangle$ and $|1\rangle$ can be the states of any two-state system, such as two atomic levels, two states of polarization, etc. There is a two-state system located at C at Bob's end. Just like in the television serial Star Trek, we would like to destroy the unknown quantum state (10.45) of the system at Alice's end and create the same quantum state on the system at Bob's end.

To be more specific, suppose we have an atom in an unknown coherent superposition of the ground state $|g\rangle$ and an excited state $|e\rangle$ of the form at Alice's end

$$|\psi(A)\rangle = c_g |g\rangle + c_e |e\rangle \tag{10.47a}$$

---

[1] Einstein's comment on Schrödinger's cat in a letter to Schrödinger dated December 22, 1950 is in K. Przibram (ed), *Letters on Wave Mechanics* (Philosophical Library, New York, 1967),p. 39.

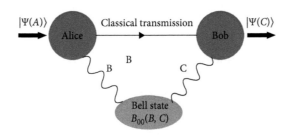

Fig. 10.5 Teleportation of a state $|\psi(A)\rangle$ at Alice's end to Bob's end. An entangled state $|B_{00}(B, C)\rangle$ is generated, particle B goes to Alice and particle C goes to Bob. Alice announces the outcome of the Bell basis measurement for states A and B to Bob on a classical channel. Bob makes local adjustments on his particle C accordingly and the state $|\psi(A)\rangle$ is teleported to Bob.

and a two-level atom with Bob. Can we teleport the state $|\psi(A)\rangle$ to the atom with Bob so that its state becomes

$$|\psi(C)\rangle = c_g |g\rangle + c_e |e\rangle? \tag{10.47b}$$

The key to quantum teleportation is the presence of two channels for exchanging information between Alice and Bob. This is shown in Fig. 10.5. The first is what we call a quantum channel, which allows two objects B and C, prepared in an entangled state, to be sent to Alice and Bob (object B to Alice and object C to Bob). The second is a classical channel such as a telephone line which Alice uses to send the outcome of her measurement to Bob.

Quantum teleportation of a quantum state of the form (10.46) can be accomplished in three steps:

Step I: We prepare two objects B and C in an entangled state:

$$|B_{00}(B, C)\rangle = \frac{1}{\sqrt{2}} [|0_B, 0_C\rangle + |1_B, 1_C\rangle]. \tag{10.48}$$

This is one of the Bell states, B_{00}, discussed in Section 10.2. The particle B is sent to Alice who has in her possession the particle A. The particle C is sent to Bob. So now particles A and B are at Alice's end and particle C is at Bob's end. The combined state of the A, B, C system is:

$$|\psi(A, B, C)\rangle = \frac{1}{\sqrt{2}} [c_0 (|0_A, 0_B, 0_C\rangle + |0_A, 1_B, 1_C\rangle) + c_1 (|1_A, 0_B, 0_C\rangle + |1_A, 1_B, 1_C\rangle)]. \tag{10.49}$$

We can rewrite $|\psi(A, B, C)\rangle$ in terms of the Bell basis states (10.40)–(10.43) for the AB system by first noting that

$$|0_A, 0_B\rangle = \frac{1}{\sqrt{2}} [|B_{00}(A, B)\rangle + |B_{10}(A, B)\rangle], \tag{10.50}$$

$$|0_A, 1_B\rangle = \frac{1}{\sqrt{2}} [|B_{01}(A, B)\rangle + |B_{11}(A, B)\rangle], \tag{10.51}$$

$$|1_A, 0_B\rangle = \frac{1}{\sqrt{2}} [|B_{01}(A, B)\rangle - |B_{11}(A, B)\rangle], \tag{10.52}$$

$$|1_A, 1_B\rangle = \frac{1}{\sqrt{2}} \left[|B_{00}(A, B)\rangle - |B_{10}(A, B)\rangle \right].$$ (10.53)

Then, substituting these expressions in Eq. (10.49), we have

$$
\begin{aligned}
|\psi(A, B, C)\rangle = \frac{1}{2} \big[&|B_{00}(A, B)\rangle (c_0|0_C\rangle + c_1|1_C\rangle) \\
+ &|B_{01}(A, B)\rangle (c_0|1_C\rangle + c_1|0_C\rangle) \\
+ &|B_{10}(A, B)\rangle (c_0|0_C\rangle - c_1|1_C\rangle) \\
+ &|B_{11}(A, B)\rangle (c_0|1_C\rangle - c_1|0_C\rangle) \big].
\end{aligned}
$$ (10.54)

So far the state of system A at Alice's end is not touched.

Step II: In this step, Alice, who has both systems A and B, makes a joint measurement on the AB system in the Bell basis. She can get one of the outcomes $|B_{00}(A, B)\rangle$, $|B_{01}(A, B)\rangle$, $|B_{10}(A, B)\rangle$, or $|B_{11}(A, B)\rangle$. Now the interesting result! It is clear from Eq. (10.54) that, if her outcome is $|B_{00}(A, B)\rangle$, then the system at Bob's end, C, reduces to $c_0|0_C\rangle + c_1|1_C\rangle$, i.e.,

$$|B_{00}(A, B)\rangle \rightarrow c_0|0_C\rangle + c_1|1_C\rangle.$$ (10.55)

Similarly,

$$|B_{01}(A, B)\rangle \rightarrow c_0|1_C\rangle + c_1|0_C\rangle,$$ (10.56)

$$|B_{10}(A, B)\rangle \rightarrow c_0|0_C\rangle - c_1|1_C\rangle,$$ (10.57)

$$|B_{11}(A, B)\rangle \rightarrow c_0|1_C\rangle - c_1|0_C\rangle.$$ (10.58)

This is remarkable as the details of the unknown state contained in the coefficients c_0 and c_1 have been "teleported" to Bob who may be, in principle, very far from Alice. However, at this stage, Bob has no knowledge about which state he has as he is unaware of the outcome of the Bell basis measurement done by Alice.

Step III: In this step, Alice informs Bob about her outcome through a classical channel, such as a telephone line. If the measurement outcome is $|B_{00}(A, B)\rangle$, then Bob knows that the state of the system C at his end is $c_0|0_C\rangle + c_1|1_C\rangle$. This is exactly the state that Alice wanted to teleport to Bob. So Bob knows that the state of the system C is the desired state and he does nothing.

If Alice tells Bob that her measurement outcome is $|B_{01}(A, B)\rangle$, then Bob knows that the state of system C is $c_0|1_C\rangle + c_1|0_C\rangle$. In this case he makes a quantum amplitude transformation $|0_C\rangle \rightarrow |1_C\rangle$ and $|1_C\rangle \rightarrow |0_C\rangle$ and the state of the system C reduces to the desired state, $c_0|0_C\rangle + c_1|1_C\rangle$.

If Alice tells Bob that her measurement outcome is $|B_{10}(A, B)\rangle$, then Bob knows that the state of system C is $c_0|0_C\rangle - c_1|1_C\rangle$. In this case he makes the quantum phase transformation $|0_C\rangle \rightarrow |0_C\rangle$ and $|1_C\rangle \rightarrow -|1_C\rangle$ reducing the state of the system C to the desired state.

Finally, if Alice tells Bob that her measurement outcome is $|B_{11}(A, B)\rangle$, then Bob knows that the state of C is $c_0|1_C\rangle - c_1|0_C\rangle$. In this case Bob makes both quantum amplitude transformation $|0_C\rangle \rightarrow |1_C\rangle$ and $|1_C\rangle \rightarrow |0_C\rangle$ as well as quantum phase transformation $|0_C\rangle \rightarrow |0_C\rangle$ and $|1_C\rangle \rightarrow -|1_C\rangle$ to bring the state of the system C to the desired state.

This finishes the teleportation protocol—the state $|\psi(A)\rangle = c_0|0_A\rangle + c_1|1_A\rangle$ has been teleported from Alice to Bob. The important resource used in the process of teleportation is quantum entanglement.

There are a couple of points worth mentioning. First we note that the process of teleportation cannot be used to transfer information faster than the speed of light. If it did, it would violate the tenet of another important theory of physics—Einstein's theory of relativity. At first sight it would appear that Alice can send information superluminally by encoding it in the state $|\psi(A)\rangle$. As soon as Alice completes the Bell basis measurement at her end the information about $|\psi(A)\rangle$ is teleported to Bob instantaneously. However this is not true as, until Alice tells Bob the outcome of her measurement on a classical channel where information is transferred at a speed less than or equal to the speed of light, Bob is not able to reproduce Alice's state.

The second point is that, in the process of teleportation, the state $|\psi(A)\rangle = c_0|0_A\rangle + c_1|1_A\rangle$ is destroyed at Alice's end before it is created at Bob's end. After Alice makes the Bell state measurement, any information about $|\psi(A)\rangle$ (which is contained in coefficients c_0 and c_1) is destroyed. Thus there is no way that Alice's state can be copied or cloned in the teleportation process.

10.5 Entanglement Swapping

Let us suppose Alice and Bob are very close friends. Similarly Cathy and David are also very close friends. Bob and Cathy are able to interact but Alice and David have never met each other. Indeed they (Alice and David) live very far away from each other, never have communicated, and do not even have a capacity to interact with each other in any way whatsoever. Is it possible that an interaction between Bob and Cathy can lead to a very close friendship between Alice and David?

An equivalent quantum problem can be stated as follows: Alice and Bob have an entangled pair of objects, such as photons, in a state

$$|\Psi_{AB}\rangle = |B_{00}(A,B)\rangle = \frac{1}{\sqrt{2}}\left[|0_A, 0_B\rangle + |1_A, 1_B\rangle\right]. \tag{10.59}$$

Similarly Cathy and David have also an entangled pair of photons in state

$$|\Psi_{CD}\rangle = |B_{00}(C,D)\rangle = \frac{1}{\sqrt{2}}\left[|0_C, 0_D\rangle + |1_C, 1_D\rangle\right]. \tag{10.60}$$

Alice and David are so far away from each other that they cannot communicate. Can Bob and Cathy make a joint measurement on the photons in their possession such that an entanglement is generated between the photons in Alice's and David's possession? As shown in the following, the answer is Yes!

Another interesting thing happens in the process: In the end there is entanglement between Bob and Cathy and between Alice and David, but no more entanglement between Alice and

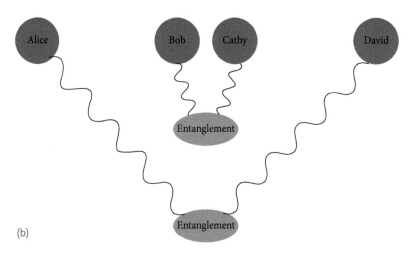

Fig. 10.6 Entanglement swapping: (*a*) Entangled states are prepared between Alice and Bob's photons as well as between Cathy and David's photons. A Bell state measurement is made between Bob and Cathy. (*b*) The result is that, after the measurement is made, the previous entanglements disappear and new entangled states are formed between Alice and David and between Bob and Cathy.

Bob and between Cathy and David. It is as if the friendships between Alice and Bob and between Cathy and David have been swapped with friendships between Alice and David and between Bob and Cathy.

The full state of the *ABCD* system is the product of entangled states between Alice and Bob and between Cathy and David (Fig. 10.6*a*):

$$
\begin{aligned}
|\Psi_{ABCD}\rangle &= |\Psi_{AB}\rangle|\Psi_{CD}\rangle \\
&= \frac{1}{2}\big[|0_A\rangle|0_B, 0_C\rangle|0_D\rangle + |0_A\rangle|0_B, 1_C\rangle|1_D\rangle + |1_A\rangle|1_B, 0_C\rangle|0_D\rangle \\
&\quad + |1_A\rangle|1_B, 1_C\rangle|1_D\rangle\big].
\end{aligned} \tag{10.61}
$$

Next we rewrite this state in the Bell basis for the BC system via the transformations

$$|0_B, 0_C\rangle = \frac{1}{\sqrt{2}} \left[|B_{00}(B, C)\rangle + |B_{10}(B, C)\rangle \right], \tag{10.62}$$

$$|0_B, 1_C\rangle = \frac{1}{\sqrt{2}} \left[|B_{01}(B, C)\rangle + |B_{11}(B, C)\rangle \right], \tag{10.63}$$

$$|1_B, 0_C\rangle = \frac{1}{\sqrt{2}} \left[|B_{01}(B, C)\rangle - |B_{11}(B, C)\rangle \right], \tag{10.64}$$

$$|1_B, 1_C\rangle = \frac{1}{\sqrt{2}} \left[|B_{00}(B, C)\rangle - |B_{10}(B, C)\rangle \right]. \tag{10.65}$$

The resulting expression for $|\Psi_{ABCD}\rangle$ is

$$
\begin{aligned}
|\Psi_{ABCD} = \frac{1}{2\sqrt{2}} \{ & |0_A\rangle [|B_{00}(B, C)\rangle + |B_{10}(B, C)\rangle] |0_D\rangle \\
+ & |0_A\rangle [|B_{01}(B, C)\rangle + |B_{11}(B, C)\rangle] |1_D\rangle \\
+ & |1_A\rangle [|B_{01}(B, C)\rangle - |B_{11}(B, C)\rangle] |0_D\rangle \\
+ & |1_A\rangle [|B_{00}(B, C)\rangle - |B_{10}(B, C)\rangle] |1_D\rangle \} \\
= & |B_{00}(A, D)\rangle |B_{00}(B, C)\rangle + |B_{01}(A, D)\rangle |B_{01}(B, C)\rangle \\
+ & |B_{10}(A, D)\rangle |B_{10}(B, C)\rangle + |B_{11}(A, D)\rangle |B_{11}(B, C)\rangle.
\end{aligned} \tag{10.66}
$$

At this point Bob and Cathy make a joint measurement in the Bell basis. There are four possible outcomes, namely $|B_{00}(B, C)\rangle$, $|B_{01}(B, C)\rangle$, $|B_{10}(B, C)\rangle$, and $|B_{11}(B, C)\rangle$, with equal probability. It can be seen from Eq. (10.66) that a measurement outcome of $|B_{00}(B, C)\rangle$ by Bob and Cathy reduces the state of Alice and David to an entangled state $|B_{00}(A, D)\rangle$, i.e.,

$$|B_{00}(B, C)\rangle \rightarrow \frac{1}{\sqrt{2}} \left[|0_A, 0_D\rangle + |1_A, 1_D\rangle \right] = |B_{00}(A, D)\rangle. \tag{10.67}$$

Similarly,

$$|B_{01}(B, C)\rangle \rightarrow \frac{1}{\sqrt{2}} \left[|0_A, 1_D\rangle + |1_A, 0_D\rangle \right] = |B_{01}(A, D)\rangle, \tag{10.68}$$

$$|B_{10}(B, C)\rangle \rightarrow \frac{1}{\sqrt{2}} \left[|0_A, 0_D\rangle - |1_A, 1_D\rangle \right] = |B_{10}(A, D)\rangle, \tag{10.69}$$

$$|B_{11}(B, C)\rangle \rightarrow \frac{1}{\sqrt{2}} \left[|0_A, 1_D\rangle - |1_A, 0_D\rangle \right] = |B_{11}(A, D)\rangle. \tag{10.70}$$

The net result is that the entanglement between Alice and Bob and between Cathy and David has been swapped to between Alice and David and Bob and Cathy, as shown in Fig. 10.6b.

Quantum entanglement swapping has applications in quantum communication when entanglement of distant objects is created by first breaking up the transmission distance into smaller distances and then making Bell state measurements at the intermediate locations.

Problems

10.1 Consider the following two-photon states:

$$|\psi_1\rangle = \frac{1}{2}\left[|\rightarrow_1, \rightarrow_2\rangle + |\rightarrow_1, \uparrow_2\rangle + |\uparrow_1, \rightarrow_2\rangle + |\uparrow_1, \uparrow_2\rangle\right],$$
$$|\psi_2\rangle = \frac{1}{2}\left[|\rightarrow_1, \rightarrow_2\rangle + |\rightarrow_1, \uparrow_2\rangle + |\uparrow_1, \rightarrow_2\rangle - |\uparrow_1, \uparrow_2\rangle\right],$$
$$|\psi_3\rangle = \cos\theta\, |\rightarrow_1, \rightarrow_2\rangle + \sin\theta\, |\uparrow_1, \uparrow_2\rangle],$$
$$|\psi_4\rangle = \frac{1}{2}\left[|\rightarrow_1, \rightarrow_2\rangle - |\rightarrow_1, \uparrow_2\rangle - |\uparrow_1, \rightarrow_2\rangle + |\uparrow_1, \uparrow_2\rangle\right].$$

Verify that all the states are correctly normalized. Which ones are entangled? Write those that are not entangled as an explicit product of single-photon states.

10.2 Show that the states

$$|C_{00}\rangle = \frac{1}{2}\left[-|0_1, 0_2\rangle + |0_1, 1_2\rangle + |1_1, 0_2\rangle + |1_1, 1_2\rangle\right],$$
$$|C_{01}\rangle = \frac{1}{2}\left[|0_1, 0_2\rangle - |0_1, 1_2\rangle + |1_1, 0_2\rangle + |1_1, 1_2\rangle\right],$$
$$|C_{10}\rangle = \frac{1}{2}\left[|0_1, 0_2\rangle + |0_1, 1_2\rangle - |1_1, 0_2\rangle + |1_1, 1_2\rangle\right],$$
$$|C_{11}\rangle = \frac{1}{2}\left[|0_1, 0_2\rangle + |0_1, 1_2\rangle + |1_1, 0_2\rangle - |1_1, 1_2\rangle\right],$$

are normalized, mutually orthogonal, and maximally entangled (concurrence, $C = 1$)

10.3 Consider the following entangled state:

$$|\psi\rangle = \frac{1}{\sqrt{54}}\left[2|\rightarrow_1, \rightarrow_2\rangle + 3|\rightarrow_1, \uparrow_2\rangle + 4|\uparrow_1, \rightarrow_2\rangle - 5|\uparrow_1, \uparrow_2\rangle\right].$$

Find the probability that the first photon is found in state $|\rightarrow\rangle$. Also find the probability that the second photon is found in the state $|\uparrow\rangle$.

10.4 Express the general two-particle state

$$|\psi_{12}\rangle = c_{00}|0_1, 0_2\rangle + c_{01}|0_1, 1_2\rangle + c_{10}|1_1, 0_2\rangle + c_{11}|1_1, 1_2\rangle$$

as a linear superposition of Bell states:

$$|\psi_{12}\rangle = d_{00}|B_{00}\rangle + d_{01}|B_{01}\rangle + d_{10}|B_{10}\rangle + d_{11}|B_{11}\rangle.$$

Find the coefficients d_{00}, d_{01}, d_{10}, and d_{11} in terms of the coefficients c_{00}, c_{01}, c_{10}, and c_{11}. What is the concurrence, C, in terms of the new coefficients?

10.5 Discuss the quantum teleportation of the state

$$|\psi(A)\rangle = c_0\,|0_A\rangle + c_1\,|1_A\rangle$$

when the initial entangled state between B and C is

$$|B_{01}(B, C)\rangle = \frac{1}{\sqrt{2}}\left(|0_B, 1_C\rangle + |1_B, 0_C\rangle\right).$$

BIBLIOGRAPHY

M. Ray, *Quantum Physics: Illusion or Reality?* (Cambridge University Press 2012).

S. Barnett, *Quantum Informatics* (Oxford University Press 2009).

J. Gribbin, *In Search of Schrödinger's Cat: Quantum Physics and Reality* (Bantam 1984).

J. A. Wheeler and W.H. Zurek (eds.), *Quantum Theory and Measurement* (Princeton University Press 1983).

C. H. Bennett, G. Brassard, C. Crépeau, R. Jozsa, A. Peres, and W. K. Wootters, *Teleporting an unknown quantum state via dual classical and Einstein–Podolsky–Rosen channels*, Physical Review Letters 70, 1895 (1993).

11 No-cloning Theorem and Quantum Copying

Heisenberg's uncertainty relation and Bohr's principle of complementarity form the foundations of quantum mechanics. If these are violated then the edifice of quantum mechanics can come crashing down. Thus, any process that can potentially lead to a violation of these *sacred* principles must be examined with utmost care. One such process is the cloning of quantum states. The question of interest is whether it is possible to make a perfect copy or a clone of an unknown quantum state $|\psi\rangle$ without destroying the original state? If this becomes possible then we could make as many copies of the state $|\psi\rangle$ as we like. We can then make measurement of any variable with arbitrary precision, leading to a violation of both the Heisenberg uncertainty relation and the principle of complementarity.

In order to illustrate this point, let us go back to the example of the measurement of the polarization of photons in different bases as discussed in Chapter 10. We noted that a photon, when observed in the $\{|\rightarrow\rangle,|\uparrow\rangle\}$ basis, could either be in the state $|\uparrow\rangle$ or in the state $|\rightarrow\rangle$. However the same photon, when observed in the $\{|\nearrow\rangle,|\nwarrow\rangle\}$ basis could either be in the state $|\nearrow\rangle$ or in the state $|\nwarrow\rangle$. The principle of complementarity does not allow us to measure the polarization of the photon in the two bases simultaneously. However, if cloning becomes possible, we can make identical copies of the state $|\psi\rangle$. We can then measure the polarization of half of them in the $\{|\rightarrow\rangle,|\uparrow\rangle\}$ basis and the other half in the $\{|\nearrow\rangle,|\nwarrow\rangle\}$ basis. These measurements can give precise values of the complementary variables, thus violating the principle of complementarity. In a similar way, we can show that, if cloning of quantum states is allowed, Heisenberg's uncertainty relation can be violated as well.

In this chapter, we show that cloning of an arbitrary quantum state is not allowed. The foundation of quantum mechanics is therefore protected. The no-cloning theorem that we discuss in Section 11.2 was formulated in a classic paper by William Wootters and Wojciech Zurek in 1982. The motivation for this paper came when a paper by Nick Herbert was published in 1982 showing that photon cloning can lead to superluminal (faster than light) communication, thus violating another tenet of modern physics, namely, Einstein's theory of relativity. According to the theory of relativity, no information can be sent faster than the speed of light.

In the next sections, we discuss how quantum cloning can lead to superluminal communication before discussing the no-cloning theorem. We also discuss that, if making a perfect copy of a quantum state is forbidden, how best a copy of a state can be made.

11.1 Cloning and Superluminal Communication

It is an everyday observance that identical copies can be made of objects. For example, a page from this book can be copied—the copy can be as close to the original as we like depending

Quantum Mechanics for Beginners: With Applications to Quantum Communication and Quantum Computing. M. Suhail Zubairy.
© M. Suhail Zubairy 2020. Published in 2020 by Oxford University Press. DOI: 10.1093/oso/9780198854227.001.0001

upon the quality of the copying machine. In principle, it should be possible to make perfect copies with a high quality copier.

However this observation is true for macroscopic objects which are governed by classical laws of physics. Here, we address the question: Can we make identical copies of an unknown quantum state? As an example, we consider a single photon in the polarization state

$$|\psi\rangle = \cos\theta\,|\rightarrow\rangle + \sin\theta\,|\uparrow\rangle, \tag{11.1}$$

where $|\rightarrow\rangle$ is the state of the photon with horizontal polarization and $|\uparrow\rangle$ is the state of the photon with vertical polarization. The coefficients $\cos\theta$ and $\sin\theta$ may be unknown. Can we make identical copies of such photons? The answer is No! We cannot make perfect copies of the quantum state $|\psi\rangle$. We thus have a no-cloning theorem.

Before we prove the no-cloning theorem, we show that, if we could make identical copies of a quantum state, then we could have superluminal communication—communication faster than the speed of light. As mentioned above, no information can be transmitted faster than the speed of light according to Einstein's theory of relativity. So this should not be possible!!! A no-cloning theorem therefore saves Einstein's theory of relativity!!!

We consider the transmission of information via photons. The information is supposed to be binary, "0" or "1." A key to our argument is two conjugate bases: a $\{|\rightarrow\rangle,|\uparrow\rangle\}$ basis and a $\{|\nearrow\rangle,|\nwarrow\rangle\}$ basis where

$$|\nearrow\rangle = \frac{1}{\sqrt{2}}(|\rightarrow\rangle + |\uparrow\rangle), \tag{11.2}$$

$$|\nwarrow\rangle = \frac{1}{\sqrt{2}}(|\uparrow\rangle - |\rightarrow\rangle), \tag{11.3}$$

are polarization states in a frame rotated with the $\{|\rightarrow\rangle,|\uparrow\rangle\}$ basis by 45°.

The key to the communication protocol is that an entangled pair of photons is shared by Alice and Bob, who could be a long distance apart. The entangled state of the photon can be of the form

$$|\Psi_{AB}\rangle = \frac{1}{\sqrt{2}}(|\rightarrow_A, \uparrow_B\rangle + |\uparrow_A, \rightarrow_B\rangle), \tag{11.4}$$

i.e., if Alice's photon is polarized along the horizontal direction, then Bob's photon is polarized along the vertical direction and vice versa. It follows from Eqs. (11.2) and (11.3) that $\{|\rightarrow\rangle,|\uparrow\rangle\}$ can be written in terms of $\{|\nearrow\rangle,|\nwarrow\rangle\}$ as

$$|\uparrow\rangle = \frac{1}{\sqrt{2}}(|\nearrow\rangle + |\nwarrow\rangle), \tag{11.5}$$

$$|\rightarrow\rangle = \frac{1}{\sqrt{2}}(|\nearrow\rangle - |\nwarrow\rangle). \tag{11.6}$$

Therefore, on substituting for $|\rightarrow\rangle$ and $|\uparrow\rangle$ from Eqs. (11.5) and (11.6) into Eq. (11.4), the entangled state (11.4) can be shown to be equivalent to

$$|\Psi_{AB}\rangle = \frac{1}{\sqrt{2}}(|\nearrow_A, \nearrow_B\rangle - |\nwarrow_A, \nwarrow_B\rangle). \tag{11.7}$$

Thus Alice–Bob's entangled state is:

$$|\Psi_{AB}\rangle = \frac{1}{\sqrt{2}}\left(|\rightarrow_A, \uparrow_B\rangle + |\uparrow_A, \rightarrow_B\rangle\right)$$

or

$$|\Psi_{AB} = \frac{1}{\sqrt{2}}\left(|\nearrow_A, \nearrow_B\rangle - |\nwarrow_A, \nwarrow_B\rangle\right)$$

depending on our choice of bases. Which basis we choose depends on the experiment we carry out. If our polarizer is such that it is vertically polarized in the $\{|\rightarrow\rangle,|\uparrow\rangle\}$ or $\oplus$ basis, then the photon polarization is measured as either in state $|\rightarrow\rangle$ or in state $|\uparrow\rangle$. On the other hand, if the polarizer is in the rotated $\{|\nearrow\rangle,|\nwarrow\rangle\}$ or $\otimes$ basis then the photon polarization is measured as either in state $|\nearrow\rangle$ or in state $|\nwarrow\rangle$.

The communication protocol proceeds as follows: If Alice wants to transmit a "0", she measures her photon in the $\{|\rightarrow\rangle,|\uparrow\rangle\}$ basis. In case her outcome is,$|\rightarrow_A\rangle$, then Bob's state collapses to $\langle\rightarrow_A|\Psi_{AB}\rangle = |\uparrow_B\rangle$ and if her outcome is, $|\uparrow_A\rangle$, then Bob's state collapses to $\langle\uparrow_A|\Psi_{AB}\rangle = |\rightarrow_B\rangle$. Here we substituted the form in Eq. (11.4) for $|\Psi_{AB}\rangle$. The probabilities of obtaining the '$|\rightarrow_A\rangle$' or '$|\uparrow_A\rangle$' state of photon by Alice are 50% each. If Alice wants to transmit a "1", she measures her photon in the $\{|\nearrow\rangle,|\nwarrow\rangle\}$ basis, collapsing Bob's state to either $\langle\nearrow_A|\Psi_{AB}\rangle = |\nearrow_B\rangle$ or $\langle\nwarrow_A|\Psi_{AB}\rangle = |\nwarrow_B\rangle$ depending upon whether Alice's outcome is $|\nearrow_A\rangle$ or $|\nwarrow_A\rangle$, respectively.

If Bob cannot clone Alice's photon, then superluminal communication cannot take place. For example, if Alice decides to transmit a "0", she measures her photon in the $\{|\rightarrow\rangle,|\uparrow\rangle\}$ basis, and let the outcome be "$|\rightarrow_A\rangle$". Now Bob can make his measurement with the basis $\{|\rightarrow\rangle,|\uparrow\rangle\}$, and the outcome is "$|\uparrow_B\rangle$". However if Bob makes his measurement in the basis $\{|\nearrow\rangle,|\nwarrow\rangle\}$, then there is a 50% chance that his outcome is "$|\nearrow_B\rangle$" and a 50% chance that his outcome is "$|\nwarrow_B\rangle$". Thus there is no way for Bob to find out what information, "0" or "1", Alice sent.

But what happens if cloning is possible? A cloning machine, if it existed, would take some auxiliary systems in some well-defined state $|0\rangle$ (where $|0\rangle$ can be $|\rightarrow\rangle$ or $|\uparrow\rangle$) in addition to the input state $|\psi\rangle$ and convert them both into $|\psi\rangle$. A cloning machine therefore acts as follows:

$$U_{\text{clone}}|\psi\rangle|0\rangle = |\psi\rangle|\psi\rangle \tag{11.8}$$

for any state $|\psi\rangle$. The cloning machine U_{clone} can thus make many identical copies of $|\psi\rangle$ as shown in Fig. 11.1.

After Alice has measured the polarization of her photon, Bob's state reduces to a state depending on the basis that Alice chose. Bob's strategy would be to first clone his photon and make a large number of identical copies. In the second step he would make his measurements in the $\{|\rightarrow\rangle,|\uparrow\rangle\}$ basis on all the cloned photons. Next we see how Bob can find out whether Alice sent a "0" or a "1".

Fig. 11.1 A cloning machine makes an identical copy of an unknown state $|\psi\rangle$.

Let us first see what happens when Alice sent a "0". In this case, she measures her photon in the $\{|\rightarrow\rangle, |\uparrow\rangle\}$ basis. If her outcome is $|\rightarrow_A\rangle$, then Bob's state collapses to $|\uparrow_B\rangle$. If Bob measures all his cloned photons in the $\{|\rightarrow\rangle, |\uparrow\rangle\}$ basis, the outcome for all of them will be a $|\uparrow_B\rangle$ state. Bob would immediately know that Alice sent a "0".

Next consider the case when Alice sends a "1" by measuring her photon in the $\{|\nearrow\rangle, |\searrow\rangle\}$ basis. If Alice's outcome is $|\nearrow_A\rangle$, then the state of Bob's photon collapses to $|\nearrow_B\rangle$ as well. All Bob's cloned photons will also be in the $|\nearrow_B\rangle$ state. Since

$$|\nearrow\rangle = \frac{1}{\sqrt{2}}(|\rightarrow\rangle + |\uparrow\rangle),$$

Bob's measurements in the $\{|\rightarrow\rangle, |\uparrow\rangle\}$ will split evenly between $|\rightarrow\rangle$ and $|\uparrow\rangle$. Thus for sufficiently large number of cloned photons, Bob can find out that Alice sent a "1".

Thus cloning allows communication faster than the speed of light which is not allowed by Einstein's theory of relativity.

11.2 No-cloning Theorem

Let $|\psi\rangle$ be an unknown quantum state. Unknown means that, for a state of the type,

$$|\psi = \cos\theta| \rightarrow\rangle + \sin\theta|\uparrow\rangle, \tag{11.9}$$

we do not know the value of θ. Can we make a cloning machine that performs the operation:

$$U_{\text{clone}}|\psi\rangle|0\rangle = |\psi\rangle|\psi\rangle$$

for any state $|\psi\rangle$? Such a machine cannot exist!! This is the no-cloning theorem. We can prove it by proving the contrary as false, i.e., if we assume that a cloning machine exists, the consequence is a result we know to be false.

In order to prove the no-cloning theorem, we first assume that there exists a transformation U_{clone} such that, for any two states $|\psi\rangle$ and $|\phi\rangle$ (such that $|\psi \neq |\varphi\rangle$) of the system, we have

$$U_{\text{clone}}|\psi\rangle|0\rangle = |\psi\rangle|\psi\rangle, \tag{11.10}$$

$$U_{\text{clone}}|\varphi\rangle|0\rangle = |\varphi\rangle|\varphi\rangle. \tag{11.11}$$

where $|0\rangle$ denotes some well-defined initial state of the target system. Next we consider the state

$$|\sigma\rangle = \frac{1}{\sqrt{2}}(|\psi\rangle + |\varphi\rangle) \tag{11.12}$$

and apply the U_{clone} operator. It follows from Eqs. (11.10) and (11.11) that

$$\begin{aligned}
U_{\text{clone}}|\sigma\rangle|0\rangle &= U_{\text{clone}}\frac{1}{\sqrt{2}}(|\psi\rangle + |\varphi\rangle)|0\rangle \\
&= \frac{1}{\sqrt{2}}(|\psi\rangle|\psi\rangle + |\varphi\rangle|\varphi\rangle).
\end{aligned} \tag{11.13}$$

However, a true cloning machine should make the clone of the state $|\sigma\rangle$, i.e., we expect

$$U_{\text{clone}}|\sigma\rangle|0\rangle = |\sigma\rangle|\sigma\rangle$$
$$= \frac{1}{\sqrt{2}}\left(|\psi\rangle|\psi\rangle + |\psi\rangle|\varphi\rangle + |\varphi\rangle|\psi\rangle + |\varphi\rangle|\varphi\rangle\right). \tag{11.14}$$

But this is not the same as Eq. (11.13). This shows that if the cloning machine works for the states $|\psi\rangle$ and $|\phi\rangle$, it cannot work for the state $|\sigma\rangle$ as given by Eq. (11.12). Thus a universal cloning machine that clones all the states cannot exist. This proves the no-cloning theorem.

11.3 Quantum Copier

We have seen that a perfect quantum copying machine cannot exist. A question of interest is that if we cannot copy quantum states exactly, how well can we copy an arbitrary state? In order to answer this question, we should define a quantity that is a measure of how good a copy is.

A suitable measure for a quantum state is the fidelity that we can define as follows. If the initial state is $|\psi\rangle$ and the final state is $|\phi\rangle$, then the fidelity is defined as

$$F = |\langle\psi|\phi\rangle|^2. \tag{11.15}$$

For a perfect copy $|\psi\rangle = |\phi\rangle$ and $F = |\langle\psi|\phi\rangle|^2 = |\langle\psi|\psi\rangle|^2 = 1$. We can illustrate how fidelity is a good measure of the quality of the copy of a quantum state vector by using an analogy with vectors.

We first recall from Section 9.2 that $\langle\psi|\phi\rangle$ is analogous to the dot product between two vectors. With this in mind, we consider a unit vector A and try to make an identical copy. For this purpose we draw another vector B of unit magnitude. If the drawn vector B is parallel to the vector A, it will be a perfect copy of A (except for the direction). When this happens, the dot product between A and B is unity, i.e.,

$$A \cdot B = \pm 1 \text{ for } |A| = |B|.$$

However the worst copy is made when the unit vector B is perpendicular to A and the dot product between A and B is zero, i.e.,

$$A \cdot B = 0 \text{ for } A \perp B.$$

In general, if there is an angle θ between the unit vectors A and B, then the dot product between the two vectors is $\cos\theta$, i.e.,

$$A \cdot B = \cos\theta.$$

We thus have a range of "similarity" between the vectors A and B with dot product ranging from 1 to 0 as θ varies from 0 to $\pi/2$ radians. However $A \cdot B$ can be negative in the range $\pi/2 < \theta < 3\pi/2$. If we do not care about the direction of the copied vector, we can modify our definition of how similar the two vectors are by considering the quantity $(A \cdot B)^2$ instead of $A \cdot B$.

Analogously if we want to copy the state $|0\rangle$ onto another state $|\phi\rangle$, then $|\langle\phi|0\rangle|^2$ is a measure of the quality of the copy. Here we take the modulus square because, in general, the inner product $\langle\phi|0\rangle$ can be complex but $|\langle\phi|0\rangle|^2$ is always real and positive and ranges from 0 to 1. Thus if $|\phi\rangle = |0\rangle$, then the fidelity $F = |\langle\phi|0\rangle|^2 = |\langle 0|0\rangle|^2 = 1$ and $|\phi\rangle$ is a perfect copy of $|0\rangle$. However if $|\phi\rangle = |1\rangle$, then $F = |\langle\phi|0\rangle|^2 = |\langle 1|0\rangle|^2 = 0$ and $|\phi\rangle$ is the worst possible copy of $|0\rangle$. In general, when

$$|\phi\rangle = \cos\theta|0\rangle + \sin\theta|1\rangle, \tag{11.16}$$

the fidelity $F = |\langle\phi|0\rangle|^2 = \cos^2\theta$. This ranges from 1 for $\theta = 0$ to 0 for $\theta = \pi/2$.

We now consider the possibility of copying the state of one system (such as a photon) in a well-defined state $|\psi\rangle$ onto another system. In the copying process, the quality of the original state decreases such that the quality (as measured by fidelity F) is the same for the two systems. This is equivalent to making a copy of an original onto a blank page. The copying machine prepares two identical copies such that the quality of both copies is less than that of the original. The quantum copying machine as discussed here was first proposed by Vladimir Buzek and Mark Hillery in 1996.

Let us consider the state in system 1:

$$|\phi_1\rangle = \cos\theta|0_1\rangle + \sin\theta|1_1\rangle. \tag{11.17}$$

We want to faithfully transfer this state to system 2 initially in state $|0_2\rangle$. We would like the quantum copying machine to carry out the operation

$$U_{copy}|\phi_1\rangle|0_2\rangle = |\psi_1\rangle|\psi_2\rangle, \tag{11.18}$$

where, due to the no-cloning theorem, the fidelity $F = |\langle\phi|\psi\rangle|^2$ is expected to be less than 1. It turns out that a copying machine with just two states may not provide maximum fidelity.

In order to achieve maximum fidelity, we need an auxiliary system 3, called *ancilla*, initially also in state $|0_3\rangle$. The role of the *ancilla* system 3 will be only to facilitate in making a high fidelity copy on system 2, and we will not be concerned about the state of the *ancilla* in the process. We thus have a situation where the system 1 can be either in state $|0_1\rangle$ or state $|1_1\rangle$ or a linear superposition of states $|0_1\rangle$ and $|1_1\rangle$. The systems 2 and 3 are initially in states $|0_2\rangle$ and $|0_3\rangle$. The initial state is thus

$$|\Phi_{123}\rangle = |\phi_1\rangle|0_2\rangle|0_3\rangle. \tag{11.19}$$

The copying machine with maximum fidelity can be obtained by the following quantum transformations:

$$U_{copy}|0_1\rangle|0_2\rangle|0_3\rangle = \sqrt{\frac{2}{3}}|0_1\rangle|0_2\rangle|0_3\rangle - \frac{1}{\sqrt{6}}(|0_1\rangle|1_2\rangle|1_3\rangle + |1_1\rangle|0_2\rangle|1_3\rangle), \tag{11.20}$$

$$U_{copy}|1_1\rangle|0_2\rangle|0_3\rangle = -\sqrt{\frac{2}{3}}|1_1\rangle|1_2\rangle|1_3\rangle + \frac{1}{\sqrt{6}}(|0_1\rangle|1_2\rangle|0_3\rangle + |1_1\rangle|0_2\rangle|0_3\rangle). \tag{11.21}$$

These transformations map the orthogonal states $|0_1\rangle|0_2\rangle|0_3\rangle$ and $|1_1\rangle|0_2\rangle|0_3\rangle$ of the initial tripartite system onto orthogonal states. These transformations can be realized experimentally in different quantum systems. In general, the unknown quantum state (11.17) undergoes a transformation by using the individual transformations (11.20) and (11.21) as (with $c \equiv \cos\theta; s \equiv \sin\theta$)

$$
\begin{aligned}
|\Psi_{123}\rangle &= U_{copy}|\Phi_{123}\rangle = U_{copy}|\phi_1\rangle|0_2\rangle|0_3\rangle \\
&= \cos\theta\, U_{copy}|0_1\rangle|0_2\rangle|0_3\rangle + \sin\theta\, U_{copy}|1_1\rangle|0_2\rangle|0_3\rangle \\
&= \sqrt{\tfrac{2}{3}}\,(\cos\theta|000\rangle - \sin\theta|111\rangle) - \tfrac{\cos\theta}{\sqrt{6}}\,(|011\rangle + |101\rangle) + \tfrac{\sin\theta}{\sqrt{6}}\,(|010\rangle + |100\rangle).
\end{aligned}
$$

$$(11.22)$$

Here, for convenience's sake, we use the shorthand notation $|000\rangle$ to represent the three-system state $|0_1\rangle|0_2\rangle|0_3\rangle$, notation $|111\rangle$ to represent the state $|1_1\rangle|1_2\rangle|1_3\rangle$ and so on. We want to find how the copying machine described by the mapping (11.22) leads to a high-quality copy of the state (11.17) onto state 2 with a large fidelity.

In order to illustrate how the mapping (11.22) can lead to a copy with high fidelity, we first consider the simple case when $\theta = \pi/2$. This simple example should help to elucidate the copying procedure. The original state that needs to be copied is $|\phi_1\rangle = |1_1\rangle$ and the initial state for the three systems is

$$
|\Phi_{123}\rangle = |\phi_1\rangle|0_2\rangle|0_3\rangle = |1_1\rangle|0_2\rangle|0_3\rangle. \tag{11.23}
$$

It follows from Eq. (11.21) that

$$
\begin{aligned}
|\Psi_{123}\rangle &= U_{copy}|\Phi_{123}\rangle = U_{copy}|\phi_1\rangle|0_2\rangle|0_3\rangle \\
&= U_{copy}|1_1\rangle|0_2\rangle|0_3\rangle \\
&= -\sqrt{\tfrac{2}{3}}|111\rangle + \tfrac{1}{\sqrt{6}}(|010\rangle + |100\rangle).
\end{aligned}
$$

$$(11.24)$$

It therefore follows that the only states that can exist are $|1_1\rangle|1_2\rangle|1_3\rangle$, $|0_1\rangle|1_2\rangle|0_3\rangle$, and $|1_1\rangle|0_2\rangle|0_3\rangle$. Following is the table that shows the probability of the outcome of all possible states after U_{copy} has been applied on the initial state. For example the probability of getting the state $|0_1\rangle|1_2\rangle|0_3\rangle$ is $|\langle 010|\Psi_{123}\rangle|^2 = 1/6$.

Outcome	Probability
$|000\rangle$	0
$|001\rangle$	0
$|010\rangle$	1/6
$|011\rangle$	0
$|100\rangle$	1/6
$|101\rangle$	0
$|110\rangle$	0
$|111\rangle$	2/3

We are interested in the first system being found in state $|1_1\rangle$. This can happen in two different ways. As we see from the table, the total probability of finding the system 1 in state $|1_1\rangle$ is the sum of the probabilities of finding the total system $|\Psi_{123}\rangle$ in states $|100\rangle$ and $|111\rangle$ with probabilities equal to 1/6 and 2/3, respectively. The total probability of getting

$$|\phi_1\rangle = |1_1\rangle$$

for the first system is therefore equal to

$$\frac{1}{6} + \frac{2}{3} = \frac{5}{6}.$$

Similarly, the probability of finding the system 2 in state $|1_2\rangle$ is the sum of the probabilities of finding the total system $|\Psi_{123}\rangle$ in states $|010\rangle$ and $|111\rangle$ with probabilities equal to 1/6 and 2/3, respectively. Thus the probability of the outcome

$$|\phi_2\rangle = |1_2\rangle$$

is also equal to 5/6.

The probability of the systems 1 and 2 being in states $|1_1\rangle$ and $|1_2\rangle$, respectively, is equal to 5/6 for each one of them. Thus the original state $|1_1\rangle$ is no longer in the state $|1_1\rangle$ with certainty. Instead it can be found in this state with a fidelity equal to $5/6 = 0.83$. The fidelity has thus gone down from 1 to 0.83. What we have gained as a result of applying the copying operation U_{copy} is that the state of the second system which was originally in state $|0_2\rangle$ with fidelity $F=|\langle\phi_2|0_2\rangle|^2 = |\langle 1_2|0_2\rangle|^2 = 0$ is now found in state $|1_2\rangle$ with 83% probability as well. The fidelity for the second system has gone up from 0 to 0.83. This is the essence of quantum copying.

Next we discuss a formal procedure of calculating fidelity for the first two systems in the above example. We can obtain the fidelity for the first system by considering all possibilities where the first system is in state $|1_1\rangle$ regardless of the states of the second and third systems. There are four possible combination of states for the second and third systems, namely, $|0_2\rangle|0_3\rangle$, $|0_2\rangle|1_3\rangle$, $|1_2\rangle|0_3\rangle$, and $|1_2\rangle|1_3\rangle$. Thus, the fidelity for the first system is the sum

$$|\langle 100|\Psi_{123}\rangle|^2 + |\langle 101|\Psi_{123}\rangle|^2 + |\langle 110|\Psi_{123}\rangle|^2 + |\langle 111|\Psi_{123}\rangle|^2 = \frac{5}{6}. \tag{11.25}$$

Here we substituted for $|\Psi_{123}\rangle$ from Eq. (11.24). Similarly, we can obtain the fidelity for the second system:

$$|\langle 010|\Psi_{123}\rangle|^2 + |\langle 011|\Psi_{123}\rangle|^2 + |\langle 110|\Psi_{123}\rangle|^2 + |\langle 111|\Psi_{123}\rangle|^2 = \frac{5}{6}. \tag{11.26}$$

These are the same values for the fidelity for the two systems that we obtained before.

We now turn to the general state of the following form for the system 1 that we want to copy:

$$|\phi_1\rangle = \cos\theta|0_1\rangle + \sin\theta|1_1\rangle. \tag{11.27}$$

The fidelity for the first system is

$$F_1 = |\langle\phi 00|\Psi_{123}\rangle|^2 + |\langle\phi 01|\Psi_{123}\rangle|^2 + |\langle\phi 10|\Psi_{123}\rangle|^2 + |\langle\phi 11|\Psi_{123}\rangle|^2, \tag{11.28}$$

where the state of the three systems $|\Psi_{123}\rangle$ is given by Eq. (11.22). Similarly, the fidelity for the second system is

$$F_2 = |\langle 0\phi 0|\Psi_{123}\rangle|^2 + |\langle 0\phi 1|\Psi_{123}\rangle|^2 + |\langle 1\phi 0|\Psi_{123}\rangle|^2 + |\langle 1\phi 1|\Psi_{123}\rangle|^2. \tag{11.29}$$

We can calculate various terms in these equations. It follows from the expression for $|\Psi_{123}\rangle$ in Eq. (11.22) that

$$\langle \phi 00|\Psi_{123}\rangle = \cos\theta \, \langle 000|\Psi_{123}\rangle + \sin\theta \, \langle 100|\Psi_{123}\rangle = \sqrt{\frac{2}{3}}\cos^2\theta + \frac{1}{\sqrt{6}}\sin^2\theta, \tag{11.30}$$

$$\langle \phi 01|\Psi_{123}\rangle = \cos\theta \, \langle 001|\Psi_{123}\rangle + \sin\theta \, \langle 101|\Psi_{123}\rangle = -\frac{1}{\sqrt{6}} \sin\theta \cos\theta, \tag{11.31}$$

$$\langle \phi 10|\Psi_{123}\rangle = \cos\theta \, \langle 010|\Psi_{123}\rangle + \sin\theta \, \langle 110|\Psi_{123}\rangle = \frac{1}{\sqrt{6}} \sin\theta \cos\theta, \tag{11.32}$$

$$\langle \phi 11|\Psi_{123}\rangle = \cos\theta \, \langle 011|\Psi_{123}\rangle + \sin\theta \, \langle 111|\Psi_{123}\rangle = -\frac{1}{\sqrt{6}}\cos^2\theta - \sqrt{\frac{2}{3}}\sin^2\theta. \tag{11.33}$$

It follows, on substituting Eqs. (11.30)–(11.33) in Eq. (11.28), that

$$F_1 = \frac{5}{6}. \tag{11.34}$$

In a similar manner, we can show that

$$F_2 = \frac{5}{6}. \tag{11.35}$$

Thus the copying machine prepares systems 1 and 2 with 83.3 percent fidelity each. The fidelity is independent of θ in state $|\phi_1\rangle$. It can be shown that this is the maximum fidelity that can be obtained in the copying process.

Problems

11.1 Consider the quantum copying machine with the following transformations:

$$U_{copy}|0_1\rangle|0_2\rangle|0_3\rangle = \sqrt{\frac{2}{3}}|0_1\rangle|0_2\rangle|0_3\rangle - \frac{1}{\sqrt{6}}\left(|0_1\rangle|1_2\rangle|1_3\rangle + |1_1\rangle|0_2\rangle|1_3\rangle\right),$$

$$U_{copy}|1_1\rangle|0_2\rangle|0_3\rangle = -\sqrt{\frac{2}{3}}|1_1\rangle|1_2\rangle|1_3\rangle + \frac{1}{\sqrt{6}}\left(|0_1\rangle|1_2\rangle|0_3\rangle + |1_1\rangle|0_2\rangle|0_3\rangle\right).$$

We consider the copying the state of system 1

$$|\phi_1\rangle = \cos\theta|0_1\rangle + \sin\theta|1_1\rangle$$

onto system 2. In the text, we have explicitly calculated the fidelity of system 1 and found it to be equal to 5/6. Show that the fidelity of the copied state of system 2 is also equal to 5/6. What is the fidelity of the auxiliary system 3?

11.2 Consider a copying machine that copies the state

$$|\phi_1\rangle = \cos\theta|0_1\rangle + \sin\theta|1_1\rangle$$

of system 1 onto system 2 via the transformation

$$U_{copy}|0_1\rangle|0_2\rangle = \cos\alpha|0_1\rangle|0_2\rangle + \sin\alpha|1_1\rangle|1_2\rangle,$$
$$U_{copy}|1_1\rangle|0_2\rangle = \cos\alpha|1_1\rangle|0_2\rangle + \sin\alpha|0_1\rangle|1_2\rangle.$$

Calculate the fidelities of systems 1 and 2. For what value of α, the two fidelities are equal?

BIBLIOGRAPHY

N. Herbert, FLASH–A superluminal communicator based upon a new kind of quantum measurement, Foundations of Physics 12, 1171 (1982).

W. K. Wooters and W. H. Zurek, A single photon cannot be cloned, Nature 299, 802 (1982).

V. Buzek and M. Hillery, Quantum copying: beyond the no-cloning theorem, Physical Review A 54, 1844 (1996).

S. Stenholm and K.-A. Suominen, Quantum Approach to Informatics (John Wiley, 2005).

V. Scarani, L. Chua, and S. Y. Liu, Six Quantum Pieces: A First Course in Quantum Mechanics (World Scientific 2010).

12 EPR and Bell Theorem

It is ironical that both Albert Einstein and Niels Bohr played crucial roles in laying the foundation of quantum mechanics but strongly disagreed on the interpretation and limitations of the theory. In Chapter 8, we discussed the first round of the debate between Einstein and Bohr when Einstein challenged Bohr's principle of complementarity at the Solvay conference in 1927 and Bohr successfully defended it. The most serious challenge, however, came in 1935 through a paper by Einstein, Podolsky, and Rosen (EPR) entitled 'Can Quantum-Mechanical Description of Physical Reality Be Considered Complete?' when they argued the incompleteness of quantum mechanics through a *gedanken* experiment.

In this chapter, we present EPR's arguments about the incompleteness of quantum mechanics as embodied in the EPR paper and Bohr's reply to it. The ultimate answer came almost 30 years later, almost ten years after Einstein's death, and the answer was nothing that Einstein would have expected.

12.1 Hidden Variables

What Einstein felt uncomfortable about in quantum mechanics was the probabilistic nature of quantum mechanical predictions. In his viewpoint, a "complete" theory should be able to give definite predictions. However, as quantum mechanics can only tell us whether an event will happen or not probabilistically, quantum mechanics should be incomplete. In order to illustrate this point, we consider a simple example.

Two photons identical in every respect are prepared in the polarized state $|\nearrow\rangle$. In the $\{|\rightarrow\rangle, |\uparrow\rangle\}$ basis, these photons are described as:

$$|\nearrow\rangle = \frac{1}{\sqrt{2}}(|\uparrow\rangle + |\rightarrow\rangle), \tag{12.1}$$

As discussed in Section 10.1, if a photon in state (12.1) passes through a polarizing beam splitter oriented to measure photons in the $\{|\rightarrow\rangle, |\uparrow\rangle\}$ basis, there is equal probability that, after passing through the beam splitter, it will be found in state $|\uparrow\rangle$ or in state $|\rightarrow\rangle$. We thus have four possible outcomes with equal probabilities for the two photons: both are found in state $|\uparrow\rangle$, the first is found in state $|\rightarrow\rangle$ and the second is found in state $|\uparrow\rangle$, the first is found in state $|\uparrow\rangle$ and the second is found in state $|\rightarrow\rangle$, and both are found in state $|\rightarrow\rangle$. This is the probabilistic aspect of quantum theory that Einstein never accepted, proclaiming that "God does not play dice."

Let us suppose that the first photon is found in state $|\uparrow\rangle$ and the other in state $|\rightarrow\rangle$ as shown in Fig. 12.1. We ask the obvious question: What is the difference between the two photons? The

Quantum Mechanics for Beginners: With Applications to Quantum Communication and Quantum Computing. M. Suhail Zubairy.
© M. Suhail Zubairy 2020. Published in 2020 by Oxford University Press. DOI: 10.1093/oso/9780198854227.001.0001

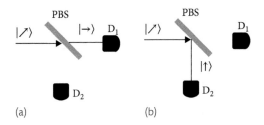

Fig. 12.1 Two photons in identical state $|\nearrow\rangle$ pass through a polarization beam splitter. One photon is found in state $|\uparrow\rangle$ and the other in state $|\rightarrow\rangle$.

equally obvious answer is that the first photon after passing through the polarizing beam splitter is found in state $|\uparrow\rangle$ and the other in state $|\rightarrow\rangle$. But then we ask the difficult question: What was the difference between the two photons before they passed through the beam splitter? If, somehow we knew the difference, we could understand why the two photons behaved differently. However, according to quantum mechanics, there was no difference between the two photons, yet one of them ended up in state $|\uparrow\rangle$ and the other in state $|\rightarrow\rangle$.

One might think that this situation is similar to flipping two coins where we have 25% probability that we get the first coin with head up and the other coin with tail up. However, in this case, if we knew all the forces and the initial orientations of the coins, we could predict with certainty whether we will get head up or tail up. This is not the case with quantum objects like the polarization state of a single photon.

One way out of this conundrum is to assume that the two photons, being apparently identical in all respects, such as the frequency and the polarization, were indeed different even before they entered the beam splitter. These photons had some "hidden" properties that we do not know and we cannot measure in the laboratory but which distinguished them from each other. If, some day, we are able to identify these "hidden variables" and incorporate them in the theory, then we will be able to make definite predictions like in classical mechanics. Quantum mechanics would then become "complete."

In Section 12.3, we discuss how this debate was resolved via certain inequalities called Bell's inequalities that could be tested in the laboratory. The results ruled out any or all theories based on "hidden variables."

12.2 The Einstein–Podolsky–Rosen (EPR) Paradox

There are certain postulates that can be described as "self-evident truths" which lie at the foundation of physics. Challenging them would amount to challenging common sense. Two such "truths" are reality and locality.

According to the postulate of reality, regularities in observed phenomena are caused by some physical reality whose existence is independent of human observers. For example, as Einstein put it, we all believe that the moon exists even when none of us looks at it. It is hard to imagine that the reality of an object like the moon depends on us directly observing it.

The other postulate about locality comes directly from the work of Einstein himself on the theory of relativity. One of the postulates of the theory of relativity is that no information of any kind can propagate faster than the speed of light in vacuum. If information can be transferred faster than the speed of light, then we would be violating another foundational principle of science, causality. According to causality, the present is determined by what happened in the past and what is happening in the present but NOT by what will happen in the future.

One of the most startling results of the twentieth century is that quantum mechanics is not consistent with at least one of these "truths" and experiments agree with the predictions of quantum mechanics. This is what we discuss in the rest of this chapter.

We begin by presenting EPR's concept of reality which, in their own words, is stated as follows:

> If without in any way disturbing a system, we can predict with certainty (i.e., with probability equal to unity) the value of a physical quantity, then there is an element of physical reality corresponding to this physical quantity.

They then argue that

> ... Every element of the physical reality must have a counterpart in the physical theory.

What this implies is that, if somehow, we can attribute well-defined orientations of polarization of a photon in both $\{|\rightarrow\rangle, |\uparrow\rangle\}$ and $\{|\nearrow\rangle, |\nwarrow\rangle\}$ bases then the theory, quantum theory, should be able to predict with certainty the outcome of measurement when this photon is passed through different orientations of the polarization beam splitter. We have seen in Section 10.1 that this is not possible.

First we address the question: How do we establish the reality of an object "without in any way disturbing" or making measurement on it? How do we know that the moon exists when we do not look at it? A way to establish reality, "physical reality whose existence is independent of a human observer", is to look at events with common cause (highly correlated events). Thus the reality of the moon can be established by going to the shore of an ocean and looking at the tides. A careful analysis of the tides (their amplitude and periodicity, etc.) can help in proving the existence of the moon without ever looking at it.

As another example we consider a box containing two balls, one blue and the other red. Let Alice and Bob, blindfolded, pick one ball each and travel in opposite directions to distances so far away that they cannot influence each other in any way. Alice then removes her blindfold and discovers that the color of the ball in her possession is red. She instantaneously concludes that the color of the ball in Bob's possession must be blue. Therefore, according to EPR's definition, there is an element of reality associated with the blue color of the ball in Bob's possession.

This simple experiment does not have any mystery about it. However a version of a similar experiment when done on quantum objects like the polarization of a single photon leads to paradoxes. Quantum mechanics does not agree with our common-sense analysis and predictions. We now turn to a simplified version of the EPR *gedanken* experiment.

Consider two photons 1 and 2 at the source that are prepared in an entangled state

$$|\psi_{12}\rangle = \frac{1}{\sqrt{2}} \left(|\uparrow_1\rangle|\uparrow_2\rangle + |\rightarrow_1\rangle|\rightarrow_2\rangle \right). \tag{12.2}$$

Therefore, if photon 1 is measured in the $\{|\rightarrow\rangle, |\uparrow\rangle\}$ basis and found to be in the state $|\uparrow_1\rangle$, then the state of photon 2 reduces to $|\uparrow_2\rangle$ and, if the photon 1 is measured in state $|\rightarrow_1\rangle$, then the state of photon 2 reduces to $|\rightarrow_2\rangle$. The same state of the two-photon system, when expressed in the $\{|\nearrow\rangle, |\searrow\rangle\}$ basis, is given by

$$|\psi_{12}\rangle = \frac{1}{\sqrt{2}} \left(|\nearrow_1\rangle|\nearrow_2\rangle + |\searrow_1\rangle|\searrow_2\rangle \right) \tag{12.3}$$

after we made the transformations (Eqs. (10.7) and (10.8)):

$$|\uparrow\rangle = \frac{1}{\sqrt{2}} (|\nearrow\rangle + |\searrow\rangle), \tag{12.4}$$

$$|\rightarrow\rangle = \frac{1}{\sqrt{2}} (|\nearrow\rangle - |\searrow\rangle). \tag{12.5}$$

These photons are separated such that photon 1 flies to the right to Alice and photon 2 flies to the left to Bob. Again Alice and Bob may be so far away that they cannot influence each other's photon in any way.

Now EPR's argument runs as follows:

Alice makes a measurement on photon 1 in the $\{|\rightarrow\rangle, |\uparrow\rangle\}$ basis by passing the photon through a polarizer whose polarization axis is along the y-axis (Fig. 12.2a). A click at the detector implies that the polarization of her photon is $|\uparrow_1\rangle$. She can then conclude instantaneously (based on Eq. (12.2)) that photon 2's polarization is $|\uparrow_2\rangle$. Alice made this conclusion "without in any way disturbing" photon 2. Thus, according to EPR, there should be an element of reality associated with the polarization $|\uparrow_2\rangle$ of photon 2.

Let, on the other hand, Alice choose to make a measurement on photon 1 in the $\{|\nearrow\rangle, |\searrow\rangle\}$ basis and let her outcome be $|\nearrow_1\rangle$. She then concludes (based on Eq. (12.3)) that photon 2's polarization should be $|\nearrow_2\rangle$. As before, Alice makes this conclusion "without in any way

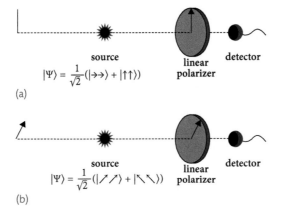

(a)

$$|\Psi\rangle = \frac{1}{\sqrt{2}}(|\rightarrow\rightarrow\rangle + |\uparrow\uparrow\rangle)$$

source   linear polarizer   detector

(b)

$$|\Psi\rangle = \frac{1}{\sqrt{2}}(|\nearrow\nearrow\rangle + |\searrow\searrow\rangle)$$

source   linear polarizer   detector

Fig. 12.2 Two photons initially prepared in an entangled state are sent in opposite directions. In (a) Alice chooses the polarization orientation along the y- axis and in (b) along an axis making and angle 45° with the y-axis.

disturbing" photon 2 and according to EPR, there should be an element of reality associated with the polarization $|\nearrow_2\rangle$ of photon 2.

We thus have a situation where it is possible to determine the state of photon 2 in the two bases $\{|\rightarrow\rangle, |\uparrow\rangle\}$ and $\{|\nearrow\rangle, |\nwarrow\rangle\}$ and these polarization states are obtained "without in any way disturbing" photon 2. Then, EPR argue that, in any complete theory, "every element of the physical reality must have a counterpart." Therefore the states of the polarization in the two bases for photon 2 should be obtainable from quantum mechanics if it is a complete theory. However, quantum mechanics is unable to determine the states of polarization of a photon in the two bases $\{|\rightarrow\rangle, |\uparrow\rangle\}$ and $\{|\nearrow\rangle, |\nwarrow\rangle\}$ simultaneously as it would violate Bohr's principle of complementarity.

This result has come to be known as the "EPR paradox." The inability of quantum mechanics to make definite predictions for the outcome of certain measurements led EPR to conclude that quantum mechanics is an incomplete theory. They postulated that the theory may be made complete if we include certain "hidden variables" that are not known and perhaps not measurable. It was hoped that an inclusion of these "hidden variables" would restore the completeness and determinism to quantum mechanics.

12.3 **Bohr's Reply**

The EPR paper was published in 1935. Leon Rosenfeld has recorded Niels Bohr's reaction to the EPR paper in these words:

> ... This onslaught came down upon us as a bolt from the blue ... Its effect on Bohr was remarkable. As soon as Bohr had heard my report of Einstein's argument, everything else was abandoned, we had to clear up the misunderstanding at once ... Bohr, greatly excited, immediately began to dictate a reply to Einstein. He found, however, that this was no easy matter. He'd start off on one track, then change his mind, backtrack, and start again. He couldn't put his finger where the problem was. "What can they mean? Do you understand it?", he would ask ... After some six weeks of work, Bohr had an answer ...

In his reply (also published in 1935 and with the same title as the EPR paper), Bohr argued that quantum mechanics does not deal with the history of objects from an earlier time when two particles were correlated to a later time when they are free. Quantum mechanics only provides us with a set of rules regarding the outcome of the measurements on the physical properties of the object when the measurements are carried out. In his words,

> The extent to which an unambiguous meaning can be attributed to such an expression as "physical reality" cannot of course be deduced from a priori philosophical conceptions, but ... must be founded on a direct appeal to experiments and measurements.

Bohr's case against EPR rests on the premise that a measurement inevitably disturbs the system. The key point of Bohr's reply is that, in the EPR argument, the two photons should not be treated as independent until a measurement is made. It is therefore incorrect to say that the photon at Bob's end is not disturbed when Alice makes her measurement. It is Alice's measurement that causes a separation between the photons at Alice's and Bob's ends. Until Alice's measurement the two photons are in the entangled state (12.2), i.e.,

$$\frac{1}{\sqrt{2}} \left(|\uparrow_1\rangle|\uparrow_2\rangle + |\rightarrow_1\rangle|\rightarrow_2\rangle \right).$$

Her measurement in the $\{|\rightarrow\rangle, |\uparrow\rangle\}$ basis reduces the state of the two-photon system to

$$|\uparrow_1\rangle|\uparrow_2\rangle.$$

Before making a measurement by Alice of her photon we could not associate the reality of the polarization state of Bob's photon. It is only Alice's measurement of her photon in the $\{|\rightarrow\rangle, |\uparrow\rangle\}$ basis that allows us to determine the polarization of Bob's photon in the same basis, namely $|\uparrow_2\rangle$.

When Alice makes a measurement in the $\{|\nearrow\rangle, |\nwarrow\rangle\}$ basis, the two photons are no longer in the entangled state (12.2) or (12.3). They have been disturbed by the first measurement and the state of the two photons is $|\uparrow_1\rangle|\uparrow_2\rangle$. Alice's second measurement in the $\{|\nearrow\rangle, |\nwarrow\rangle\}$ basis further disturbs the system. The prediction of the outcome of the next measurement cannot be made in advance and we cannot associate an element of reality with the component of polarization in the $\{|\nearrow\rangle, |\nwarrow\rangle\}$ basis. The principle of complementarity is therefore intact.

In this interpretation, the reality of an object does not exist until either the object has been measured or is in a position to be measured with a predictable result. So, in the EPR *gedanken* experiment discussed above, the reality of the polarization states $|\uparrow_2\rangle$ of Bob's photon is established only after Alice made a measurement in the $\{|\rightarrow\rangle, |\uparrow\rangle\}$ basis of her photon. In order to establish the reality of the polarization state of Bob's photon in the $\{|\nearrow\rangle, |\nwarrow\rangle\}$ basis another measurement in the $\{|\nearrow\rangle, |\nwarrow\rangle\}$ basis is needed. However this measurement is made on the disturbed system.

Bohr argues that the inevitable disturbance during the measurement process leads to "*final renunciation of the classical ideal of causality and a radical revision of our attitude towards the problem of physical reality.*" The principle of complementarity explains the quantum-mechanical description of physical phenomena and this fulfills, "*within its scope, all rational demands of completeness.*" The reality at the quantum level does not exist until the object is measured. According to Bohr, quantum mechanics is therefore not incomplete.

There is therefore a fundamental difference in the definitions of reality by Einstein and Bohr. According to Einstein, an object is real if it exists independent of a human observer. On the other hand, according to Bohr, we cannot assign reality to an object until either it is measured or is in a position to be measured with a predictable result.

12.4 Bell's Inequality

In the absence of a concrete experimental situation to test the *reality* and *locality* aspects of quantum mechanics, the debate concerning the foundations of quantum mechanics remained philosophical. It was also not clear whether any inclusion of "hidden variables" in quantum mechanics could make it *complete*—a theory with deterministic description like Newtonian mechanics.

The situation changed dramatically in 1964 when John Bell proposed an inequality that would be satisfied by any theory whose main ingredients were locality and reality. This inequality involved experimentally measurable quantities and thus afforded an opportunity to experimentally test whether quantum mechanics always satisfied this inequality and thus ensure that our common-sense belief in locality and reality is justified. We now derive a form of Bell's inequality that was first given by J. F. Clauser and M. A. Horne in 1974.

We consider the EPR *gedanken* experiment illustrated in Fig. 12.3. As before, two photons 1 and 2 are prepared at the source in an entangled state

$$|\psi_{12}\rangle = \frac{1}{\sqrt{2}} \left(|\uparrow_1\rangle|\uparrow_2\rangle + |\rightarrow_1\rangle|\rightarrow_2\rangle \right). \tag{12.6}$$

Photon 1 travels to the right and photon 2 travels to the left. For the purpose of proving Bell's theorem we are interested in the probability of photon 1 passing through a polarizer whose polarization axis is oriented at an angle a with the vertical direction and that photon 2 passes through a polarizer whose polarization axis is oriented at an angle b to the vertical.

Just to make sure that our conclusions are valid for all theories that include hidden variables as well, we introduce quantities $p_1(a, \mathcal{H})$ and $p_2(b, \mathcal{H})$ where $p_1(a, \mathcal{H})$ is the probability of detecting photon 1 for setting a of the polarizer and the setting $\mathcal{H}$ of the hidden variable and $p_2(b, \mathcal{H})$ is the probability of detecting photon 2 for setting b of the polarizer and the setting $\mathcal{H}$ of the hidden variable. By definition, hidden variables are variables which we have no way of knowing about or measuring. Measurable quantities are therefore obtained by taking an average over all the hidden variables $\{\mathcal{H}\}$, i.e.,

$$p_1(a) = \sum_{\mathcal{H}} p_1(a, \mathcal{H}) \rho(\mathcal{H}); p_2(b) = \sum_{\mathcal{H}} p_2(b, \mathcal{H}) \rho(\mathcal{H}). \tag{12.7}$$

Here $\rho(\mathcal{H})$ is the weight function of the hidden variable $\mathcal{H}$ and contains all the information about the hidden variable. Therefore measurable quantities in this set-up are:

$p_1(a)$ − Probability of detecting photon 1 for setting a of the polarizer
$p_2(b)$ − Probability of detecting photon 1 for setting b of the polarizer.

The condition for locality requires that the two polarizers, 1 and 2, are so far away that they cannot communicate with each other. Therefore the joint probability, $p_{12}(a, b, \mathcal{H})$, of detecting photon 1 for setting a and photon 2 for setting b of the polarizers and the setting $\mathcal{H}$ of the unknown hidden variables is the product of the probabilities $p_1(a, \mathcal{H})$ and $p_2(b, \mathcal{H})$, i.e.,

$$p_{12}(a, b, \mathcal{H}) = p_1(a, \mathcal{H}) p_2(b, \mathcal{H}). \tag{12.8}$$

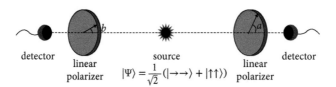

detector   linear   source   linear   detector
polarizer $|\Psi\rangle = \frac{1}{\sqrt{2}}(|\rightarrow\rightarrow\rangle + |\uparrow\uparrow\rangle)$ polarizer

Fig. 12.3 The set-up for the test of Bell's inequality. Two photons initially prepared in an entangled state are sent in opposite directions. The photon at Alice's end is detected with a polarizer oriented with an angle a with the vertical and the photon at Bob's end is detected with a polarizer oriented along angle b with the vertical.

Here we used the property that the joint probability of independent events is the product of individual properties. The measured joint probability can, however, be obtained by taking an average over the hidden variables as before:

$$p_{12}(a, b) = \sum_{\mathcal{H}} p_{12}(a, b, \mathcal{H}) \rho(\mathcal{H}) = \sum_{h} p_1(a, \mathcal{H}) p_2(b, \mathcal{H}) \rho(\mathcal{H}). \tag{12.9}$$

Thus, hidden variables may (or may not) introduce correlation, i.e., $p_{12}(a, b)$ may no longer be a product of $p_1(a)$ and $p_2(b)$. The probabilities can be numbers that lie between 0 and 1. Thus, in general,

$$0 \leq p_1(a, \mathcal{H}), p_2(b, \mathcal{H}) \leq 1. \tag{12.10}$$

So, before deriving the Bell inequality, we reiterate that locality is enforced by our assumption (12.8) that the probability of detecting the two photons is independent of each other and the joint probability $p_{12}(a, b, \mathcal{H})$ factorizes.

The reality is incorporated by noting that even if we cannot make simultaneous measurements on particle 1 along both a_1 and a_2 directions and on particle 2 along both b_1 and b_2 directions, we can speak about the outcome of measurements in the two directions of the analyzers. These polarization directions can be chosen using a more elaborate set-up as shown in Fig. 12.6 when we discuss the real experimental arrangements to test Bell's inequality.

The Bell inequality can now be derived by first noting that, for any four numbers that lie between 0 and 1, i.e.,

$$0 \leq X_1, X_2, Y_1, Y_2 \leq 1, \tag{12.11}$$

the following inequality holds

$$U = (X_1 Y_1 - X_1 Y_2 + X_2 Y_1 + X_2 Y_2 - X_2 - Y_1) \leq 0. \tag{12.12}$$

We can prove this inequality in two steps. First let us assume $X_1 \geq X_2$. In this case $U = (X_1 - 1) Y_1 + (Y_1 - 1) X_2 + (X_2 - X_1) Y_2$. Since all the terms are less than or equal to zero, $U \leq 0$. Next we assume that $X_2 > X_1$. In this case, $U = X_1 Y_1 + X_2 Y_1 + X_2 Y_2 - Y_1 + X_2(Y_2 - 1) \leq X_1 Y_1 + X_2 Y_1 + X_2 Y_2 - Y_1 + X_1(Y_2 - 1) = X_1(Y_1 - 1) + (X_2 - 1) Y_1$. Again as both terms are less than or equal to zero, $U \leq 0$. Thus the inequality (12.12) is proved for all possible cases.

Next we substitute for X_1, X_2, Y_1, Y_2 in Eq. (12.12) the following probabilities

$$X_1 \equiv p_1(a_1, \mathcal{H}), \tag{12.13}$$

$$X_2 \equiv p_1(a_2, \mathcal{H}), \tag{12.14}$$

$$Y_1 \equiv p_2(b_1, \mathcal{H}), \tag{12.15}$$

$$Y_2 \equiv p_2(b_2, \mathcal{H}), \tag{12.16}$$

The resulting inequality is

$$\begin{aligned} p_1(a_1, \mathcal{H}) p_2(b_1, \mathcal{H}) - p_1(a_1, \mathcal{H}) p_2(b_2, \mathcal{H}) + p_1(a_2, \mathcal{H}) p_2(b_1, \mathcal{H}) \\ + p_1(a_2, \mathcal{H}) p_2(b_2, \mathcal{H}) - p_1(a_2, \mathcal{H}) - p_2(b_1, \mathcal{H}) \leq 0. \end{aligned} \tag{12.17}$$

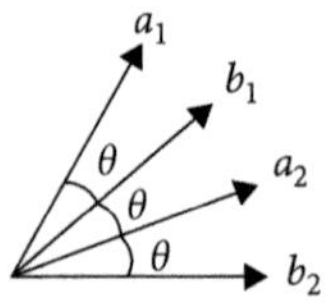

Fig. 12.4 The polarization directions a_1, b_1, a_2, and b_2.

We recall that

$$p_{12}(a, b, \mathcal{H}) = p_1(a, \mathcal{H}) p_2(b, \mathcal{H}). \tag{12.18}$$

The inequality (12.17) becomes

$$p_{12}(a_1, b_1, \mathcal{H}) - p_{12}(a_1, b_2, \mathcal{H}) + p_{12}(a_2, b_1, \mathcal{H}) + p_{12}(a_2, b_2, \mathcal{H})$$
$$- p_1(a_2, \mathcal{H}) - p_2(b_1, \mathcal{H}) \le 0. \tag{12.19}$$

Finally, taking an average over the hidden variables we obtain

$$\sum_{\mathcal{H}} \rho(\mathcal{H}) \, d\mathcal{H} \, [] \to \frac{p_{12}(a_1, b_1) - p_{12}(a_1, b_2) + p_{12}(a_2, b_1) + p_{12}(a_2, b_2)}{p_1(a_2) + p_2(b_1)} \le 1. \tag{12.20}$$

This is Bell's inequality that can be tested in an experiment. This inequality has been obtained with only two assumptions: locality and reality. A violation of this inequality would therefore imply that locality and reality, our cherished self-evident truths, may not co-exist.

The Bell's inequality

$$\frac{p_{12}(a_1, b_1) - p_{12}(a_1, b_2) + p_{12}(a_2, b_1) + p_{12}(a_2, b_2)}{p_1(a_2) + p_2(b_1)} \le 1$$

can be further simplified. In all the experiments to date, $p_1(a)$, $p_2(b)$ are independent of the direction and the joint probability $p_{12}(a, b)$ depends only on the angle between a and b, i.e.,

$$p_{12}(a, b) = p_{12}(\theta), \tag{12.21}$$

where $\theta = a - b$. If we assume the angles between a_1 and b_1, b_1 and a_2, and a_2 and b_2 to be θ then the angle between a_1 and b_2 is equal to 3θ as shown in Fig. 12.4. Bell's inequality (12.20) then reduces to

$$S(\theta) = \frac{3p_{12}(\theta) - p_{12}(3\theta)}{p_1 + p_2} \le 1. \tag{12.22}$$

12.5 Quantum Mechanical Prediction

Next we calculate the value S for the experimental set-up in Fig. 12.3 based on quantum mechanics. We stress that a violation of the inequality (12.22) even for a single value of θ would mean that locality and reality do not co-exist in quantum mechanics.

The quantum state for the two photons 1 and 2 is given by

$$|\psi_{12}\rangle = \frac{1}{\sqrt{2}}\left(|\uparrow_1\rangle|\uparrow_2\rangle + |\rightarrow_1\rangle|\rightarrow_2\rangle\right). \tag{12.23}$$

The joint probability of detection of photons 1 and 2 with polarizer orientations a and b with the horizontal, respectively, is given by

$$P_{12}(a,b) = |\langle a|\langle b|\psi_{12}\rangle|^2 = \frac{1}{2}\,|\langle a|\uparrow_1\rangle\langle b|\uparrow_2\rangle + \langle a|\rightarrow_1\rangle\langle b|\rightarrow_2\rangle|^2$$
$$= \frac{1}{2}[\sin a \sin b + \cos a \cos b]^2 = \frac{1}{2}\cos^2(a-b) = \frac{1}{2}\cos^2\theta. \tag{12.24}$$

We therefore obtain

$$P_{12}(a,b) \equiv P_{12}(\theta) = \frac{1}{2}\cos^2\theta. \tag{12.25}$$

For each individual photon, 1 and 2, there is no well-defined polarization direction if they are prepared in the entangled state (12.23). The situation is therefore similar to the passage of an unpolarized photon as discussed in Section 9.1. The probability $p_1(a)$ can be calculated as follows:

$$p_1(a) = |\langle a|\langle \uparrow_2|\psi_{12}\rangle|^2 + |\langle a|\langle \rightarrow_2|\psi_{12}\rangle|^2 = \frac{1}{2}\left(|\langle a|\uparrow_1\rangle\langle \uparrow_2|\uparrow_2\rangle|^2 + |\langle a|\rightarrow_1\rangle\langle \rightarrow_2|\rightarrow_2\rangle|^2\right)$$
$$= \frac{1}{2}(\sin^2 a + \cos^2 a) = \frac{1}{2}. \tag{12.26}$$

Similarly $p_2(b) = 1/2$.

On substituting these values in the expression for S, as given by Eq. (12.22), we obtain

$$S(\theta) = \frac{1}{2}\left(3\cos^2\theta - \cos^2 3\theta\right).$$

In Fig. 12.5, we plot $S(\theta)$ as a function of θ. The maximum value of $S(\theta)$ is obtained for $\theta = \pi/8$ which is given by

$$S(\pi/8) = 1.207. \tag{12.27}$$

Thus S can be greater than 1 and the Bell's inequality is violated by quantum mechanics. Therefore locality and reality do not co-exist in quantum mechanics. This is a startling result.

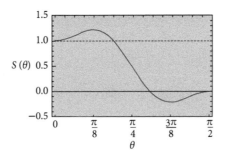

Fig. 12.5 A plot of $S(\theta)$ vs θ as predicted by quantum mechanics. In the region where $S(\theta) > 1$, Bell's inequality (12.22) is violated.

12.6 **Experiments to Test Bell's Inequality**

The experimental activity to test Bell's inequality has an interesting and rich history. The first experiment was done by S. J. Freedman and J. F. Clauser at the University of California at Berkeley in 1972. The outcome was in violation of Bell's inequality and agreed with the predictions of quantum mechanics. In 1974, R. A. Holt and F. M. Pipkin carried out a similar experiment at Harvard University. This experiment was in agreement with Bell's inequality but in disagreement with quantum mechanics. This work was never formally published and remains in reprint form. So, in 1974, we had two experiments at the two coasts of the United States, both disagreeing with each other. A decisive experiment was carried out by E. S. Fry and R. C. Thompson at Texas A&M University two years later in 1976. This experiment proved Freedman–Clauser correct and demonstrated the violation of Bell's inequality. Another experiment that explicitly addressed the issue of locality was done by A. Aspect, J. Dalibard, and G. Roger in 1982.

Most experiments to test Bell's inequality have been a variation of an experiment of the type shown in Fig. 12.6. The source of the entangled pair of photons is a three-level atom initially excited state which can emit two photons of slightly different frequencies ν_1 and ν_2. The atomic levels are chosen in such a way that either both photons are vertically polarized or they are horizontally polarized. The resulting state of the emitted photons is an entangled state of the form (12.2), i.e.,

$$|\psi_{12}\rangle = \frac{1}{\sqrt{2}}\left(|\uparrow_1\rangle|\uparrow_2\rangle + |\rightarrow_1\rangle|\rightarrow_2\rangle\right).$$

The emitted photons move in opposite directions as shown in Fig. 12.6. An optical filter on the right-hand side allows only photons with frequency ν_1 to be transmitted and a filter on the left-hand side transmits only photons of frequency ν_2. On each side there are two path selectors c_1 and c_2. The role of the path selectors is to direct the incoming photons to the appropriate polarizers. The path selector c_1 can direct the photon of frequency ν_1 to the polarizer oriented in the direction a_1 or a_2 with the horizontal. Similarly the path selector c_2 directs the photon of frequency ν_2 to the polarizer oriented in the directions b_1 or b_2 with the horizontal. The photons that pass through the polarizers are detected at D_{a_1}, D_{a_2} and D_{b_1}, D_{b_2}. We carry out

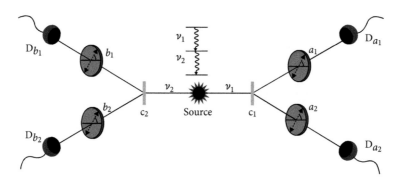

Fig. 12.6 Schematics for an experiment to test Bell's inequality.

this experiment a large number of times with random orientations. Thus, we have four sets of data: (i) when polarizer 1 is oriented along a_1 and polarizer 2 is oriented along b_1, (ii) when polarizer 1 is oriented along a_1 and polarizer 2 is oriented along b_2, (iii) when polarizer 1 is oriented along a_2 and polarizer 2 is oriented along b_1, and (iv) when polarizer 1 is oriented along a_2 and polarizer 2 is oriented along b_2.

If we want to measure the joint probability, such as $p_{12}(a_1, b_1)$, we consider only those events where polarizer 1 is oriented along a_1 and polarizer 2 is oriented along b_1. If the total number of such events is $N_{a_1 b_1}$, we find the number of events $n_{a_1 b_1}$ when both D_{a_1} and D_{b_1} click. Then

$$p_{12}(a_1, b_1) = \frac{n_{a_1 b_1}}{N_{a_1 b_1}}. \tag{12.28}$$

Similarly

$$p_{12}(a_1, b_2) = \frac{n_{a_1 b_2}}{N_{a_1 b_2}}, \tag{12.29}$$

$$p_{12}(a_2, b_1) = \frac{n_{a_2 b_1}}{N_{a_2 b_1}}, \tag{12.30}$$

$$p_{12}(a_2, b_2) = \frac{n_{a_2 b_2}}{N_{a_2 b_2}}. \tag{12.31}$$

These measurement results are then substituted in the expression (12.20) or (12.22) for S to test Bell's inequality.

All the experiments done so far to test it have violated Bell's inequality and agreed with the predictions of quantum mechanics.

12.7 Bell–CHSH Inequality

In Section 12.2, we derived the Bell's inequality that involved polarization measurements along one axis only. In this section we derive another Bell's inequality which was first proposed by J. F. Clauser, M. Horne, A. Shimony, and R. Holt, and is usually referred as the Bell–CHSH inequality. This inequality involves polarization measurements along both orthogonal axes.

The experimental set-up is as shown in Fig. 12.7. Again we consider a source that emits two entangled photons of slightly different frequencies ν_1 and ν_2. They are initially prepared in the entangled state

$$|\psi_{12}\rangle = \frac{1}{\sqrt{2}} \left(|\uparrow_1\rangle|\uparrow_2\rangle + |\rightarrow_1\rangle|\rightarrow_2\rangle \right). \tag{12.32}$$

The first photon travels to Alice to the right and the second photon to Bob to the left; each photon passes through the respective path selectors c_1 and c_2. The path selector c_1 directs the photon of frequency ν_1 to the Pockel cells that rotate the polarization of the incoming photon by the angles $-a_1$ or $-a_2$ with the horizontal followed by the polarization beam splitter and the detectors. As discussed in Section 9.5 (see Fig. 9.14), this set-up allows us to measure both the horizontal and vertical components of the photons, with the possible orientations a_1 or a_2

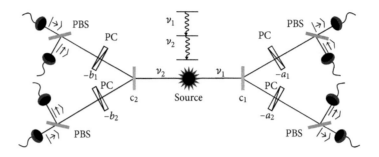

Fig. 12.7 Schematics for an experiment to test the CHSH inequality.

for the first photon. Similarly, the path selector c_2 directs the photon of frequency ν_2 to the Pockel cells that rotate the polarization of incoming photon by the angles $-b_1$ or $-b_2$ with the horizontal followed by the polarization beam splitter and the detectors. This set-up allows us to measure both the horizontal and vertical components of the photons with the possible orientations b_1 or b_2 for the second photon.

The net result is that, at Alice's end, the photon can be detected in states $|\uparrow\rangle$ and $|\rightarrow\rangle$ along the directions a_1 or a_2 and similarly, at Bob's end, in states $|\uparrow\rangle$ and $|\rightarrow\rangle$ along b_1 or b_2.

Let $r_1(a)$ and $r_2(b)$ be the measurement outcomes observed by Alice and Bob for measurements along angles a and b. If Alice measures $|\uparrow\rangle$ along the angle a then $r_1(a) = +1$. Similarly if Bob measures $|\uparrow\rangle$ along the angle b then $r_2(b) = +1$. On the other hand, if Alice or Bob measure the polarization directions to be horizontal $|\rightarrow\rangle$ along the angles a and b then $r_1(a) = -1$ and $r_2(b) = -1$. Here, as before, we consider two possible orientations for a (a_1 and a_2) as well as for b (b_1 and b_2).

Next we define the probabilities $P_{++}(a_i, b_j)$, $P_{+-}(a_i, b_j)$, $P_{-+}(a_i, b_j)$, and $P_{--}(a_i, b_j)$ with $i = 1, 2$. Here $P_{++}(a_i, b_j)$ is the probability that we get $r_1(a_i) = +1$ and $r_2(b_j) = +1$. Similarly $P_{+-}(a_i, b_j)$ is the probability that we get $r_1(a_i) = +1$ and $r_2(b_j) = -1$, $P_{-+}(a_i, b_j)$ is the probability that we get $r_1(a_i) = -1$ and $r_2(b_j) = +1$, and $P_{--}(a_i, b_j)$ is the probability that we get both $r_1(a) = -1$ and $r_2(a) = -1$.

These probabilities are measured in the experiment. For example, we carry out the experiment N times with the beam splitter orientations along a_1 and b_2 where N can be very large. We measure the outcomes at Alice's and Bob's ends for the polarization of photons. Suppose n_{++} is the number of times the measurement outcomes are $|\uparrow\rangle$ for Alice and $|\uparrow\rangle$ for Bob; then

$$P_{++}(a_1, b_2) = \frac{n_{++}}{N}. \tag{12.33}$$

Similarly, for n_{+-} measurement outcomes of $|\uparrow\rangle$ for Alice and $|\rightarrow\rangle$ for Bob,

$$P_{+-}(a_1, b_2) = \frac{n_{+-}}{N}, \tag{12.34}$$

for n_{-+} measurement outcomes of $|\rightarrow\rangle$ for Alice and $|\uparrow\rangle$ for Bob,

$$P_{-+}(a_1, b_2) = \frac{n_{-+}}{N}, \tag{12.35}$$

and for n_{--} measurement outcomes of $|\rightarrow\rangle$ for Alice and $|\rightarrow\rangle$ for Bob,

$$P_{--}(a_1, b_2) = \frac{n_{--}}{N}. \tag{12.36}$$

Here

$$n_{++} + n_{+-} + n_{-+} + n_{--} = N. \tag{12.37}$$

In the same way, we can measure the joint probabilities $P_{++}(a_i, b_j)$, $P_{+-}(a_i, b_j)$, $P_{-+}(a_i, b_j)$, and $P_{--}(a_i, b_j)$ for other orientations of the polarizers at Alice's and Bob's ends.

If Alice and Bob are very far away so that one cannot influence the other in any way, we expect the joint probabilities to be factorized, i.e., $P_{++}(a_i, b_j) = P_+(a_i)P_+(b_j)$. The source of the correlation can be the hidden variables. Again we average over a distribution of hidden variables the way we did it in Section 10.2.

We define the correlation function $\langle r_1(a_i)r_2(b_j)\rangle$ between the measurements at Alice's and Bob's ends as follows. There are four possibilities. We can have $r_1(a_i) = +1$ and $r_2(b_j) = +1$. The probability of such events happening is $P_{++}(a_i, b_j)$. Thus the contribution of such events to $\langle r_1(a_i)r_2(b_j)\rangle$ is $(+1)(+1)P_{++}(a_i, b_j)$. Similarly the contribution of the events $r_1(a_i) = +1$ and $r_2(b_j) = -1$ is $(+1)(-1)P_{+-}(a_i, b_j)$, of the events $r_1(a_i) = -1$ and $r_2(b_j) = +1$ is $(-1)(+1)P_{-+}(a_i, b_j)$, and of the events $r_1(a_i) = -1$ and $r_2(b_j) = -1$ is $(-1)(-1)P_{--}(a_i, b_j)$. If we add all these contributions, we get

$$\begin{aligned}\langle r_1(a_i)r_2(b_j)\rangle = &(+1)(P_{++}(a_i, b_j) + P_{--}(a_i, b_j)) \\ &+(-1)(P_{+-}(a_i, b_j) + P_{-+}(a_i, b_j)).\end{aligned} \tag{12.38}$$

We label this quantity as $E(a_i, b_j)$, i.e.,

$$\begin{aligned}E(a_i, b_j) &= \langle r_1(a_i)r_2(b_j)\rangle \\ &= P_{++}(a_i, b_j) + P_{--}(a_i, b_j) - P_{+-}(a_i, b_j) - P_{-+}(a_i, b_j)\end{aligned} \tag{12.39}$$

It is clear that the correlation function $E(a_i, b_j)$ can be constructed from the experimentally measured quantities $P_{++}(a_i, b_j)$, $P_{+-}(a_i, b_j)$, $P_{-+}(a_i, b_j)$, and $P_{--}(a_i, b_j)$.

We now form a quantity S defined as

$$\begin{aligned}S &= E(a_1, b_1) - E(a_1, b_2) + E(a_2, b_1) + E(a_2, b_2) \\ &= \langle r_1(a_1)r_2(b_2)\rangle - \langle r_1(a_1)r_2(b_2)\rangle + \langle r_1(a_2)r_2(b_1)\rangle + \langle r_1(a_2)r_2(b_2)\rangle \\ &= \langle r_1(a_1)(r_2(b_1) - r_2(b_2))\rangle + \langle r_1(a_2)(r_2(b_1) + r_2(b_2))\rangle.\end{aligned} \tag{12.40}$$

Since both $r_1(\alpha)$ and $r_2(\beta)$ can have only two values, $+1$ and -1, S can have a value in the range -2 to $+2$, i.e.,

$$|S| \le 2. \tag{12.41}$$

This is another version of Bell's inequality. Just like the earlier inequality discussed in Section 12.2, this inequality is also based only on the postulates of locality and reality.

What is the prediction of quantum mechanics for this inequality? To answer this question, we have to calculate the probabilities $P_{++}(a_i, b_j)$, $P_{+-}(a_i, b_j)$, $P_{-+}(a_i, b_j)$, and $P_{--}(a_i, b_j)$. We recall that the initially prepared state is an entangled state

$$|\psi_{12}\rangle = \frac{1}{\sqrt{2}}\left(|\uparrow_1\rangle|\uparrow_2\rangle + |\rightarrow_1\rangle|\rightarrow_2\rangle\right).\tag{12.42}$$

In the set-up shown in Fig. 12.7, for the setting a, the measurement basis is

$$|+a\rangle \equiv |a\rangle = \cos a|\rightarrow\rangle + \sin a\,|\uparrow\rangle,\tag{12.43}$$

$$|-a\rangle \equiv |a+\pi/2\rangle = \cos a\,|\uparrow\rangle - \sin a|\rightarrow\rangle.\tag{12.44}$$

This follows from our discussion in Section 9.5 (see Eqs. (9.56) and (9.57)). Here $|+a\rangle$ is the polarization state when the polarizer is rotated by an angle a with respect to the horizontal. Similarly $|-a\rangle$ is the polarization state when the polarizer is rotated by an angle $a + \pi/2a$ with respect to the horizontal. As an example, if $a = 0°$, then $|+a\rangle = |\rightarrow\rangle$ and $|-a\rangle = |\uparrow\rangle$. This corresponds to the photon passing through a simple polarization beam splitter. However, if $a = 45°$, then $|+a\rangle = (|\rightarrow\rangle + |\uparrow\rangle)/\sqrt{2} = |\nearrow\rangle$ and $|-a\rangle = (|\uparrow\rangle - |\rightarrow\rangle)/\sqrt{2} = |\nwarrow\rangle$. This corresponds to the photon passing through the polarization beam splitter after passing through a Pockel cell that rotates the polarization of the incoming photon by an angle $-45°$ (see Fig. 9.14).

Similarly the measurement basis with setting b, i.e., when the polarization axis is oriented at an angle b with the horizontal direction, is

$$|+b\rangle = \cos b|\rightarrow\rangle + \sin b|\uparrow\rangle,\tag{12.45}$$

$$|-b\rangle = \cos b|\uparrow\rangle - \sin b|\rightarrow\rangle.\tag{12.46}$$

The probability of an outcome with polarization $|\rightarrow\rangle$ for both photons with orientations of the polarizers along a_i for photon 1 and along b_j for photon 2 is thus

$$P_{++}(a_i, b_j) = |\langle+a_i|\langle+b_j|\psi_{12}\rangle|^2 = \frac{1}{2}\cos^2(a_i - b_j).\tag{12.47}$$

Here $i = 1, 2$ and $j = 1, 2$. Similarly

$$P_{+-}(a_i, b_j) = |\langle+a_i|\langle-b_j|\psi_{12}\rangle|^2 = \frac{1}{2}\sin^2(a_i - b_j),\tag{12.48}$$

$$P_{-+}(a_i, b_j) = |\langle-a_i|\langle+b_j|\psi_{12}\rangle|^2 = \frac{1}{2}\sin^2(a_i - b_j),\tag{12.49}$$

$$P_{--}(a_i, b_j) = |\langle-a_i|\langle-b_j|\psi_{12}\rangle|^2 = \frac{1}{2}\cos^2(a_i - b_j).\tag{12.50}$$

It follows, on substituting from Eqs. (12.47)–(12.50) into Eq. (12.39), that the correlation function

$$\begin{aligned}E(a_i, b_j) &= P_{++}(a_i, b_j) + P_{--}(a_i, b_j) - P_{+-}(a_i, b_j) - P_{-+}(a_i, b_j)\\ &= \cos^2(a_i - b_j) - \sin^2(a_i - b_j)\\ &= \cos(2(a_i - b_j))\end{aligned}\tag{12.51}$$

If we choose $a_1 - b_1 = b_1 - a_2 = a_2 - b_2 = \theta$, then $a_1 - b_2 = 3\theta$ and

$$S = E(a_1, b_1) - E(a_1, b_2) + E(a_2, b_1) + E(a_2, b_2)$$
$$= 3\cos(2\theta) - \cos(6\theta). \tag{12.52}$$

The maximum value for S is obtained for $\theta = \pi/8$,

$$S = 2\sqrt{2}. \tag{12.53}$$

This represents violation of the Bell–CHSH inequality (12.41).

Problems

12.1 Show that, for any four numbers, X_1, X_2, Y_1, Y_2, such that

$$0 \le X_1, X_2 \le A \text{ and } 0 \le Y_1, Y_2 \le B$$

the following inequality holds

$$-AB \le (X_1 Y_1 - X_1 Y_2 + X_2 Y_1 + X_2 Y_2 - X_2 - Y_1) \le 0.$$

[Hint: See Appendix A of the paper J. F. Clauser and M. A. Horne, Phys. Rev. D 10, 526 (1974)]

12.2 Consider the experimental set-up in Fig. 12.3. The initial quantum state for the two photons 1 and 2 is given by

$$|\psi_{12}\rangle = \frac{1}{\sqrt{2}}(|\rightarrow_1\rangle|\uparrow_2\rangle - |\uparrow_1\rangle|\rightarrow_2\rangle).$$

Find the joint probability of detection of photons 1 and 2, $p_{12}(a, b)$, with polarizer orientations a and b with the horizontal, respectively. Show that the Bell inequality (12.20) is violated.

BIBLIOGRAPHY

A. Einstein, B. Podolsky, and N. Rosen, *Can quantum-mechanical description of physical reality be considered complete?*, Physical Review 47, 777 (1935).

N. Bohr, *Can quantum-mechanical description of physical reality be considered complete?*, Physical Review 48, 696 (1935).

J. Bell, *On the Einstein Podolsky Rosen paradox*, Physics 1, 195 (1965).

J. F. Clauser, M. Horne, A. Shimony, and R. Holt, *Proposed experiment to test local hidden-variable theories*. Physical Review Letters 23, 880 (1969).

J. F. Clauser and M. A. Horne, *Experimental consequences of objective local theories*, Physical Review D 10, 526 (1974).

S. J. Freedman and J. F. Clauser, *Experimental test of local hidden-variable theories*, Physical Review Letters 28, 938 (1972).

E. S. Fry and R. C. Thompson, *Experimental test of local hidden-variable theories*, Physical Review Letters 37, 465 (1976).

A. Aspect, J. Dalibard, and G. Roger, *Experimental test of Bell's inequalities using time-varying analyzers*, Physical Review Letters 49, 1804 (1982).

PART

Quantum Communication

13 Quantum Secure Communication

Since ancient times, a topic of great interest has been the exchange of information over long distances with complete secrecy and security. The sender of the information (say Alice) should be confident that her message is received by the receiver (say Bob) in such a way that no one else has access to this message. This topic has many applications ranging from commercial transactions where the information transferred is kept secret from a potential eavesdropper to military applications where the security and confidentiality of information can make the difference between victory and defeat.

Cryptography is a method of secure communication between two or more parties. Here, the sender, Alice, and receiver, Bob, first exchange a key in a secure manner. Alice then encodes her message with the key, which is known only to Alice and Bob. The message gets encoded in a way that makes it scrambled. This encoded message is then sent to Bob who can decode the message using the key.

As an example of this kind of old-fashioned cryptography, let us assume that Alice and Bob exchange a key such that each letter in alphabet is shifted by one, i.e.,

$$A \rightarrow Z, B \rightarrow A, C \rightarrow B \cdots \cdots Z \rightarrow Y$$

This is perhaps the simplest of the keys. Typically, the keys could be much more complex. Now let us suppose that Alice wants to send the message:

I AM HAPPY TO BE READING THIS BOOK

The message is encoded with the above key by displacing each letter by one ("I" becomes "J", "AM" becomes "BN", etc.). The sent message would therefore be

J BN IBQQZ UP CF SFBEJOH UIJT CPPL

The message is then sent through a public channel like a telephone line or via the internet. If, during transmission, an eavesdropper, Eve, is able to intercept this message, it would not make any sense to her unless she knows the key. When the message reaches Bob, he can decode the message by shifting each letter by one in the opposite direction ("J" becomes "I", "BN" becomes "AM", etc.) and the full message is recovered.

There are however two problems with this type of cryptography. First Alice and Bob should exchange the key through highly reliable and secure channels. For example, such a key cannot be exchanged on a telephone line that can be intercepted rather easily by a resourceful eavesdropper. A secure channel could be through a physical contact between Alice and Bob or their reliable representatives who can then travel to their destinations before starting the exchange of information. The second problem is that a clever eavesdropper can, by a careful analysis of the sent information, reconstruct the key. Both Alice and Bob may continue

Quantum Mechanics for Beginners: With Applications to Quantum Communication and Quantum Computing. M. Suhail Zubairy.
© M. Suhail Zubairy 2020. Published in 2020 by Oxford University Press. DOI: 10.1093/oso/9780198854227.001.0001

communicating without knowing that their key is known to a clever Eve and consequently their communication security is compromised. Of course many clever schemes for conventional cryptography have been proposed and implemented, but ultimately all of them are plagued with these problems.

In this chapter, we discuss how to overcome these problems. A scheme for exchanging a key over public channels, the so-called RSA algorithm, is presented, followed by a discussion of quantum cryptographic techniques. However, first we discuss the role of numbers that are the tools for communication in the present-day world

13.1 Binary Numbers

We are most familiar with the decimal numbers in base-10, which require 10 integers, 0, 1, 2, 3, 4, 5, 6, 7, 8, 9, to represent any number. For example, the number 3568 represents

$$3568 \equiv (3568)_{10} = (3 \times 10^3) + (5 \times 10^2) + (6 \times 10^1) + (8 \times 10^0)$$

Modern communication and computing, however, is done in binary numbers that require only 0 and 1. The reason for using binary numbers is that the processing becomes very simple with just two options, a high voltage may refer to "1" and a low voltage to "0." Each 0 and 1 is referred to as a bit. The price we pay is that a decimal number when written as a binary number may have many more digits. For example the decimal number 13 with only two digits requires four digits to write it in binary form, as follows:

$$1101 \equiv (1101)_2 = (1 \times 2^3) + (1 \times 2^2) + (0 \times 2^1) + (1 \times 2^0) = (13)_{10}.$$

In a similar manner, we can show that $(111001)_2 = (57)_{10}$, i.e.,

$$111001 = (111001)_2 = (1 \times 2^5) + (1 \times 2^4) + (1 \times 2^3) + (1 \times 2^0) = (57)_{10}.$$

The first nine numbers are shown in the following conversion table between the decimal (base-10) and binary (base-2):

Decimal	0	1	2	3	4	5	6	7	8
Binary	0000	0001	0010	0011	0100	0101	0110	0111	1000

For most communication purposes, an ASCII code is followed where each letter (lower and upper case), symbol, space, etc. is characterized by a sequence of binary numbers, as in the following table.

Thus any message can be represented by a long sequence of 0s and 1s. For example, the word "Book" corresponds to the sequence

$$\overbrace{0100\ 0010}^{B}\ \overbrace{0110\ 1111}^{o}\ \overbrace{0110\ 1111}^{o}\ \overbrace{0110\ 1011}^{k}$$

ASCII Code: Character to Binary

0	0011	0000	I	0100	1001	a	0110	0001	s	0111	0011
1	0011	0001	J	0100	1010	b	0110	0010	t	0111	0100
2	0011	0010	K	0100	1011	c	0110	0011	u	0111	0101
3	0011	0011	L	0100	1100	d	0110	0100	v	0111	0110
4	0011	0100	M	0100	1101	e	0110	0101	w	0111	0111
5	0011	0101	N	0100	1110	f	0110	0110	x	0111	1000
6	0011	0110	O	0100	1111	g	0110	0111	y	0111	1001
7	0011	0111	P	0101	0000	h	0110	1000	z	0111	1010
8	0011	1000	Q	0101	0001	i	0110	1001	space	0010	0000
9	0011	1001	R	0101	0010	j	0110	1010	!	0010	0001
A	0100	0001	S	0101	0011	k	0110	1011	"	0010	0010
B	0100	0010	T	0101	0100	l	0110	1100	'	0010	0111
C	0100	0011	U	0101	0101	m	0110	1101	(	0010	1000
D	0100	0100	V	0101	0110	n	0110	1110	)	0010	1001
E	0100	0101	W	0101	0111	o	0110	1111	'	0010	1100
F	0100	0110	X	0101	1000	P	0111	0000	.	0010	1110
G	0100	0111	Y	0101	1001	q	0111	0001	:	0011	1010
H	0100	1000	Z	0101	1010	r	0111	0010	;	0011	1011
									?	0011	1111

This example shows that any message or information can be communicated through numbers. Thus a secure way of sending numbers would ensure that any message can be sent in a secure manner.

13.2 Public Key Distribution, RSA

Conventional cryptography was used for secure communication for millennia until around 1970, when the advent of public key algorithms such as the RSA protocol changed the nature of cryptography dramatically. Even before the internet became a widely used means of communication, there was a need for a key exchange through a public channel. If a bank wants

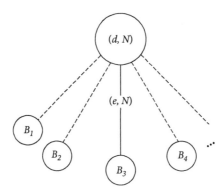

Fig. 13.1 Schematics of RSA key distribution. Alice announces the encryption key (e, N) to everyone including her friends as well as potential eavesdroppers. She keeps the decryption key d secret.

to communicate sensitive financial information with branches scattered all over the world, it would be too impractical and expensive to exchange keys via physical contact. The key has to be exchanged on the communication channels that are readily available, such as a telephone line or the internet. With the arrival of the internet in the 1990s and its usefulness in e-commerce, it became essential that data and information transfer should be exclusively on public channels, and, at the same time, should be secure.

Three mathematicians, Ronald Rivest, Adi Shamir, and Leonard Adleman, devised the so-called RSA protocol in 1978 that allows us to exchange keys through public channels with very high level of security. The RSA protocol is a beautiful (and useful) example of the purest branch of mathematics, called number theory.

Now the RSA public key system is described. It is shown how a key can be exchanged on a public channel and a message in the form of a decimal number can be sent in an almost secure manner. The RSA algorithm is schematically shown in Fig. 13.1. Here Alice wants to exchange information with many friends Bob1, Bob2,⋯ scattered around the world, in complete security. She would like to do it in such a way that if Bob1 sends some information to Alice, then Bob2, Bob 3, etc. should not be able to figure out the message even if they have access to the information that Bob1 sent to Alice.

The way this is done is that Alice forms a key consisting of two numbers e and N that she announces to all her friends on some public channel like the internet, a cell phone, or a telephone line. This encryption key is accessible to any eavesdropper as well. Alice retains a decryption key d that is kept secret. Now if any of Alice's friends wants to send a message (a decimal number) he/she can encode the message with the key e and N in a prescribed manner and send to Alice. Alice, at her end, uses the decoding key d to decipher the message. The amazing aspect of the RSA protocol is that, although all the friends and eavesdroppers have access to both the encoded message and the encoding key (e and N), it is almost impossible for them to decipher the message. This seems almost too good to be true—announcing the key to everyone with no one being able to decipher the sent message. How this is done is discussed in the following.

Anyone with a pocket calculator knows it is trivial to multiply two prime numbers, however large they may be. However, carrying out the reverse—finding the unknown prime factors given a large number—is extremely difficult. To appreciate the difficulty, try to find the prime factors of the number 21583.[1] The difficulty in finding the prime factors of a number lies at the heart of the RSA encryption system.

In order to understand the RSA algorithm, we need a mathematical tool that is used very frequently in number theory—the modulo operation. For two integers a and b, the modulo operation is defined via the relation

$$a \bmod b = r, \tag{13.1}$$

where a is the divided, b is the divisor (or modulus), and r is the remainder. As examples, 11 mod 4 = 3 because 11 divided by 4 (twice) with 3 remaining, 25 mod 5 = 0 because 25 divided by 5 (five times) with 0 remaining; 3 mod 2 = 1 because 3 divided by 2 (once) with 1 remaining.

The steps involved in RSA encryption and decryption are as follows:

1. The sender (Alice) chooses two large prime numbers p and q. Alice multiplies these prime numbers to obtain another number N. In RSA, N is typically a 256 digit number.

2. From N, p, and q, Alice generates two additional numbers e and d. Here e stands for encryption and d stands for decryption. The encryption key e is chosen randomly such that it is co-prime (no common factors) with $(p-1)(q-1)$ and the decryption key is made via

$$1 = d \cdot e \bmod ((p-1)(q-1)), \tag{13.2}$$

that is, when the product $d \cdot e$ is divided by $(p-1)(q-1)$, the remainder is 1, i.e., $d \cdot e = 1 + k(p-1)(q-1)$ where k is an integer. Finding d according to this equation may be a bit complicated but there are known ways of doing it. Alice announces the encryption key (e, N) to everyone through a public channel. She can even put this encryption key on her website for everyone to see. She keeps the decryption key d secret and it is only known to herself.

3. If anyone, (say) Bob, wants to send a message, it should be encoded with the encryption key (e, N). Typically the message is a decimal number m. The encrypted message then is given by

$$c = m^e \bmod N. \tag{13.3}$$

Bob sends the encrypted message c to Alice. In principle, everyone is able to see that Bob has sent c to Alice.

4. Next comes the interesting and crucial step: Alice can decrypt the message with her decryption key (d, N) via

$$c^d \bmod N = m \tag{13.4}$$

and the message is recovered. This equation follows from a theorem of number theory and is central to the RSA algorithm.

---

[1] The answer can be found on page 246.

Here we consider an example.

Let Alice choose her prime numbers to be $p = 47, q = 71$ so that $N = p \cdot q = 47 \times 71 = 3337$. We then obtain

$$(p-1)(q-1) = 46 \times 70 = 3220$$

Alice chooses the encryption key $e = 79$ and verifies that 79 has no common factors with 3220. The decryption key is then obtained via

$$1 = d \cdot e \bmod ((p-1)(q-1)) = 79 \cdot d \pmod{3220}$$

The resulting value of d is 1019. We can verify this result by noting that $1019 \times 79 = 80501$ is equal to $3220 \times 25 + 1$, i.e., when $d \cdot e = 80501$ is divided by $(p-1)(q-1) = 3220$, the remainder is 1.

Alice then sends her encryption key $(e, N) \equiv (79, 3337)$ to everyone including Bob, but keeps the decryption key $d = 1019$ secret.

Next we suppose that Bob wants to send a numerical message

$$m = 688.$$

He encodes the message m with the encryption key $(e, N) \equiv (79, 3337)$ and generates the number

$$c = m^e \bmod N = 688^{79} \bmod 3337 = 1570.$$

This is the encrypted message that is sent to Alice on a public channel which is accessible to everyone.

When Alice receives this encrypted message, she applies her decryption key $(d, N) \equiv (1019, 3337)$ to recover the actual message:

$$c^d \bmod N = 1570^{1019} \bmod 3337 = 688.$$

The message sent by Bob is finally recovered by Alice.

It seems very surprising that a key can be exchanged between Alice and Bob on a public channel. More remarkable is the fact that the key is shared with everyone so that anyone can use the same key to send a message to Alice through a public channel without being fearful that his/her message will be compromised.

Central to the security of the RSA protocol is the difficulty of factorizing the number N. To see this, we note that what is public knowledge in the RSA protocol are the encryption key e and N. In order to break the security one needs to know the decryption key d and that, in turn, requires a knowledge of the prime numbers p and q. Thus the difficulty of factorizing N ensures the security of the RSA protocol.

In order to appreciate the difficulty of factorizing a large number, we note that it would take decades to find prime factors of a 256-decimal-digit number on one of the fastest computers available today. A 1000-decimal-digit number would take about ten billion years (10^{10} years). This is indeed a long time considering that the present estimate of the age of the universe is 13.8 billion years. Thus by increasing the size of the prime numbers, the amount of difficulty in breaking the security can be made extremely large.

The RSA protocol may be a secure way to communicate in today's world. However, with the exponential growth of technology, it is not inconceivable that sometime soon it will be possible to find prime factors in an efficient way. There is, therefore, a need to develop foolproof methods to communicate with security. It turns out that cryptographic systems based on quantum concepts like Bohr's principle of complementarity can make it possible. We turn to these quantum cryptography protocols next.

13.3 Bennett–Brassard 84 (BB-84) Protocol

We have seen how it is possible to exchange a key on a public channel using the RSA protocol. However, the ultimate security of communication is not guaranteed as a clever eavesdropper can, via efficient factorization algorithms, find the decoding key and jeopardize the security of the protocol.

As discussed earlier, there are two requirements to be met for a practical secure communication channel: Firstly, the communication should take place on a public channel, and, secondly, if there is an eavesdropper trying to intercept the communication, he/she should be readily detected. We have seen that the RSA algorithm can accomplish the first task, i.e., a key is exchanged between Alice and Bob on a public channel. However, anyone who can find an efficient way of factoring large numbers in a short time can eavesdrop and the multi-billion dollar field of e-commerce will be put into serious danger. Remarkably, an algorithm, which is potentially very strong in factorizing numbers, is based on the principles of quantum physics, the so-called Shor's factorization algorithm. This algorithm is an important application of the emerging field of quantum computing and is discussed in Chapter 16.

In recent years, new and novel approaches to secure communication have been proposed that are based upon the fundamental principles of quantum mechanics. These quantum mechanical protocols for quantum cryptography are based on novel aspects of quantum entanglement and the inherent probabilistic nature of quantum measurement. Quantum cryptography makes it possible to exchange a key on a public channel and to do the impossible task of detecting an eavesdropper with certainty.

Before discussing quantum cryptography, we discuss the procedure for encoding and decoding the message via a random key made up of binary numbers, schematically shown in Fig. 13.2.

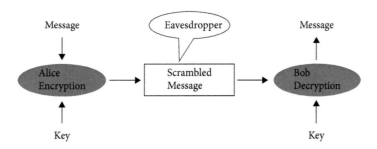

Fig. 13.2 Schematics for secure communication.

Typically, transmission via a random key takes place as follows: The objective is to send the data (a sequence of bits 1 and 0):

Message: 1 1 0 1 0 0 0 1 1 1 1 0 1 0 0 0 0 1 1 0 0 1 0

from Alice to Bob. These data are combined with a random sequence of bits, called the key, and then sent through a communication channel. The random key is another sequence of bits known only to the sender (Alice) and the receiver (Bob):

Key: 1 1 0 0 0 1 1 1 1 1 0 0 1 1 1 0 0 0 0 1 1 0 1

The transmitted sequence is obtained by adding the two sequences using the rules of binary addition, namely, $0 + 0 = 0, 0 + 1 = 1 + 0 = 1$, and $1 + 1 = 0$. Using these addition rules, the transmitted sequence is thus a scrambled sequence of 1s and 0s:

Srambled message: 0 0 0 1 0 1 1 0 0 0 1 0 0 1 1 0 0 1 1 1 1 1 1

The scrambled message is sent by Alice to Bob on a public channel. An important point is that the key is known only to Alice and Bob. Once Bob receives the scrambled message encoded with the secret key, he unscrambles the message by adding the sequence representing the key, i.e.,

Key: 1 1 0 0 0 1 1 1 1 1 0 0 1 1 1 0 0 0 0 1 1 0 1

thus recovering the original message

Message: 1 1 0 1 0 0 0 1 1 1 1 0 1 0 0 0 0 1 1 0 0 1 0.

The randomness of the key ensures that the transmitted data is also random and is inaccessible to a potential eavesdropper who does not have the key. The safety of the channel therefore depends critically on the secrecy of the key. A problem with a classical channel is that, in principle, eavesdropping can take place without the sender or the receiver knowing. This is not true of quantum cryptography, in which (as we see below) eavesdropping disturbs the transmitted sequence in a detectable way.

The major objective of quantum cryptography is therefore exchanging the key (which is a random sequence of binary numbers) on a public channel in such a way that an eavesdropper can be traced immediately.

The first protocol for quantum cryptography was proposed by Charles Bennett and Gilles Brassard in 1984 and is referred as the BB-84 protocol. In the next section, we discuss another protocol which was proposed by Bennett in 1992 and is referred as the B-92 protocol.

The main idea behind the BB-84 protocol is Bohr's principle of complementarity, i.e., two observables are *complementary* if precise knowledge of one of them implies that all possible outcomes of measuring the other one are equally probable (see Fig. 13.3).

As we discussed earlier, the measurement of a quantum system in general causes a disturbance. In quantum cryptography, this aspect of quantum mechanics is used to allow two parties, Alice and Bob, to communicate in absolute secrecy, even in the presence of an eavesdropper, Eve.

There are two channels of communication in the BB-84 protocol: one is the quantum transmission channel through which an array of polarized photons is sent from Alice to Bob and the other is a classical channel like a telephone line or the internet where Alice and Bob can

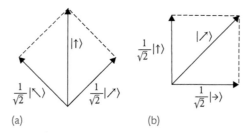

Fig. 13.3 Complementarity for a polarized photon. A vertically polarized photon in state $|\uparrow\rangle$, when measured in the $\{|\nearrow\rangle, |\nwarrow\rangle,\}$ or $\otimes$ basis is found in state $|\nearrow\rangle$ or $|\nwarrow\rangle$ with equal probability. Similarly a photon in state $|\nearrow\rangle$, when measured in the $\{|\rightarrow\rangle, |\uparrow\rangle\}$ or $\oplus$ basis is found in state $|\uparrow\rangle$ or$|\rightarrow\rangle$ with equal probability.

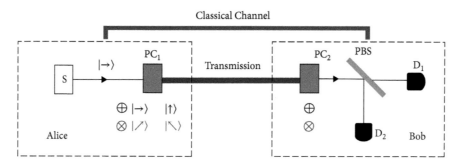

Fig. 13.4 Schematics for the BB-84 protocol.

exchange information about the preparation and measurement of the photons. It is essential that the eavesdropper, Eve, should not be able to block the classical channel and impersonate Alice to Bob and Bob to Alice. It is, however, assumed that Eve has unlimited resources to manipulate the photon in the transmission or quantum channel. The schematics of the BB-84 protocol are shown in Fig. 13.4.

Let us assume that Alice has a source of horizontally polarized photons. Such a source can, for example, be formed by sending unpolarized light through a polarizer whose polarization axis is along the horizontal direction. In the BB-84 protocol, Alice sends a stream of single photons with four possible polarizations along 0°, 45°, 90°, or 135° with the horizontal corresponding to the states $|\rightarrow\rangle$, $|\nearrow\rangle$, $|\uparrow\rangle$, and $|\nwarrow\rangle$, respectively. This can be accomplished by using a Pockel cell. In a Pockel cell, as discussed in Section 9.5, the polarization of the incoming photon can be rotated by applying an appropriate voltage. In the BB-84 experimental set-up, as shown in Fig. 13.4, the Pockel cell PC_1 rotates the polarization vector by angles of 0°, 45°, 90°, or 135° for each photon at Alice's choice. Thus Alice is able to transmit a beam of photons with each photon's polarization orientation being 0°, 45°, 90°, or 135° with the horizontal direction corresponding to states $|\rightarrow\rangle$, $|\nearrow\rangle$, $|\uparrow\rangle$, or $|\nwarrow\rangle$, respectively.

The 0° and 90° orientations corresponds to $\{|\rightarrow\rangle, |\uparrow\rangle\}$ or $\oplus$ basis whereas 45° and 135° orientations correspond to $\{|\nearrow\rangle, |\nwarrow\rangle\}$ or $\otimes$ basis. The two polarization states, say along 0° and 45°, stand for the bit "0" while the other two, along 90°, and 135°, stand for the bit "1." Alice encodes her sequence of data bits, switching randomly between $\oplus$ and $\otimes$ bases. She then transmits the photons to Bob with regular time intervals between them.

Bob's apparatus includes a second Pockel cell PC_2 which can rotate the polarization vector of the incoming photon by an angle of either 0° or −45° at Bob's choice, followed by a polarization beam splitter that can detect photons in the $\{|\rightarrow\rangle, |\uparrow\rangle\}$ or $\oplus$ basis. The two choices, 0° and −45° for the rotation by the Pockel cell, are therefore equivalent to detecting in the $\oplus$ and $\otimes$ bases, respectively. This follows from our discussion in Section 9.5. After deciding the basis in which to detect, the incoming photon, after passing through the Pockel cell PC_2 and the polarization beam splitter PBS, is eventually detected by either detector D_1 or detector D_2. For the choice 0° for PC_2, a click at D_1 corresponds to the photon state $|\rightarrow\rangle$ and a click at D_2 corresponds to the photon state $|\uparrow\rangle$. Similarly, for the choice −45° for PC_2, a click at D_1 corresponds to the photon state $|\nearrow\rangle$ and a click at D_2 corresponds to the photon state $|\nwarrow\rangle$.

Bob receives the photons and records the results using a random choice of $\oplus$ and $\otimes$ detection bases as determined by the rotation angle of his Pockel cell and the outcomes of the detection at detectors D_1 and D_2. When the basis chosen by Bob is the same as that of Alice, the polarization of the received photon is perfectly correlated with Alice's photon. Thus, a photon polarized along 90° received through the $\{|\rightarrow\rangle, |\uparrow\rangle\}$ or $\oplus$ basis will be found polarized along 90° and so on. However, no such correlation exists when the basis chosen by the receiver is different from that of the sender, i.e., a photon polarized along 90° will be found polarized either along 45°, or 135° with equal probability if received through the $\{|\nearrow\rangle, |\nwarrow\rangle,\}$ or $\otimes$ basis. This is a consequence of Bohr's principle of complementarity.

As an example, let Alice send a stream of 9 photons with the sequence of polarizations along

$$0°, 90°, 135°, 0°, 45°, 135°, 45°, 45°, 90°$$

This sequence is sent by Alice choosing the bases

$$\oplus, \oplus, \otimes, \oplus, \otimes, \otimes, \otimes, \otimes, \oplus$$

This stream of photons, when received by Bob through a sequence of basis

$$\otimes, \oplus, \oplus, \otimes, \otimes, \otimes, \oplus, \otimes, \oplus$$

yields the outcome sequence with polarizations along

$$(45° \text{ or } 135°), 90°, (0° \text{ or } 90°), (45° \text{ or } 135°), 45°, 135°, (0° \text{ or } 90°), 45°, 90°.$$

Bob records the outcome of his measurements in secrecy. Here notice that Bob's outcome is identical to that of Alice if he measures the photon in the same basis $\oplus$ or $\otimes$ as that of Alice. However, when the basis chosen by Bob is the opposite to that of Alice, then there is an equal probability for the two possible outcomes in his basis.

Next, Alice and Bob compare their sequences of basis through a classical channel (such as a telephone line) without revealing the results. They retain the instances where they use the same basis and discard the rest. Thus, in the above example, outcomes at 2, 5, 6, 8, and 9 are retained and the outcomes 1, 3, 4, and 7 are discarded. When translated into bits 0 or 1 (1 0 1 0 1 in the above example), the key is obtained.

Next we study what happens if there is an eavesdropper, Eve. In the BB-84 protocol the choice of Alice's and Bob's bases is completely hidden from Eve. Passive eavesdropping in this protocol is not possible as any attempt at eavesdropping would lead to discrepancies between the sequences.

As an example, consider the case when Alice sent a photon by choosing the $\{|\rightarrow\rangle, |\uparrow\rangle\}$ or $\oplus$ basis with the polarization of the photon oriented along 90°. Also suppose that Bob received the photon in the same basis and his outcome for the polarization is along 90°. It is important to note that the $\{|\rightarrow\rangle, |\uparrow\rangle\}$ or $\oplus$ basis is selected by Alice and Bob before they compared their sequence of bases through a public channel which may be available to a potential eavesdropper. The eavesdropper, Eve, located between Alice and Bob, can intercept the photon and can make any measurement on it when it passes through her. She has no way of knowing what basis Alice chose to send her photon and what basis Bob chose to detect the photon when it arrives at his end. Therefore, Eve has no way of knowing what basis Alice and Bob chose.

Eve thus has two choices to infer about what is being transmitted. She can either use the same basis, $\oplus$, that Alice and Bob used or the conjugate basis $\otimes$. We now discuss these cases:

In the case Eve chooses $\oplus$ basis, and this may happen about 50% of the time, she gets the same outcome, 90°, that Alice chose and which Bob also detects.

However, in those roughly 50% instances when Eve chose the wrong basis $\otimes$, she may get 45° or 135° orientations of the polarization with equal probability. If she gets a 45° orientation of the photon that she resends to Bob, the outcome at Bob's end will no longer be 90°, but, as a consequence of Bohr's principle of complementarity as discussed above, will be 0° and 90° with equal probability. The same happens if Eve finds the photon polarization orientation along 135°.

Thus, when Eve tries to eavesdrop with random orientation of her basis, roughly 25% of the outcomes at Bob's end will be different from those of Alice. Thus Alice and Bob can try to infer the presence of an eavesdropper by comparing part of their data. If about 25% discrepancies are found, they can then be confident about the presence of an Eve who is trying to eavesdrop. They can then reject their data and start over.

Thus the BB-84 protocol achieves the impossible: Exchanging a key on a public channel in such a way that an eavesdropper can be traced with almost 100% accuracy. This is, thus, a fascinating application of quantum mechanics and a direct consequence of Bohr's principle of complementarity.

We also note that the security of BB-84 is due to the no-cloning theorem. If cloning of the quantum state were possible, Eve could make a large number of clones of the transmitted photon. She could then send the original photon to Bob and measure the cloned photons, half in $\oplus$ basis and the other half in $\otimes$ basis. When Alice and Bob compare their sequences of basis on the public channel, Eve can also listen in and find the correct basis ($\oplus$ in the above example) and keep the outcome corresponding to this basis. Thus she has the same information that Alice and Bob have. In this process, Bob's measurement is not affected as Eve sent the original photon to Bob. So when Alice and Bob compare part of their data, they would not see any discrepancy and will not be able to trace Eve. The no-cloning theorem does not allow this scenario and an absolutely secure communication is made possible.

13.4 Bennett-92 (B-92) Protocol

We have seen that, in the BB-84 protocol, Alice can send binary information "0" and "1" through polarized photons in the two bases, the $\oplus$ basis or the $\otimes$ basis. In 1992, Charles Bennett

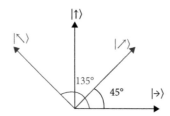

Fig. 13.5 In the B-92 protocol, if Alice sends the bit "0" encoded with a polarized photon in the state $|\rightarrow\rangle$, Bob can unambiguously determine this state if his outcome is $|\nwarrow\rangle$ if measured in the $\{|\nearrow\rangle, |\nwarrow\rangle\}$ basis. Similarly if Alice sends the bit "1" encoded with a polarized photon in the state $|\nearrow\rangle$, Bob can unambiguously determine this state if his outcome is $|\uparrow\rangle$ if measured in the $\{|\rightarrow\rangle, |\uparrow\rangle\}$ basis.

proposed another protocol, the so-called Bennett-92 or B-92 protocol, in which, instead of four pairwise orthogonal states, Alice employs only two non-orthogonal states $|\rightarrow\rangle$ and $|\nearrow\rangle$, corresponding to values 0 and 1 of her random bit. Bob, upon receiving the photons, measures them in a randomly chosen basis $\{|\rightarrow\rangle, |\uparrow\rangle\}$ or $\{|\nearrow\rangle, |\nwarrow\rangle\}$, as in the BB-84 protocol.

The B-92 protocol works on the following premise: If Alice sends photons only in states $|\rightarrow\rangle$ and $|\nearrow\rangle$ to Bob, then Bob can determine the $|\rightarrow\rangle$ state unambiguously only in those events where he happens to choose the $\{|\nearrow\rangle, |\nwarrow\rangle\}$ basis and the outcome is $|\nwarrow\rangle$. Only when he finds the received photon in the $|\nwarrow\rangle$ state can he be certain that it is not in the $|\nearrow\rangle$ state as the two states are mutually orthogonal and, by default, it should be in $|\rightarrow\rangle$ state. Similarly he can determine the $|\nearrow\rangle$ state unambiguously only in those instances where he uses the $\{|\rightarrow\rangle, |\uparrow\rangle\}$ basis and the outcome is $|\uparrow\rangle$. In this case, he can be certain that the incoming photon is not in the state $|\rightarrow\rangle$, and therefore, by default, it should be in the $|\nearrow\rangle$ state. In all other cases he can never be sure about the state of Alice's photon. This can be understood from Fig. 13.5.

Now we present the steps in the B-92 protocol:

- Alice sends a stream of bits in such a way that the bit "0" is encoded with a polarized photon in state $|\rightarrow\rangle$ and the bit "1" encoded in state $|\nearrow\rangle$.

- Bob assigns the value "0" to his bit if he chooses the $\{|\nearrow\rangle, |\nwarrow\rangle\}$ basis and assigns the value "1" if he chooses the $\{|\rightarrow\rangle, |\uparrow\rangle\}$ basis. Thus he forms a random string of bits.

- Bob then measures the polarization of the received photon in the chosen basis. Based on the measurement outcome, Bob forms another string of bits which he calls "control bits" by assigning the value "0" for the states $|\rightarrow\rangle$ or $|\nearrow\rangle$ and the value "1" for the states $|\uparrow\rangle$ and $|\nwarrow\rangle$.

- At this point, Alice has a stream of bits with "0" corresponding to $|\rightarrow\rangle$ and a "1" corresponding to $|\nearrow\rangle$ whereas Bob has two strings: a stream of bits with "0" corresponding to the basis $\{|\nearrow\rangle, |\nwarrow\rangle\}$ and "1" corresponding to the basis $\{|\rightarrow\rangle, |\uparrow\rangle\}$ that he chooses, as well as a stream of control bits with "0" corresponding to the outcomes $|\rightarrow\rangle$ or $|\nearrow\rangle$ and "1" for the outcomes $|\uparrow\rangle$ and $|\nwarrow\rangle$.

- Bob informs Alice the values of the control bits without telling her the bases that he used. They decide to keep only those values of the bits for which the values of the control bit are "1." These values of the bits form the key.

Now consider an example that illustrates how this protocol works. Suppose Alice wants to send to Bob a "0." She sends a photon polarized in the horizontal direction in the state $|\rightarrow\rangle$. Bob measures the polarization by choosing either the $\{|\rightarrow\rangle, |\uparrow\rangle\}$ basis or the $\{|\nearrow\rangle, |\nwarrow\rangle\}$ basis.

Let us first assume that Bob chooses the $\{|\rightarrow\rangle, |\uparrow\rangle\}$ basis and he assigns the value "1" to his bit. Since the polarization of the photon is $|\rightarrow\rangle$, Bob's outcome is definitely going to be the state $|\rightarrow\rangle$ and he assigns the value "0" for the control bit. But can Bob conclude that Alice sent him the photon in the state $|\rightarrow\rangle$? The answer is No because if Alice had sent a "1" via the $|\nearrow\rangle$ state then due to the decomposition

$$|\nearrow\rangle = \frac{1}{\sqrt{2}} (|\rightarrow\rangle + |\uparrow\rangle), \tag{13.5}$$

there is a 50% probability that the outcome will also be $|\rightarrow\rangle$. Thus a detection in the state $|\rightarrow\rangle$ means that Alice could have sent either $|\rightarrow\rangle$ or $|\nearrow\rangle$.

Next we assume that Bob choses the $\{|\nearrow\rangle, |\nwarrow\rangle\}$ basis and he assigns the value "0" to his bit. Since

$$|\rightarrow\rangle = \frac{1}{\sqrt{2}} (|\nearrow\rangle - |\nwarrow\rangle), \tag{13.6}$$

there is equal probability that his outcome will be $|\nearrow\rangle$ or $|\nwarrow\rangle$. If his outcome is $|\nearrow\rangle$, Bob assigns the value "0" for the control bit. In this case he again cannot be sure what polarization Alice sent, $|\rightarrow\rangle$ or $|\nearrow\rangle$, as both possibilities are allowed.

Lastly, we examine the remaining possibility that his outcome is $|\nwarrow\rangle$, for which he assigns the value "1" for the control bit. For this outcome, Bob can be sure about one thing: Alice did not send the state $|\nearrow\rangle$ as it is orthogonal to $|\nwarrow\rangle$. Since Alice was allowed to send the photon in either state $|\rightarrow\rangle$ or state $|\nearrow\rangle$, Bob can now conclude with certainty that Alice sent the state $|\rightarrow\rangle$.

Finally, Bob tells Alice the value of the control bit. If it happens to be "0" corresponding to the outcomes $|\rightarrow\rangle$ or $|\nearrow\rangle$, they conclude that Bob has not been able to decipher what Alice sent. However, if the value of the control bit is "1" corresponding to the outcomes $|\uparrow\rangle$ or $|\nwarrow\rangle$, they decide to keep the value of the bit, which happens to be "0" in our example.

What happens if an eavesdropper, Eve, is present? As, in the BB-84 protocol, Eve has a 50/50 probability of choosing the correct and incorrect bases. Eve can be intercepted in the B-92 protocol on the same ground as in BB-84, the basic premise being that a quantum measurement disturbs the system.

Both Alice and Bob compare the state of the bit "0" or "1" of some of the successful events, events for which the control bit is "1", corresponding to the outcomes $|\uparrow\rangle$ or $|\nwarrow\rangle$. The presence of Eve is inferred if Eve's measurements lead to a wrong outcome for Bob. This can happen in two ways.

1. If Alice sent a "0" with a photon in state $|\rightarrow\rangle$ and Bob detects it in the state $|\uparrow\rangle$, thus concluding that Eve sent "1" with state $|\nearrow\rangle$. Such an outcome can happen if Eve measures the incoming photon in the $\{|\nearrow\rangle, |\nwarrow\rangle\}$ basis and Bob makes the measurement in the $\{|\rightarrow\rangle, |\uparrow\rangle\}$ basis. The reader can see that the probability of such an outcome is 12.5%.

2. Similarly, if Alice sent a "1" with a photon in state $|\nearrow\rangle$ and Bob detects it in the state $|\nwarrow\rangle$, thus wrongly concluding that Alice sent a "0" with the photon in state $|\rightarrow\rangle$. This can

happen if Eve measures the incoming photon in the $\{|\rightarrow\rangle, |\uparrow\rangle\}$ basis and Bob makes the measurement in the $\{|\rightarrow\rangle, |\uparrow\rangle\}$ basis. The probability of such an outcome is again 12.5%.

Therefore, if Alice and Bob find 25% discrepancy in their outcomes, they can infer that an eavesdropper is present and start all over again.

13.5 Quantum Money

An interesting and potentially important application of quantum security is quantum money. This idea, due to Stephen Wiesner, was first presented in 1983, and was a precursor to the BB-84 protocol for communication security. The problem is how to design a foolproof system to avoid counterfeiting of currency notes. More specifically, if we have a currency note, a counterfeiter could try to make an identical copy. Is it possible that, regardless of how good a forgery, the copy is identified as such with certainty?

In the present world, tremendous efforts have been made to make the task of the counterfeiter very difficult. Each currency note has embedded strips, holograms, special inks, or microprinting to safeguard its integrity. However, from a classical perspective, no matter how sophisticated our techniques, it is impossible to absolutely guarantee that a counterfeiter cannot succeed for the simple reason that any printing device a good guy can build, a determined bad guy can also build an identical copy.

Quantum aspects like no-cloning and complementarity can however make it possible to potentially create a currency note that cannot be copied and it should be possible to identify a forged copy with certainty.

Typically a currency note is identified only by a serial number. But, in a quantum currency, each bill can have, in addition to a classical serial number, n trapped photons secretly prepared in one of the four BB-84 states $|\rightarrow\rangle, |\uparrow\rangle, |\nearrow\rangle, |\nwarrow\rangle$. At the bank, there is a record of all the polarizations and the corresponding serial number as shown in Fig. 13.6. On the bank note, the serial number is printed, but the polarizations are kept secret. The bank can always verify the polarizations by measuring the polarization of each photon in the correct basis without introducing any disturbance.

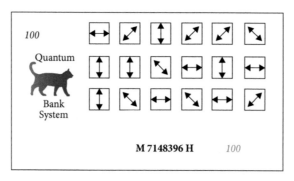

Fig. 13.6 A quantum currency note has a serial number and a number of trapped photons each stored in one of the four BB-84 states $|\rightarrow\rangle, |\nearrow\rangle, |\uparrow\rangle, |\nwarrow\rangle$.

For a potential forger, the serial number is known, but the challenge is how to copy the polarization states. He needs to know the bases for each photon to copy the photon polarization states. He faces the same dilemma that Eve faced in the BB-84 protocol. If the forger uses the correct basis (and this may happen 50% of the time), he can find the secret polarization state and copy it faithfully on his forged copy. However if he chooses the wrong basis (and this happens also about 50% of the time), it will change the polarization of the photon in the trap. The forged banknote created will be with this wrong polarization.

Let us illustrate this with an example. We assume that the polarization state of the photon in the currency note is $|\rightarrow\rangle$. If the forger measures it with the $\{|\rightarrow\rangle, |\uparrow\rangle\}$ basis, he gets the correct polarization state $|\rightarrow\rangle$ and his forgery may not be detected. However, if he measures it in the $\{|\nearrow\rangle, |\nwarrow\rangle\}$ basis, then, since

$$|\rightarrow\rangle = \frac{1}{\sqrt{2}} (|\nearrow\rangle - |\nwarrow\rangle),$$

he gets either the state $|\nearrow\rangle$ or the state $|\nwarrow\rangle$ with equal probability. Let us assume his outcome is $|\nearrow\rangle$. Now, at the bank, when polarization is checked with the correct $\{|\rightarrow\rangle, |\uparrow\rangle\}$ basis, there is a 50% chance that the outcome will be $|\rightarrow\rangle$, but a 50% chance that the outcome will be $|\uparrow\rangle$. Thus for each photon the success probability is 3/4 in duplicating it correctly but a failure probability of 1/4. If the total number of photons on the bank note is N, a duplicate will have a probability $(3/4)^N$ of passing the bank's verification test. If N is large, this probability can become extremely small. For $N = 20$, the success probability is only about 0.3 percent. The fact that a quantum state cannot be copied is ultimately guaranteed by the no-cloning theorem, which underlies the security of this system.

Problems

13.1 Express the following decimal numbers as binary numbers: 688, 573, 894, and 974.

13.2 Express the following binary numbers as decimal numbers: 10010111, 11011010, 10010001, and 11110011.

13.3 In the RSA protocol, let us choose two prime numbers $p = 47$ and $= 59$. Find an encryption key e and a decryption key d. If the numerical message is 537, what would be the encrypted message? Show that the decryption key can decode the encrypted message.

13.4 Consider an elementary code in which the letters are represented by five-bit binary numbers from 1 to 26 according to the sequence of the alphabet. Thus A is 00001, B is 00010, C is 00011, and so on until Z is 11010. The "space" is represented by 00000. Develop a table for all the letters from A to Z. The encryption process is addition modulo 2, and the random key is

10011 01010 00101 11000 10101 00010 10101 10101 10100 11100 01010 00110 10101 11110 10101 00100 10010 01111 01010 11001 11100 01001 10011 00110.

Decode the message

11010 01010 01001 10111 00011 00111 10101 00100 00001 11101 00100 10010 00000 10011 10101 01001 10111 01100 00010 11000 10010 00000 10000 10101.

13.5 Assume the letters A to Z are encoded as in Problem 13.1. Let the exchanged key between Alice and Bob be

   "10100 11100 01101 01100 01000 10001 01000 10010"

   Alice sends the following message encoded with the key:

   "10010 01110 01000 11111 00000 11100 01001 11100."

   Find the message.

13.6 Suppose that Alice sends the message "001011001011" according to the BB-84 protocol with the following sequence of bases:

   $\oplus\ \oplus\ \otimes\ \oplus\ \otimes\ \otimes\ \otimes\ \oplus\ \oplus\ \otimes\ \oplus\ \otimes.$

   What is the state of the polarization of Alice's photons? Bob uses the following sequence of detection bases:

   $\otimes\ \oplus\ \oplus\ \otimes\ \otimes\ \oplus\ \oplus\ \oplus\ \otimes\ \oplus\ \otimes\ \otimes.$

   Find the exchanged key. Explain how Alice and Bob can figure out that there is an eavesdropper.

13.7 Consider the B-92 protocol for quantum key distribution. Suppose Alice wants to send to Bob the bit "1" encoded in state $|\nearrow\rangle$. Discuss all the possibilities with Bob measuring the photon in the $\{|\rightarrow\rangle, |\uparrow\rangle\}$ basis or the $\{|\nearrow\rangle, |\nwarrow\rangle\}$ basis. For what outcome can Bob be confident that Alice sent the bit "1" to him?

BIBLIOGRAPHY

R. L. Rivest, A. Shamir, and L. M. Adleman, *A method of obtaining digital signatures and public-key cryptosystems*, Communications of the ACM 21, 120 (1978).

C. H. Bennett and G. Brassard, *Quantum cryptography: Public key distribution and coin tossing*, in Proceedings of IEEE International Conference on Computers, Systems, and Signal Processing, Bangalore, India (1984), pp 175–9.

C. H. Bennett, *Quantum cryptography using any two nonorthogonal states*, Physical Review Letters 68, 3121 (1992).

A. K. Ekert, *Quantum cryptography based on Bell's theorem*, Physical Review Letters 67, 661 (1991).

C. H. Bennett, G. Brassard, and A. K. Eckert, *Quantum cryptography*, Scientific American 267, 50 (1992).

S. Wiesner, *Conjugate coding*, SIGACT News 15, 77 (1983).

M. A. Nielson and I. L. Chuang, *Quantum Computation and Quantum Information* (Cambridge University Press 2000).

P. Lambropoulos and D. Petrosyan, *Fundamentals of Quantum Optics and Quantum Information* (Springer 2007).

It has always been a self-evident feature of any kind of communication that there should be an exchange of objects between the sender and the receiver to convey any information. For example, light pulses or photons are carriers of information in optical communication. For a binary information transfer (0 or 1), the information can be coded as a horizontally polarized photon as "0," and a vertically polarized photon can be coded as "1." Thus, if, in a communication between (say) Alice and Bob, Alice wants to send "0," she sends a horizontally polarized photon and if she wants to send a "1," she sends a vertically polarized photon. Another example is that Alice may decide to send no photon if she wants to send "0" and send a photon if she wants to send a "1." In these and any other examples of communication, an exchange of an object, like a photon, always takes place even if it is for one half of the communication (as in the second example). It is inconceivable that a communication between Alice and Bob can take place in such a way that there is no photon present in the channel of transmission (such as empty space or fiber optic cables) in both cases, when the sender wants to send a "0" or wants to send a "1."

It was shown in 2013 by the author in collaboration with his colleagues M. Alamri, Z.-H. Li, and H. Salih that quantum mechanics can be used to achieve this apparently impossible objective. Such a communication protocol is called counterfactual communication. Counterfactual communication is thus optical communication with invisible photons. The protocol for counterfactual communication is schematically shown in Fig. 14.1.

The objective is for Bob to send binary information, "0" or "1," to Alice. As shown in Fig. 14.1a, Alice has a source of single photons, an optical set-up involving optical devices such as mirrors and beam splitters (represented by the box) as well as two photon detectors D_1 and D_2. Bob only has a mirror. Alice sends a photon to Bob such that it interacts with optical devices inside the box in a specified manner before proceeding to Bob via a transmission channel that can be free space or optical fiber. Bob has now two choices: He can either block the photon from reaching the mirror, for example, by placing his hand in front of the mirror (Fig. 14.1b) or he does nothing and allows the photon to reflect from the mirror (Fig. 14.1c). In the case when Bob blocks the photon intending to send a "0," the detector D_1 clicks at Alice's end (Fig. 14.1b). This signals Alice that Bob sent a "0" to Alice. On the other hand if Bob allows the photon to get reflected, intending to send a "1," the detector D_2 clicks at Alice's end (Fig. 14.1c), signaling to Alice that Bob sent a "1." The amazing and highly counterintuitive result is that, in both instances, when the photon is supposed to be bouncing back and forth between Alice and Bob, the probability of finding the photon in the transmission channel is zero. Thus the photon bounces back and forth from mirrors and other optical instruments at Alice's end placed inside the box and give us a click at D_1 or D_2 without traveling to Bob but sensing what Bob has done, either put his hand in front of his mirror or done nothing.

Quantum Mechanics for Beginners: With Applications to Quantum Communication and Quantum Computing. M. Suhail Zubairy.
© M. Suhail Zubairy 2020. Published in 2020 by Oxford University Press. DOI: 10.1093/oso/9780198854227.001.0001

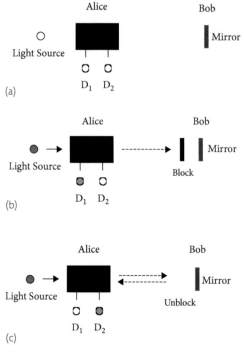

Fig. 14.1 Schematics for counterfactual communication.

How can this be done? This looks almost magical or supernatural like psychic communication. In order to understand the protocol for counterfactual communication, we first consider some simple interferometric configurations before we explain a system of mirrors and beam splitters that can exhibit counterfactual communication.

14.1 Mach–Zehnder Interferometer

First we consider an interferometer with two beam splitters and two mirrors as shown in Fig. 14.2*a*. This is called a Mach–Zehnder interferometer.

The main optical components involved in our analysis are mirrors and beam splitters. The input is a single photon in one port of the beam splitter and none at the other. We make use of the transformation properties of these devices that we discussed in Sections 9.3 and 9.4. We assume perfect reflection from a mirror. The input–output relations for a beam splitter with a single photon input are given by Eqs. (9.53) and (9.54), i.e.,

$$|10\rangle \rightarrow \cos\theta|10\rangle + \sin\theta|01\rangle, \tag{14.1}$$

$$|01\rangle \rightarrow \cos\theta|01\rangle - \sin\theta|10\rangle. \tag{14.2}$$

Here, the probability that the photon is reflected is equal to $R = \cos^2\theta$ and the probability that it is transmitted is equal to $T = \sin^2\theta$.

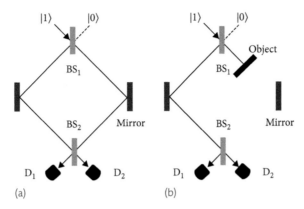

Fig. 14.2 (a) In a Mach–Zehnder interferometer, a photon incident on a beam splitter BS_1 is reflected by mirrors, recombined at the second beam splitter BS_2, and is detected either at detector D_1 or D_2. (b) The same set-up as in (a) but with a photon absorber on the right-hand side of the interferometer.

In a Mach–Zehnder interferometer (Fig. 14.2a), a single photon in state $|1\rangle$ is incident from the left on a beam splitter (BS). There is no photon on the right-hand side. The input state is therefore designated as $|10\rangle$. The input photon passes through the first beam splitter, BS_1, then reflected by the mirrors, passes through the second beam splitter, BS_2, and finally onto the detectors D_1 and D_2. We calculate the probabilities that the photon gives a click at D_1 and D_2.

According to Eq. (14.1), the photon state after passing through the first beam splitter, BS_1, is

$$|10\rangle \rightarrow \cos\theta_1|10\rangle + \sin\theta_1|01\rangle. \tag{14.3}$$

Nothing happens when the photon is reflected from either mirror. The input state at the second beam splitter, BS_2, is therefore equal to $\cos\theta_1 \,|\,10\rangle + \sin\theta_1|01\rangle$. Thus, after the second beam splitter, the output state is

$$
\begin{aligned}
\cos\theta_1|10\rangle + \sin\theta_1|01\rangle &\rightarrow \cos\theta_1\,(\cos\theta_2|10\rangle + \sin\theta_2|01\rangle) \\
&\quad + \sin\theta_1\,(\cos\theta_2|01\rangle - \sin\theta_2|10\rangle) \\
&= \cos(\theta_1 + \theta_2)|10\rangle + \sin(\theta_1 + \theta_2)|01\rangle
\end{aligned} \tag{14.4}
$$

Here $\cos\theta_i$ and $\sin\theta_i$ are the reflection and transmission coefficients of the ith beam splitter, respectively. Each term in this equation has a physical interpretation. The amplitude $\cos\theta_1 \cos\theta_2$ is the amplitude at D_1 resulting from two reflections, one at beam splitter 1 and the other at beam splitter 2, the amplitude $(-\sin\theta_1 \sin\theta_2)$ is also the amplitude at D_1 resulting from two transmissions, one at beam splitter 1 and the other at beam splitter 2. The amplitudes $\cos\theta_1 \sin\theta_2$ and $\sin\theta_1 \cos\theta_2$ are the amplitudes at D_2 resulting from one reflection and one transmission, one at beam splitter 1 and the other at beam splitter 2.

For identical beam splitters, $\theta_1 = \theta_2 = \theta$ and the state of the output state is

$$\cos(2\theta)|10\rangle + \sin(2\theta)|01\rangle. \tag{14.5}$$

For a 50/50 beam splitter, $\theta = \pi/4$, so that the probability that the photon is reflected is $R = \cos^2\theta = 1/2$ and the probability that it is transmitted is $T = \sin^2\theta = 1/2$. The two probabilities are equal. However, we see from Eq. (14.5) that the output state becomes $|01\rangle$. Thus we get a click at detector D_2 with unit probability. We also note that, for $\theta \leq \pi/4$, the photon state has an increasing amplitude for $|01\rangle$.

Next we consider the configuration as shown in Fig. 14.3b. Again a photon is sent in the Mach–Zehnder interferometer from the left, i.e., the input state is $|10\rangle$. After the first beam splitter, BS_1, the quantum state of the photon is, as before,

$$|10\rangle \rightarrow \cos\theta\,|10\rangle + \sin\theta\,|01\rangle$$

The difference with the earlier case is that if the photon is transmitted by the beam splitter and ends up on the right arm of the interferometer, it is absorbed. As the probability amplitude for the photon being in the right arm, i.e., in being state $|01\rangle$, is $\sin\theta$, the probability of the photon being absorbed or lost on the right-hand side is

$$P_{right} = \sin^2\theta. \tag{14.6}$$

The probability amplitude of the photon being on the left arm is $\cos\theta$. The photon state just before the second beam splitter, BS_2, is $\cos\theta|10\rangle$. The output state is

$$\cos\theta\,(\cos\theta|10\rangle + \sin\theta|01\rangle). \tag{14.7}$$

Thus the probability that the photon is detected at D_1 is

$$P_{D1} = \cos^4\theta. \tag{14.8}$$

If the photon is detected at D_1, it must have been reflected from the two beam splitters, each time contributing a factor equal to the reflectivity $r = \cos\theta$.

It follows from Eq. (14.7) that the probability of the photon being detected at D_2 is

$$P_{D2} = \cos^2\theta\sin^2\theta. \tag{14.9}$$

There are only three possibilities: either the photon is lost on the RHS or the photon is detected at D_1, or the photon is detected at D_2. The sum of the three probabilities should therefore be unity, i.e.,

$$P_{right} + P_{D_1} + P_{D_2} = 1. \tag{14.10}$$

14.2 Interaction-free Measurement

There is an interesting feature associated with the Mach–Zehnder interferometer shown in Figs. 14.2a and 14.2b. In the special case of a 50/50 beam splitter ($\theta = \pi/4$), we have already seen that the output state becomes $|01\rangle$ when there is no absorber present on the right-hand side of the interferometer (Fig. 14.2a). Thus, as a result of quantum interference, we get a click at D_2 with unit probability.

In the case when there is an object present on the right-hand side of the interferometer, the probability of the photon being absorbed by the object is

$$P_{right} = \sin^2\theta = \frac{1}{2}. \tag{14.11}$$

The probability of a click at detector D_1 is

$$P_{D_1} = \cos^4\theta = \frac{1}{4} \tag{14.12}$$

and the probability of a click at detector D_2 is

$$P_{D_2} = \cos^2\theta\sin^2\theta = \frac{1}{4}. \tag{14.13}$$

We have an interesting situation: If the object is not there then there is a definite click at D_2, but if there is an object present, then there is a click at D_1 and D_2 with 25% probability each and 50% probability of absorption by the object. Next we address the question: Can we get information whether the object is present or not by looking only at the output detectors D_1 or D_2?

Now note that if there is a click at D_2, we have no way of knowing whether the object is there or not because there is a non-vanishing probability of a click at D_2 in both situations, when there is an object and when there is no object on the right-hand side.

However, if there is a click at D_1, we know with certainty that there is an object present on the right-hand side of the interferometer. The amazing result is that we can determine the presence of the object in those instances in such a way that the photon does not interact with the object and follows a trajectory entirely on the left-hand side (Fig. 14.2b). This is called Interaction-free Measurement.

The concept of interaction-free measurement was first suggested by A. C. Elitzur and L. Vaidman in 1993. In a colorful example, they proposed it as a method for a bomb tester. Suppose we have a collection of bombs, some good and some bad. The bad bombs do not absorb photons and let them pass through unhindered. The good bombs however absorb the photons and explode. The question is: Can we make an optical measurement to find the good bombs without exploding them? In any conventional method, the good bombs will always explode in an attempt to find them. However an interaction-free measurement, as described above, can lead to finding the good bombs with a 25% success probability.

14.3 An Array of N Mach–Zehnder Interferometers

Next we consider an array of N Mach–Zehnder interferometers, as shown in Figs. 14.3a and 14.3b. In both cases we have an array of N beam splitters with a reflectivity $r = \cos(\pi/2N)$ and transmittivity $(t = \sin(\pi/2N)$. Thus the whole array of Mach–Zehnder interferometers is characterized by a single number N with N being the number of Mach–Zehnder interferometers as well as the reflectivity of the beam splitters,

$$r = \cos\theta; \theta = \frac{\pi}{2N}. \tag{14.14}$$

In Fig. 14.3a, there is no object on the right-hand side for any of the Mach–Zehnder interferometers, whereas, in Fig. 14.3b, there is an object or absorber in the right arm of each interferometer. The input, as before, is $|10\rangle$, a single photon is incident on the first beam splitter from the left. We calculate the probabilities of a click at D_1 and D_2 in both cases.

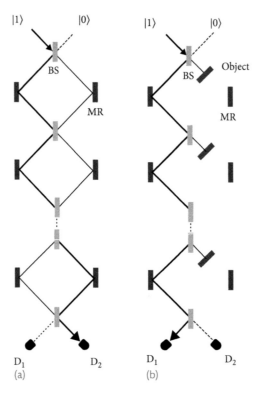

Fig. 14.3 (a) An array of N Mach–Zehnder interferometers. (b) Same set-up as in (a) but with an absorber at each step such that if the photon leaks to the right-hand side, it is absorbed.

First, we consider the case when there is no object on the right-hand side (Fig. 14.3a). We have already seen that if the photon, in the input state $|10\rangle$, passes through the two beam splitters, the output state is

$$\cos\left(\theta_1 + \theta_2\right)|10\rangle + \sin\left(\theta_1 + \theta_2\right)|01\rangle, \tag{14.15}$$

where $\cos\theta_i$ and $\sin\theta_i$ are the reflection and transmission coefficients of the *i*th beam splitter, respectively. For an N beam splitter set-up, the corresponding result is obtained by generalizing Eq. (14.15), and we obtain

$$\cos\left(\theta_1 + \theta_2 + \cdots \theta_N\right)|10\rangle + \sin\left(\theta_1 + \theta_2 + \cdots \theta_N\right)|01\rangle. \tag{14.16}$$

For equal reflectivity at the beam splitters ($\theta_1 = \theta_2 = \cdots \theta_N = \theta$), the result in the output is

$$\cos\left(N\theta\right)|10\rangle + \sin\left(N\theta\right)|01\rangle. \tag{14.17}$$

Since $N\theta = \pi/2$, the final state after the passage through N Mach–Zehnder interferometers is $|01\rangle$, i.e., the detector D_2 clicks with certainty. This result is independent of the number N of the Mach–Zehnder interferometers provided the reflectivity of each beam splitter in the set-up is such that $\theta = \pi/2N$.

Next we analyze the set-up as shown in Fig. 14.3*b*. Here there is an object that absorbs the photon if it is incident on it, at each step in the multi Mach–Zehnder interferometric set-up.

At each step the reflected amplitude is $\cos \theta$ and the transmitted amplitude is $\sin \theta$. The transmitted amplitude is lost as any photon found on the right-hand side is absorbed. After N cycles, the initial state $|10\rangle$ therefore evolves to

$$|10\rangle \to \cos^{N-1}\theta(\cos\theta|10\rangle + \sin\theta|01\rangle). \tag{14.18}$$

Here, $\cos^N \theta$ is the amplitude that is reflected N times and gives a click at D_1. The probability of a click at D_1 is therefore

$$P_{D1} = \left|\cos^N\theta\right|^2 = \cos^{2N}\theta. \tag{14.19}$$

The amplitude $\cos^{N-1}\theta \sin \theta$ is the amplitude that is reflected $(N-1)$ times until the last beam splitter and then transmitted to D_2. The probability of a click at D_2 is therefore

$$P_{D2} = \cos^{2N-2}\theta\sin^2\theta. \tag{14.20}$$

There is also the probability P_{right} that the photon may cross over to the right-hand side and get absorbed. Since the total probability is unity, i.e.,

$$P_{right} + P_{D1} + P_{D2} = 1, \tag{14.21}$$

we obtain

$$P_{right} = (1 - P_{D1} - P_{D2}) = (1 - \cos^{2N-2}\theta). \tag{14.22}$$

It turns out that, for large N, $\cos^{2N}\theta = \cos^{2N}(\pi/2N) \approx 1$.[1] The photon is almost completely reflected and the detector clicks with unit probability.

In summary, we have shown that when there is no object or absorber as shown in Fig. 14.4*a*, the detector D_1 clicks with certainty, whereas if there is an object or absorber at each step as shown in Fig. 14.4*b*, then D_2 clicks with almost unit probability. In the next section we use these results to show how counterfactual communication (no photon in the public channel) can be realized.

14.4 **Counterfactual Communication**

Next we discuss the protocol for counterfactual communication—communication with no photons present in the transmission channel.

---

[1] This follows from the series expansion (2.35) for $\cos \theta$ and the binomial expansion

$$(1 + x)^n = 1 + nx + \frac{n(n+1)}{2!}x^2 + \cdots$$

Thus

$$\cos^{2N}\left(\frac{\pi}{2N}\right) = \left(1 - \frac{1}{2!}\left(\frac{\pi}{2N}\right)^2 + \cdots\right)^{2N} = 1 - \frac{1}{2!}\frac{\pi^2}{2N} + \text{terms of the order } \left(\frac{1}{N^2}\right).$$

In the limit of large $N(N \to \infty)$, we obtain $\cos^{2N}(\pi/2N) \to 1$.

Consider a set-up as shown in Figs. 14.4a and 14.4b. Here we consider an array of M Mach–Zehnder interferometers. The reflectivity and the transmittivity of each beam splitter BS_M is given by $\cos\theta_M$ and $\sin\theta_M$, respectively, with

$$\theta_M = \frac{\pi}{2M}. \tag{14.23}$$

In the right arm of each Mach–Zehnder interferometer, there is an array of N Mach–Zehnder interferometers of the type shown in Figs. 14.3a and 14.3b and discussed in the previous section. The reflectivity and the transmittivity of all the beam splitters BS_N in these arrays of interferometers are given by $\cos\theta_N$ and $\sin\theta_N$, respectively, with

$$\theta_N = \frac{\pi}{2N}. \tag{14.24}$$

Thus only two numbers, M and N, characterize this array of M interferometers with the arrays of N interferometers in the right arms of each of these interferometers; the reflectivities and transmittivities of the corresponding beam splitters are described by θ_M and θ_N. Typically both $M, N \gg 1$.

The system is set in such a way that the source of photons, all the M beam splitters BS_M, all the $N \times M$ beam splitters BS_N, all the M mirrors on the left side of the outer interferometers, all the $N \times M$ mirrors on the left side of the inner interferometers, and the detectors D_1 and D_2 are all on Alice's side. The only objects at Bob's end are the $N \times M$ right side mirrors of the inner interferometers and the detectors D_L as shown in Figs. 14.4a and 14.4b. The region between Alice's set-up and Bob's mirrors is the transmission region, which can be made large.

We now describe the counterfactual communication protocol. In this protocol, Alice sends a single photon from the left as shown in Fig. 14.4. The input state is therefore $|10\rangle$. The information is sent by Bob: He sends a "1" when he does not place any absorbing object in front of all the mirrors, as shown in Fig. 14.4a, and a "0" by placing an object that absorbs the photon in front of all the mirrors at his end, as shown in Fig. 14.4b. We now show that, when Bob sends "1," the detector D_1 at Alice's end clicks and she knows that Bob sent a "1." However when Bob sends "0," the detector D_2 clicks and Alice knows that Bob sent a "0." The amazing result is that, in both instances, there is no photon present in the transmission channel.

First, consider the case when Bob wants to send "1" to Alice. He does not block the photon as shown in Fig 14.4a. For large M, there is a large probability that the photon is reflected from the beam splitters BS_M. However, there is a small probability of transmission. In those rare instances when photon is transmitted from BS_M to the right side, it ends up at the detectors D_L and is lost (see Fig. 14.3a). There is, thus, zero probability that the photon may enter the large interferometer back from the right side to the left side through the beam splitters BS_M. Basically the array of N Mach–Zehnder interferometers in the right arms of the M outer interferometers acts as an absorber. The situation is therefore effectively similar to that in Fig. 14.3b and the probability of a click at D_1 is

$$P_{D_1} = \left|\cos^M\theta_M\right|^2 = \cos^{2M}\theta_M. \tag{14.25}$$

For $M \gg 1, P_{D_1} \cong 1$. Thus, we have a click at D_1 with an almost unit probability. The important thing to note is that, in those instances when we get a click at D_1, the photon follows an outer trajectory, reflecting from all BS_M from the left side, and is never found in the transmission channel.

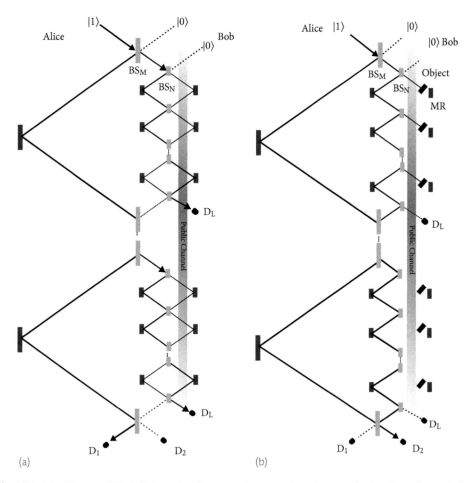

Fig. 14.4 A double array of Mach–Zehnder interferometers for counterfactual communication. Here Alice sends the photon and Bob decides whether to send (*a*) bit "1" by not blocking the mirrors and (*b*) bit "0" by absorbing the photon if it enters the Bob side of the interferometer.

Next, we consider the case when Bob wants to send "0" to Alice. He blocks the photon at each stage as shown in Fig 14.4*b*. There is an array of blocked Mach–Zehnder interferometers on the right-hand side of the M interferometers at each step. We have seen, in our discussion of such a blocked array (Fig. 14.3*b*), that the photon ends up on the left side (at detector D_1) with almost unit probability ($\cos^{2N}\theta_N = \cos^{2N}(\pi/2N) \cong 1$). Thus the configuration shown in Fig. 14.4*b* is effectively the configuration for M Mach–Zehnder interferometers (similar to Fig. 14.4*a*) with no block and the photon is detected at D_2 with almost unit probability. A click at D_2 signals Alice that Bob sent a "0." Again the important observation is that we can be sure that the photon never crossed through the transmission channel, because if it had, it would have been absorbed by one of the blockers.

We thus achieve a highly counterintuitive result: When Alice sends a photon, Bob can send a "1" by allowing the photon to be reflected from his mirrors and Alice gets a click at detector D_1 AND Bob can send a "0" by blocking the photon at each stage and Alice gets a click at D_2. In both instances, the probability that the photon exists in the transmission channel is zero.

Problems

14.1 Consider an optical set-up involving a Mach–Zehnder interferometer with another Mach–Zehnder interferometer in one arm as shown in the figure.

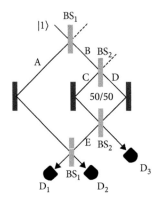

A single photon is incident as shown in the figure. Both the beam splitters BS_2 are 50/50 beam splitters whereas the beam splitters BS_1 have transmission and reflections amplitudes t and r.

(a) Show that the probability of existence of a photon at E is zero no matter what the values of t and r are.

(b) What is the probability of detecting the photon at detectors D_1, D_2, and D_3?

Hint: Beam splitter transformation for BS_1 (with $r = \cos\theta, t = \sin\theta$):

$$|10\rangle \rightarrow \cos\theta|10\rangle + \sin\theta|01\rangle$$
$$|01\rangle \rightarrow \cos\theta|01\rangle - \sin\theta|10\rangle$$

BIBLIOGRAPHY

A.C. Elitzur and L. Vaidman, *Quantum mechanical interaction-free measurements*, Foundations of Physics 23, 987 (1993).

P. G. Kwiat, H. Weinfurter, T. Herzog, A. Zeilinger, and M. A. Kasevich, *Interaction-free measurement*, Physical Review Letters 74, 4763 (1995).

H. Salih, Z.-H. Li, M. Alamri, and M. S. Zubairy, *Protocol for direct counterfactual communication*, Physical Review Letters 110, 170502 (2013).

Z.-H. Li, M. Alamri, and M. S. Zubairy, *Direct quantum communication with almost invisible photons*, Physical Review A 89, 052334 (2014).

Y. Cao, Y.-H. Li, Z. Cao, J. Yin, Y.-A. Chen, H.-L. Yin, T.-Y. Chen, X. Ma, C.-Z. Peng, J.-W. Pan, *Direct counterfactual communication via quantum Zeno effect*, Proceedings of the National Academy of Sciences 114, 4920 (2017).

PART

Quantum Computing

15 Quantum Computing I

The emergence of quantum computing as a major field of research is due to the realization that certain problems can be solved much faster on a quantum computer than on a conventional computer. The extraordinary speed-up of a quantum computer is due to some novel features of quantum mechanics, such as coherent superposition and quantum entanglement. Thus, the computational power of a quantum computer can exceed that of conventional computers.

The first ideas about quantum computing can be traced to Richard Feynman when he proposed solving complicated quantum mechanical problems by simulating them on a quantum computer around 1982. The mathematical framework of quantum computing was developed by David Deutsch in the late 1980s. There was, however, no practical problem that could be solved on a quantum computer with a substantial speed-up compared to conventional computers. The interest in the field of quantum computing remained limited to a very small number of researchers. The situation changed drastically in the mid 1990s when two major quantum computing algorithms were proposed. One related to factoring a large number into its prime factors and the second related to finding a marked object in an unsorted database. Since then there has been tremendous activity in this field. Quantum computing provides a beautiful example of how quantum mechanical concepts such as coherent superposition of states and quantum entanglement can lead to incredibly fast speed-up in the solution of certain problems.

15.1 Introduction to Quantum Computing

The basic building block of a computer is a bit that can take on two values, "0" or "1". In a conventional computer, these bits are classical objects like voltage—high voltage corresponds to "1" and low voltage corresponds to "0". These are therefore referred to as classical bits or simply as "bits."

On the other hand, a quantum bit (or a "qubit") is a system that can exist in two possible quantum states that we label as $|0\rangle$ and $|1\rangle$. The qubits, in the laboratory, can correspond to many different realizations. For example, a photon that can exist in the polarization state $|\rightarrow\rangle$ or $|\uparrow\rangle$, corresponding to "0" or "1", respectively, is a qubit. Other examples include an atom in the ground state $|g\rangle$ and the excited state $|e\rangle$ and a radiation field with states with no photon $|0\rangle$ or one photon $|1\rangle$ corresponding to "0" or "1", respectively. Any computer that carries out computation using qubits, instead of classical bits, is called a quantum computer.

Quantum Mechanics for Beginners: With Applications to Quantum Communication and Quantum Computing. M. Suhail Zubairy.
© M. Suhail Zubairy 2020. Published in 2020 by Oxford University Press. DOI: 10.1093/oso/9780198854227.001.0001

Quantum bits, being quantum states, can satisfy two novel aspects which are inaccessible to their classical counterparts.

Coherent superposition: The first is the possibility that qubits can exist in a state of coherent superposition. For example, an atom can exist in a superposition of ground state $|g\rangle$ and an excited state $|e\rangle$, i.e., $c_g|g\rangle + c_e|e\rangle$ where c_g and c_e are, in general, complex numbers that satisfy $|c_g|^2 + |c_e|^2 = 1$. In general, for any qubit, we can have the state

$$|\psi\rangle = c_0|0\rangle + c_1|1\rangle. \tag{15.1}$$

Quantum entanglement: A system consisting of two or more qubits can be found in an entangled state where the two qubits lose their independent identity and the state of one qubit depends on the state of the other. For example, when the two qubits can be found in the state

$$|\psi(1,2)\rangle = \frac{1}{\sqrt{2}}(|0_1, 1_2\rangle + |1_1, 0_2\rangle), \tag{15.2}$$

the state of the two qubits cannot be written separately in the form $|\psi(1,2)\rangle = |\psi(1)\rangle|\psi(2)\rangle$. If the first qubit is found in state $|0\rangle$ then the state of the second qubit is $|1\rangle$ corresponding to $|0_1, 1_2\rangle$ and if the first qubit is found in state $|1\rangle$, then the state of the second qubit is $|0\rangle$ corresponding to $|1_1, 0_2\rangle$. The two qubits cannot be expressed independently. In order to appreciate the potential power of a quantum computer based on quantum entanglement we consider the states generated by N qubits.

In general, the quantum state of two qubits can be written as

$$\frac{1}{2}(c_0|0,0\rangle + c_1|0,1\rangle + c_2|1,0\rangle + c_3|1,1\rangle), \tag{15.3}$$

i.e., a superposition of $2^2 = 4$ states. Here, as before, c_i $(i = 0, 1, \cdots 3)$ are complex numbers. Thus, a two-qubit computer can store 4 complex numbers.

Similarly, the most general state generated by 4 qubits is

$$\begin{aligned} \frac{1}{4}(c_0|0,0,0,0\rangle &+ c_1|0,0,0,1\rangle + c_2|0,0,1,0\rangle + c_3|0,0,1,1\rangle \\ &+ c_4|0,1,0,0\rangle + c_5|0,1,0,1\rangle + \cdots + c_{15}|1,1,1,1\rangle). \end{aligned} \tag{15.4}$$

Thus we have a superposition of $2^4 = 16$ states described by 16 complex numbers $c_0, c_1, \cdots \cdots c_{15}$. Similarly we see that N qubits can store 2^N numbers "simultaneously."

If we increase the size of the quantum computer to 256 qubits, then, following the same argument as above, we have the possibility of a superposition of 2^{256} states with 2^{256} complex numbers as amplitudes. This is an enormously large number, larger than 10^{77}. It is estimated that there are between 10^{78} to 10^{82} atoms in the known, observable universe. That works out to between one quadrillion vigintillion and ten thousand quadrillion vigintillion atoms[1]. Thus if we have the capability to manipulate just 256 qubits, such as 256 atoms, we can have a computer with a power that can be acquired by our conventional computers only if they work with almost all the atoms in the known universe as bits. This realization is scintillating, almost mind boggling. This observation also clearly shows why one feels extremely excited at the prospect of a quantum computer.

---

[1]   1 quadrillion is equal to 10^{15} and 1 vigintillion is equal to 10^{63}.

Then the obvious question: If, even with the modest number of 256 qubits, we can potentially make a computer that will match a conventional computer with all the atoms in the known universe in its memory, why is such a computer not a reality so far? The answer to this question lies in two aspects of quantum mechanics that would appear to destroy the prospect of a quantum computer. Just as the quantum mechanical concepts of coherent superposition and quantum entanglement lead us to the prospect of a quantum computer with almost unbelievable capability, the other two aspects of quantum mechanics, namely the probabilistic nature of the measurement outcome and decoherence, can prove fatal to the design of a robust and reliable computer.

The probabilistic nature of quantum mechanics has been emphasized throughout this book. In the present context, let us consider (say) a two bit computer with an initial state of the form

$$|\psi_{input}\rangle = |0, 0\rangle. \tag{15.5}$$

After some computational manipulations, let the output state of the two qubits be a superposition

$$|\psi_{output}\rangle = \frac{1}{2}(d_0|0, 0\rangle + d_1|0, 1\rangle + d_2|1, 0\rangle + d_3|1, 1\rangle). \tag{15.6}$$

If we measure the output state of the two qubits, it will be $|0, 0\rangle$, $|0, 1\rangle$, $|1, 0\rangle$, or $|1, 1\rangle$ with probabilities $|d_0|^2$, $|d_1|^2$, $|d_2|^2$, and $|d_3|^2$, respectively. The outcome of the measurement is probabilistic. Before the measurement we have no way of knowing what the outcome will be. This is very disappointing as a probabilistic answer to any computation is typically not acceptable. So this should be the end of the "quantum computer" fantasy.

The second issue relating to decoherence is more subtle. It has to do with the errors induced in the state of qubits by their interaction with the environment. In our everyday experience, we know that any interaction of a given system with the environment can lead to a change in the state of the system even when we do not intend to do so. For example, a cup of hot coffee can cool to room temperature by losing energy to the environment. The environment is huge and we do not notice any change of temperature of the environment. However the temperature of the coffee in our little cup goes down. This process is irreversible as the cooling of the coffee can and does take place by losing energy to the surrounding but it never happens that a cup of cold coffee suddenly heats up by extracting energy from the surroundings. Quantum mechanically the situation is even more complex. It has to do with the nature of vacuum. Before the advent of quantum mechanics, a vacuum was perceived as *nothing*—a place where no light existed, nothing moved, and there was no energy present. The quantum mechanical picture of a vacuum turned out to be dramatically different. According to quantum mechanics, a vacuum, even at almost absolute zero temperature when nothing moves and no energy is supposed to be present, has an infinite amount of energy associated with the electromagnetic field. In addition, there are quantum mechanical fluctuations as a result of Heisenberg's uncertainty relation that cannot be neglected. Thus when a qubit, such as an atom in the excited state, experiences fluctuating fields associated with vacuum fluctuations, it can decay to the ground state spontaneously. This becomes a source of inevitable error.

Thus, we have a situation where, on the one hand, the possibility of preparing a qubit in a coherent superposition of its states and the possibility of multiple qubits existing in an entangled state raises the possibility of processing a large (indeed extremely large) amount of data much faster, but, on the other hand, the probabilistic outcome of the measurement and the presence of inevitable decoherence due to the interaction of the qubits with the environment appear to kill the prospect of quantum computing. The challenge, therefore, is to come up with problems that can be solved by taking advantage of the tremendous potential of quantum entanglement but are not adversely affected by the probabilistic nature of the measurement outcome. Decoherence is still a stumbling block. This is a practical problem that requires searching for the systems where we can minimize the effect of decoherence or using some error correction codes and procedures. This challenge is met in a limited number of problems—the two most famous being the determination of the prime factors of a composite number and the search of an unsorted database as discussed above. These problems are discussed in the next chapter.

In the following, we discuss the conditions for quantum computing systems and, in the next section, the basic building blocks such as the logic gates for quantum computing. A logic gate converts the input states of the qubits in the output states in a prescribed manner.

A typical quantum computing system, like any computing device, consists of an input, a processor, and an output as shown in Fig. 15.1. The input consists of qubits in prescribed initial states. Inside the processor, these qubits can carry out the necessary manipulations via logic gates. The output is again a set of qubits in a prescribed quantum entangled state and a measurement is made on these qubits to achieve the result.

A quantum computer can adopt many different technologies as the choice of the form of the qubits and the processing systems can vary depending on the requirements. For example, the qubits can be photons, atoms, electrons, or some other two-state quantum systems. There is, however, a set of criteria, called Di Vincenzo criteria, that should be satisfied by any quantum computer. These are listed as follows:

- A quantum computer should be a scalable physical system with well-defined qubits. The quantum computing system should consist of qubits that can take only two possible states and no more. For example, as we have seen in our discussion of the hydrogen atom, there are many, indeed infinite, energy levels. The atom can therefore exist in any of these levels. However the atom can be treated as a qubit only if two levels (such as the ground state and the first excited state) are relevant and the atom is never found in the higher excited states. Additionally, the complexity of the computation should be directly proportional to the number of qubits required.

- A quantum computing device should be initializable to a simple state, i.e., the qubits are prepared in a prescribed initial state. Such a state, for example, can be of the form $|0, 0, 0, \ldots\rangle$, i.e, all qubits are initially in the state $|0\rangle$. Any fluctuation or uncertainty can become a source of error.

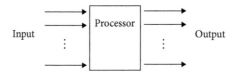

Fig. 15.1 Schematics of a quantum computer. It consists of three parts: The input consists of qubits in some well-defined states, the processor consists of some quantum logic gates, and the output gives the outcome of measurements.

- The qubits should have long decoherence times. As discussed above, decoherence arising out of the fluctuations may cause error. The system evolves randomly under the action of these fluctuations. There is a timescale associated with the decoherence process. A quantum computing system will be viable only if we can do a sufficiently large number of operations before errors are introduced.

- A quantum computer should have a universal set of quantum gates. Both in classical computing and quantum computing, we can implement an algorithm with the help of logic gates. For a quantum computer, a universal set of logic gates is a very small set of 1 and 2 qubit gates that can implement a large computing device by using just this small set of gates. It is therefore necessary to have such logic gates readily available and they are able to carry out the desired computation.

- Qubit-specific measurements should be possible. In any quantum computing device, the qubits at the input interact with each other to form an array of quantum logic gates. When the computation is done, we have an output state of the device that should be read by making the measurement on the output qubits. The measurements should be highly reliable and not introduce significant error.

It is hard to imagine that there will be an all-purpose quantum computer that is able to solve a wide range of problems like our everyday computers in the near future. However, it should be possible to design a quantum computer that can solve particular problems much faster than conventional computers.

15.2 Quantum Logic Gates

The basic building blocks of any computer are logic gates. The logic gates, in conventional computers, are electronic devices that can have two or more ports at the input as well as ports in the output. The input ports have binary inputs "0" and "1", corresponding to low and high voltages, respectively. The output port can have a prescribed binary output. A table, describing the input–output relation, is called a truth table. As examples, the truth tables for AND and exclusive-OR (XOR) gates are given as follows.

AND			XOR		
A	B	$A \cdot B$	A	B	$A \oplus B$
0	0	0	0	0	0
0	1	0	0	1	1
1	0	0	1	0	1
1	1	1	1	1	0

Here A and B are the states of the input ports. For binary numbers, there are four possibilities for the input, $\{0_A, 0_B\}, \{0_A, 1_B\}, \{1_A, 0_B\}, \{1_A, 1_B\}$. For an AND gate, represented by $A \cdot B$, the output is "1" only when both inputs are "1." In the other instances when one or both inputs are "0", the output is also "0." For an XOR gate, represented by $A \oplus B$, the output is "1" in those instances when either A or B is "1." When both inputs are "0"s or "1"s, then the output is "0." The symbols of AND and XOR gates are shown in Fig. 15.2.

A careful look at these gates indicates that whereas the XOR gate represents the sum of two binary numbers:

$$0 + 0 = 0,$$
$$0 + 1 = 1,$$
$$1 + 0 = 1,$$
$$1 + 1 = 0,$$

the AND gate represents the carry, the carry being "1" only when we add "1" and "1." As shown in Fig. 15.3, these two gates can be combined together to make a "half adder" such that, for an input A and B, the output is the sum $(A \cdot B)$ and the carry $(A \oplus B)$.

The classical logic gates are irreversible, i.e., from the output, we cannot determine the input uniquely. For example, in the XOR gate, if the output is "1," we cannot determine whether the input is $\{0_A, 1_B\}$ or $\{1_A, 0_B\}$.

Next, consider quantum logic gates. Due to the coherent nature of quantum mechanics, quantum logic gates should be reversible. The basic building blocks are the one-qubit and two-qubit gates.

First, we discuss the one-qubit gate. A general one-qubit gate is called a unitary gate, designated by U_θ. It has one input and one output such that, when it acts either on state $|0\rangle$ or on state $|1\rangle$, it generates, in general, a coherent superposition of states $|0\rangle$ and $|1\rangle$. i.e.,

$$U_\theta|0\rangle = \cos\theta|0\rangle + \sin\theta|1\rangle, \tag{15.7}$$

$$U_\theta|1\rangle = \sin\theta|0\rangle - \cos\theta|1\rangle. \tag{15.8}$$

As seen in Chapter 9, a polarizing beam splitter can lead to a similar transformation when $|0\rangle$ and $|1\rangle$ are identified with horizontally and vertically polarized photons.

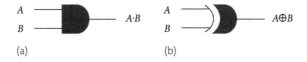

Fig. 15.2 The symbols for classical AND and Exclusive-OR or XOR gates.

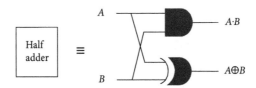

Fig. 15.3 A circuit for a half adder. The XOR gate gives the sum and the AND gate gives the carry in the binary summation.

As a special case, for $\theta = \pi/4$, the maximum coherence is obtained between $|0\rangle$ and $|1\rangle$, yielding the so-called Hadamard gate H, i.e.,

$$H|0\rangle \equiv U_{\pi/4}|0\rangle = \frac{1}{\sqrt{2}}\left(|0\rangle + |1\rangle\right), \tag{15.9}$$

$$H|1\rangle \equiv U_{\pi/4}|1\rangle = \frac{1}{\sqrt{2}}\left(|0\rangle - |1\rangle\right). \tag{15.10}$$

A bit flip gate ($|0\rangle \to |1\rangle$ and $|1\rangle \to |0\rangle$), labelled as X, is obtained for $\theta = \pi/2$, i.e.,

$$X|0\rangle \equiv U_{\pi/2}|0\rangle = |1\rangle, \tag{15.11}$$

$$X|1\rangle \equiv U_{\pi/2}|1\rangle = |0\rangle. \tag{15.12}$$

In a similar manner, a Z gate is defined as one that changes the sign only when the input is $|1\rangle$. Such a gate is obtained when $\theta = 0$, i.e.,

$$Z|0\rangle \equiv U_0|0\rangle = |0\rangle, \tag{15.13}$$

$$Z|1\rangle \equiv U_0|1\rangle = -|1\rangle. \tag{15.14}$$

In the table below, we list these important one-bit gates.

Unity	I	$\begin{aligned}&|0\rangle \to |0\rangle\\ &|1\rangle \to |1\rangle\end{aligned}$
H	H	$\begin{aligned}&|0\rangle \to \frac{1}{\sqrt{2}}(|0\rangle + |1\rangle)\\ &|1\rangle \to \frac{1}{\sqrt{2}}(|0\rangle - |1\rangle)\end{aligned}$
X	X	$\begin{aligned}&|0\rangle \to |1\rangle\\ &|1\rangle \to |0\rangle\end{aligned}$
Z	Z	$\begin{aligned}&|0\rangle \to |0\rangle\\ &|1\rangle \to -|1\rangle\end{aligned}$

It should be noted that there is no corresponding classical one-bit logic gate: an input "0" remains a "0" and a "1" remains a "1" for a classical one bit gate as identified by I.

Next, consider two-qubit gates. These gates are responsible for creating entangled states between two qubits. Unlike the classical logic gates discussed above, the two-qubit gates are two-input and two-output gates and are reversible.

There are two most prominent two-qubit gates. The first is a quantum phase gate and is designated by the operator Q_ϕ. For the two inputs, the input–output transformation of a quantum phase gate is given as follows:

$$Q_\phi|0_1, 0_2\rangle = |0_1, 0_2\rangle, \tag{15.15}$$

$$Q_\phi|0_1, 1_2\rangle = |0_1, 1_2\rangle, \tag{15.16}$$

$$Q_\phi|1_1, 0_2\rangle = |1_1, 0_2\rangle, \tag{15.17}$$

$$Q_\phi|1_1, 1_2\rangle = e^{i\phi}|1_1, 1_2\rangle. \tag{15.18}$$

Thus a quantum phase gate Q_ϕ does not change the input state when one or both qubits are in the state $|0\rangle$. Only when the input state is $|1_1, 1_2\rangle$ do we have a phase shift $e^{i\phi}$. As a special case,

$$Q_\pi|0_1, 0_2\rangle = |0_1, 0_2\rangle, \tag{15.19}$$

$$Q_\pi|0_1, 1_2\rangle = |0_1, 1_2\rangle, \tag{15.20}$$

$$Q_\pi|1_1, 0_2\rangle = |1_1, 0_2\rangle, \tag{15.21}$$

$$Q_\pi|1_1, 1_2\rangle = -|1_1, 1_2\rangle. \tag{15.22}$$

Another important two-qubit quantum gate is the Controlled-NOT gate or the CNOT gate. The truth table of the CNOT gate is given as follows:

CNOT Gate

$|A\rangle$	$|B\rangle$	$|A\rangle$	$|A \oplus B\rangle$
$|0\rangle$	$|0\rangle$	$|0\rangle$	$|0\rangle$
$|0\rangle$	$|1\rangle$	$|0\rangle$	$|1\rangle$
$|1\rangle$	$|0\rangle$	$|1\rangle$	$|1\rangle$
$|1\rangle$	$|1\rangle$	$|1\rangle$	$|0\rangle$

Here $|A\rangle$ is called the controlling bit and $|B\rangle$ is called the controlled or target bit. When the controlling qubit $|A\rangle$ is in the state $|0\rangle$, the controlled bit $|B\rangle$ remains the same. However when the controlling qubit $|A\rangle$ is in the state $|1\rangle$, the controlled bit $|B\rangle$ flips, i.e., $|0\rangle \rightarrow |1\rangle$ and $|1\rangle \rightarrow |0\rangle$. Thus a CNOT gate, designated by U_{CNOT}, leads to the following transformations:

$$U_{CNOT}|0_1, 0_2\rangle = |0_1, 0_2\rangle, \tag{15.23}$$

$$U_{CNOT}|0_1, 1_2\rangle = |0_1, 1_2\rangle, \tag{15.24}$$

$$U_{CNOT}|1_1, 0_2\rangle = |1_1, 1_2\rangle, \tag{15.25}$$

$$U_{CNOT}|1_1, 1_2\rangle = |1_1, 0_2\rangle. \tag{15.26}$$

A CNOT gate is typically represented in the circuit diagrams as shown in Fig. 15.4.

In earlier chapters we have seen that the Bell state basis

$$|B_{00}(1, 2)\rangle = \frac{1}{\sqrt{2}}(|0_1, 0_2\rangle + |1_1, 1_2\rangle), \tag{15.27}$$

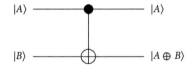

Fig. 15.4 Symbol for the CNOT gate.

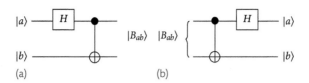

Fig. 15.5 Circuit diagrams for (a) generating Bell states via Hadamard and CNOT gates and (b) measurement of the Bell basis states.

$$|B_{01}(1,2)\rangle = \frac{1}{\sqrt{2}}(|0_1,1_2\rangle + |1_1,0_2\rangle), \tag{15.28}$$

$$|B_{10}(1,2)\rangle = \frac{1}{\sqrt{2}}(|0_1,0_2\rangle - |1_1,1_2\rangle), \tag{15.29}$$

$$|B_{11}(1,2)\rangle = \frac{1}{\sqrt{2}}(|0_1,1_2\rangle - |1_1,0_2\rangle), \tag{15.30}$$

plays a key role in many applications such as quantum teleportation and quantum swapping. As an example, we show how the Bell basis states can be created using the quantum logic gates discussed in this section, and how Bell basis states can be measured using the logic gates.

In Fig. 15.5a, the circuit involving Hadamard and CNOT gates can generate a desired Bell state. As an example, we consider the initial state $|0_1, 0_2\rangle$ and show how we can generate the Bell basis state $|B_{00}(1,2)\rangle$. First apply a Hadamard gate H_1 on the first qubit in the initial state $|0_1, 0_2\rangle$. It follows from Eq. (15.9) that the resulting state is

$$\frac{1}{\sqrt{2}}(|0_1,0_2\rangle + |1_1,0_2\rangle).$$

Next, apply a CNOT gate on the qubits. A CNOT gate keeps the state $|0_1, 0_2\rangle$ unchanged according to Eq. (15.23) and transforms the state $|1_1, 0_2\rangle$ to $|1_1, 1_2\rangle$ according to Eq. (15.25). The net result is that the output is the Bell basis state $|B_{00}\rangle$ as given by Eq. (15.27). Formally, this is

$$U_{CNOT}H_1|0_1,0_2\rangle = |B_{00}(1,2)\rangle. \tag{15.31}$$

In the same way, the states $|0_1, 1_2\rangle$, $|1_1, 0_2\rangle$, and $|1_1, 1\rangle$ are transformed into the Bell basis states $|B_{01}(1,2)\rangle$, $|B_{10}(1,2)\rangle$, and $|B_{11}(1,2)\rangle$, respectively.

The reverse process corresponds to a measurement of the Bell basis states—a Bell basis state $|B_{ab}(1,2)\rangle$ $(a, b = 0, 1)$ can generate the state $|a, b\rangle$ in the output which can then be measured as shown in Fig. 15.5b. A measurement outcome of state $|a\rangle$ for the first qubit and state $|b\rangle$ for the second qubit yields the information that the input state is the Bell state $|B_{ab}(1, 2)\rangle$ in the input.

15.3 The Deutsch Problem

In 1985, David Deutsch discussed a simple problem that indicates the power of quantum computers. It is a toy problem that may not be useful in any practical application: However, the remarkable aspect of the Deutsch problem is that, despite the probabilistic nature of quantum mechanics, we obtain a deterministic and definite answer to a quantum mechanical problem.

The problem can be stated as follows: Suppose we are given a binary function of a binary variable $f(x)$, i.e., x can take on only two values "0" and "1," and both $f(0)$ and $f(1)$ can take on two values, again "0" and "1." Thus

$f(0)$ can be 0 or 1
$f(1)$ can be 0 or 1

There are now two possibilities: Either $f(0)$ is equal to $f(1)$ (i.e., both $f(0)$ and $f(1)$ are equal to 0 or 1) or $f(0)$ is not equal to $f(1)$ (i.e., if $f(0)$ is equal to 0 then $f(1)$ is equal to 1 or vice versa). The question we ask is: By making only one measurement, can we determine with certainty whether

$$f(0) = f(1) \tag{15.32}$$

or

$$f(0) \neq f(1)? \tag{15.33}$$

Classically we always require two measurements to solve this problem. We should measure both $f(0)$ and $f(1)$. Only then can we find out whether the two functions are equal or unequal. Can quantum mechanics do better and solve the problem by making only one measurement? The answer is, remarkably, "yes" as we see in the following.

The problem is solved using two qubits. Let the initial state of the two qubits be $|x\rangle|y\rangle$ or $|x, y\rangle$ and the quantum computer then carries out the following transformation on the input qubits:

$$|x, y\rangle \rightarrow |x, y \oplus f(x)\rangle. \tag{15.34}$$

Here $|y \oplus f(x)\rangle$ is the CNOT gate such that the binary states transform as follows,

$$|0, 0\rangle \rightarrow |0, 0 \oplus f(0)\rangle = |0, f(0)\rangle, \tag{15.35}$$

$$|0, 1\rangle \rightarrow |0, 1 \oplus f(0)\rangle = |0, \overline{f(0)}\rangle, \tag{15.36}$$

$$|1, 0\rangle \rightarrow |1, 0 \oplus f(1)\rangle = |1, f(1)\rangle, \tag{15.37}$$

$$|1, 1\rangle \rightarrow |1, 1 \oplus f(1)\rangle = |1, \overline{f(1)}\rangle, \tag{15.38}$$

where the bar indicates the opposite value: $\bar{0} = 1, \bar{1} = 0$. For example, if $f(0) = 0$ then $\overline{f(0)} = 1$.

Let the two qubits be prepared initially in states $|0\rangle$ and $|1\rangle$ as shown in Fig. 15.6. Hadamard transform (15.9) and (15.10) on both qubits leads to

$$|x\rangle = H|0\rangle = \frac{1}{\sqrt{2}} (|0\rangle + |1\rangle), \tag{15.39}$$

$$|y\rangle = H|1\rangle = \frac{1}{\sqrt{2}} (|0\rangle - |1\rangle). \tag{15.40}$$

Therefore the input state of the two qubits is

$$
\begin{aligned}
|\psi_{in}\rangle &= |x\rangle|y\rangle \\
&= \frac{1}{\sqrt{2}}(|0\rangle + |1\rangle)\frac{1}{\sqrt{2}}(|0\rangle - |1\rangle) \\
&= \frac{1}{2}(|0,0\rangle - |0,1\rangle + |1,0\rangle - |1,1\rangle).
\end{aligned}
\tag{15.41}
$$

According to Eq. (15.34), the output state is

$$
\begin{aligned}
|\psi_{out}\rangle &= |x, y \oplus f(x)\rangle \\
&= \frac{1}{2}(|0, f(0)\rangle - |0, \overline{f(0)}\rangle + |1, f(1)\rangle - |1, \overline{f(1)}\rangle) \quad , \\
&= \frac{1}{2}[|0\rangle(|f(0)\rangle - |\overline{f(0)}\rangle) + |1\rangle(|f(1)\rangle - |\overline{f(1)}\rangle)]
\end{aligned}
\tag{15.42}
$$

where we used transformations (15.35)–(15.38).

We next consider the two cases: If $f(0) = f(1)$ then $\overline{f(0)} = \overline{f(1)}$, and we obtain

$$
|\psi_{out}\rangle = \frac{1}{2}(|0\rangle + |1\rangle)(|f(0)\rangle - |\overline{f(0)}\rangle).
\tag{15.43}
$$

However, if $f(0) \neq f(1)$ then $\overline{f(1)} = f(0)$ and $f(1) = \overline{f(0)}$, and the result is

$$
|\psi_{out}\rangle = \frac{1}{2}(|0\rangle - |1\rangle)(|f(0)\rangle - |\overline{f(0)}\rangle).
\tag{15.44}
$$

Next we make the Hadamard transformation

$$
\begin{aligned}
|0\rangle &\rightarrow \frac{1}{\sqrt{2}}(|0\rangle + |1\rangle), \\
|1\rangle &\rightarrow \frac{1}{\sqrt{2}}(|0\rangle - |1\rangle),
\end{aligned}
$$

on the first qubit. The output state for $f(0) = f(1)$ (Eq. (15.43) becomes

$$
|\psi_{out}\rangle = |0\rangle(|f(0)\rangle - |\overline{f(0)}\rangle)
\tag{15.45}
$$

and for $f(0) \neq f(1)$ (Eq. (15.44) becomes

$$
|\psi_{out}\rangle = |1\rangle(|f(0)\rangle - |\overline{f(0)}\rangle).
\tag{15.46}
$$

Finally, we make a measurement on the first qubit $|a\rangle$ (Fig. 15.6). It is clear from Eqs. (15.45) and (15.46) that if the state is found to be $|0\rangle$ we know that $f(0) = f(1)$, and if the outcome is $|1\rangle$ then $f(0) \neq f(1)$. Thus, by making just a single measurement, we can find with certainty whether $f(0) = f(1)$ or $f(0) \neq f(1)$.

This is a remarkable result as we cannot devise a classical algorithm which can solve this problem with just a single measurement.

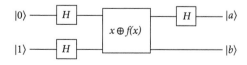

Fig. 15.6 Circuit diagram for the implementation of the Deutsch algorithm.

15.4 **Quantum Teleportation Revisited**

In Section 10.4, we learned how entanglement can be used to teleport an arbitrary unknown state

$$|\psi_C\rangle = c_0|0\rangle + c_1|1\rangle \tag{15.47}$$

from Alice to Bob. Here we revisit quantum teleportation from a quantum circuit point of view.

The circuit for quantum teleportation is shown in Fig. 15.7. Alice and Bob share a pair of qubits in a Bell state

$$|B_{00}(A,B)\rangle = \frac{1}{\sqrt{2}}(|0_A,0_B\rangle + |1_A,1_B\rangle). \tag{15.48}$$

The initial state $|\psi_3\rangle$ of the system of three qubits is

$$\begin{aligned}|\psi_3^{(0)}\rangle &= |\psi_C\rangle|B_{00}(A,B)\rangle \\ &= \frac{1}{\sqrt{2}}\left[c_0|0_C\rangle(|0_A\rangle|0_B\rangle + |1_A\rangle|1_B\rangle) + c_1|1_C\rangle(|0_A\rangle|0_B\rangle + |1_A\rangle|1_B\rangle)\right].\end{aligned} \tag{15.49}$$

The first two qubits (C and A) are at Alice's location, and the last qubit (B) is at Bob's location. Alice applies the CNOT transformation to her two qubits, with the controlling qubit being the qubit C to be teleported to Bob. Thus $|0_C\rangle|0_A\rangle \to |0_C\rangle|0_A\rangle$, $|0_C\rangle|1_A\rangle \to |0_C\rangle|1_A\rangle$, $|1_C\rangle|0_A\rangle \to |1_C\rangle|1_A\rangle$, and $|1_C\rangle|1_A\rangle \to |1_C\rangle|0_A\rangle$, and the resulting state is

$$|\psi_3^{(1)}\rangle = \frac{1}{\sqrt{2}}\left[c_0|0_C\rangle(|0_A\rangle|0_B\rangle + |1_A\rangle|1_B\rangle) + c_1|1_C\rangle(|1_A\rangle|0_B\rangle + |0_A\rangle|1_B\rangle)\right]. \tag{15.50}$$

Alice then applies the Hadamard transformation ((15.9) and (15.10)) to qubit C, with the result

$$\begin{aligned}|\psi_3^{(2)}\rangle = \frac{1}{2}[&c_0(|0_C\rangle + |1_C\rangle)(|0_A\rangle|0_B\rangle + |1_A\rangle|1_B\rangle) \\ +&c_1(|0_C\rangle - |1_C\rangle)(|1_A\rangle|0_B\rangle + |0_A\rangle|1_B\rangle)],\end{aligned} \tag{15.51}$$

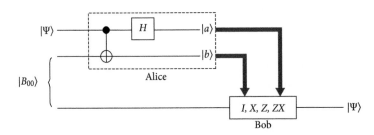

Fig. 15.7 Circuit diagram for the quantum teleportation of the state $|\psi\rangle$ from Alice to Bob. A Bell basis state $|B_{00}\rangle$ is created between Alice and Bob. Alice makes a joint Bell basis measurement on the two qubits at her end and conveys the outcome to Bob via a classical channel. Bob transforms his state by applying I, X, Z, and ZX gates depending on Alice's measurement and recovers the state $|\psi\rangle$.

which can be cast in the form

$$|\psi_3^{(2)}\rangle = \frac{1}{2}[|0_C, 0_A\rangle (c_0|0_B\rangle + c_1|1_B\rangle) + |0_C, 1_A\rangle (c_0|1_B\rangle + c_1|0_B\rangle)$$
$$+ |1_C, 0_A\rangle (c_0|0_B\rangle - c_1|1_B\rangle) + |1_C, 1_A\rangle (c_0|1_B\rangle - c_1|0_B\rangle)].$$
(15.52)

Finally, Alice measures the two qubits C and A in her possession and communicates the result to Bob with the two bits of information. The possible outcomes are four possible states $|0_C, 0_A\rangle, |0_C, 1_A\rangle, |1_C, 0_A\rangle,$ and $|1_C, 1_A\rangle$.

The measurement outcome $|0_C, 0_A\rangle$ reveals that the state of Bob's qubit is equivalent to the original state $|\psi\rangle$ that was to be teleported. So, if Bob receives from Alice a two-bit message 00, he knows that the state of his qubit B coincides with $|\psi_C\rangle$ and he does not change it. This corresponds to applying a unity I operation.

If, on the other hand, Bob receives message 01, he applies the X (NOT) transformation to his qubit, whose state then becomes $|\psi_C\rangle$.

Similarly, messages 10 or 11 instruct Bob to apply, respectively, the Z or ZX transformations to attain state $|\psi_C\rangle$.

This completes the protocol. Here we have seen how an application of the one-bit and two-bit gates can lead to the quantum teleportation of an unknown state.

15.5 Quantum Dense Coding

As discussed before, photons are ideal carriers of information. Typically a single photon can carry only one bit of information $\{0, 1\}$. For example, a photon in the horizontal polarization state $|\rightarrow\rangle$ can represent "0" and in the vertical polarization state $|\uparrow\rangle$ can represent "1." Thus Alice can send a single bit of information "0" or "1" to Bob by suitably choosing the polarization of her photon. A question of interest is whether we can send two bits of information $\{00, 01, 10, 11\}$ by exchanging just one photon. If this becomes possible, then the information capacity can be doubled. In the following we show that quantum entanglement can help us in achieving this goal.

As a first step, Alice and Bob can create an entangled pair of photons in the state

$$\frac{1}{\sqrt{2}}[|0_A, 0_B\rangle + |1_A, 1_B\rangle],$$
(15.53)

with Alice holding the first qubit and sending the second qubit to Bob. This can, for example be done at a time when communication traffic is slow. We also assume that the qubits can be stored in their respective states for a sufficiently long time before decoherence can induce any errors. Next Alice sends two bits of information by sending her qubit to Bob after making one of the four operations. This is shown in the box in Fig. 15.8.

If Alice wants to send "00" she does nothing (logic gate I) and sends her photon to Bob. In this case the state of the two photons at Bob's end is

$$|B_{00}(A, B)\rangle = \frac{1}{\sqrt{2}}(|0_A, 0_B\rangle + |1_A, 1_B\rangle).$$
(15.54)

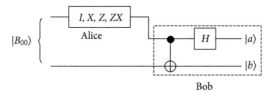

Fig. 15.8 Circuit diagram for quantum dense coding.

If Alice wants to send "01" she applies the logic gate X on her qubit transforming $|0_A\rangle \rightarrow |1_A\rangle$ and $|1_A\rangle \rightarrow |0_A\rangle$ before sending her photon to Bob. In this case the state of the two photons at Bob's end is

$$|B_{01}(A, B)\rangle = \frac{1}{\sqrt{2}}(|0_A, 1_B\rangle + |1_A, 0_B\rangle). \qquad (15.55)$$

If Alice wants to send "10" she applies the logic gate Z on her qubit transforming $|0_A\rangle \rightarrow |0_A\rangle$ but $|1_A\rangle \rightarrow -|1_A\rangle$ before sending her photon to Bob. In this case the state of the two photons at Bob's end is

$$|B_{10}(A, B)\rangle = \frac{1}{\sqrt{2}}(|0_A, 0_B\rangle - |1_A, 1_B\rangle). \qquad (15.56)$$

Finally, if Alice wants to send "11" she applies first the logic gate Z on her qubit transforming $|0_A\rangle \rightarrow |0_A\rangle$ but $|1_A\rangle \rightarrow -|1_A\rangle$ and then applies the logic gate X transforming $|0_A\rangle \rightarrow |1_A\rangle$ and $|1_A\rangle \rightarrow |0_A\rangle$ before sending her photon to Bob. In this case the state of the two photons at Bob's end is (apart from a trivial overall factor -1)

$$|B_{11}(A, B)\rangle = \frac{1}{\sqrt{2}}(|0_A, 1_B\rangle - |1_A, 0_B\rangle). \qquad (15.57)$$

Thus, the resulting entangled state at Bob's end is one of the Bell states which are mutually orthogonal. Bob can therefore uniquely determine the Bell state, and hence the two-bit information, by making the measurement on the two qubits using a Bell State Discriminator consisting of a CNOT gate and a Hadamard gate H as shown in Fig. 15.8. In the presence of a prior entangled state Alice's photon can, therefore, carry two bits of information.

Problems

15.1 A swap gate U_{SW} is characterized by the following transformations:

$$U_{SW}|0\rangle|0\rangle = |0\rangle|0\rangle,$$
$$U_{SW}|0\rangle|1\rangle = |1\rangle|0\rangle,$$
$$U_{SW}|1\rangle|0\rangle = |0\rangle|1\rangle,$$
$$U_{SW}|1\rangle|1\rangle = |1\rangle|1\rangle.$$

Show that this gate can be implemented by a combination of three CNOT gates as follows:

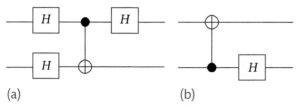

15.2 Show that the following circuits effect the same transformation

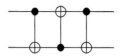

(a)   (b)

List the transformation for all the four possible input states $|0_1, 0_2\rangle$, $|0_1, 1_2\rangle$, $|1_1, 0_2\rangle$, and $|1_1, 1_2\rangle$.

15.3 Show that a CNOT gate can be accomplished via a combination of Hadamard gates and a quantum phase gate Q_π.

15.4 In the text we discussed explicitly how the Bell state $|B_{00}\rangle$ can be created by applying a Hadamard gate on the first qubit H_1 followed by a CNOT gate on the initial state $|0_1, 1_2\rangle$. Show, in a similar manner, that a Hadamard gate on the first qubit H_1 followed by a CNOT gate on the initial states $|0_1, 1_2\rangle$, $|1_1, 0_2\rangle$, and $|1_1, 1_2\rangle$ yield the Bell basis states $|B_{01}\rangle$, $|B_{10}\rangle$, and $|B_{11}\rangle$, respectively.

15.5 Show that, as depicted in Fig. 15.5b, a CNOT gate followed by a Hadamard gate on the first qubit, H_1, transforms the initial Bell basis states $|B_{00}(A, B)\rangle$, $|B_{01}(A, B)\rangle$, $|B_{10}(A, B)\rangle$, and $|B_{11}(A, B)\rangle$ into states $|0_A, 0_B\rangle$, $|0_A, 1_B\rangle$, $|1_A, 0_B\rangle$, and $|1_A, 1_B\rangle$, respectively.

 BIBLIOGRAPHY

R. P. Feynman, *Simulating physics with computers*, International Journal of Theoretical Physics 21, 467 (1982).

R. P. Feynman, *Quantum mechanical computer*, Optics News 11, 11 (1985).

D. Deutsch, *Quantum theory, the Church–Turing principle,and the universal quantum computer*, Proceedings of the Royal Society (London) 400, 97 (1985).

P. Shor, *Algorithms for Quantum Computation*, in Proceedings of 35th Annual Symposium on Foundations of Computer Science (1994), pp 124.

T. Sleator and H. Weinfurter, *Realizable universal quantum gates*, Physical Review Letters 74, 15 (1995).

D. Deutsch, *Quantum communication*, Physics World 5, 57 (1992).

C. H. Bennett, G. Brassard, C. Crepeau, R. Jozsa, A. Peres, and W. Wooters, *Teleporting an unknown quantum state via dual classical and EPR channels*, Physical Review Letters 70, 1895 (1993).

C. Bennett and S. Wiesner, *Communication via one- and two-particle operators on Einstein–Podolsky–Rosen states,* Physical Review Letters 69, 2881 (1992).

M. A. Nielson and I. L. Chuang, *Quantum Computation and Quantum Information* (Cambridge University Press 2000).

P. Lambropoulos and D. Petrosyan, *Fundamentals of Quantum Optics and Quantum Information* (Springer 2007).

C. P. Williams and S. H. Clearwater, *Explorations in Quantum Computing*, (Springer 1998).

S. Stenholm and K.-A. Suominen, *Quantum Approach to Informatics* (John Wiley 2005).

In Chapter 13, we discussed the RSA algorithm that is used to exchange a key on a public channel such as optical fiber or telephone line. This algorithm provides the foundation for the communication security in the present internet-based e-commerce. We are able to exchange confidential information like our credit card number on the internet due to the security provided by the RSA algorithm in the exchange of a key on a public channel. We also discussed how RSA is critically dependent on the difficulty of finding the prime factors of a number N that is a product of two prime numbers p and q. On a conventional computer, it takes an extremely long time to find the factors p and q, which ensures that the RSA-based exchange of information is safe. Recall that a 256-digit number that is typically used in the present-day RSA would require several decades on the fastest computer available to factorize. However, if we can find an algorithm that can accomplish this task (finding prime factors p and q of the number N) much faster, RSA would be compromised and e-commerce might be seriously endangered.

In 1994, Peter Shor proposed a quantum computing algorithm to find the prime factors p and q of the number N much more efficiently. This sent a major alarm throughout the international community. The potential of quantum mechanics to seriously attack the foundation of e-commerce sent tremors through the security community and created great interest in the emerging field of quantum computing. Shor's algorithm remains the most powerful example of the applications of quantum computing.

Another algorithm created by Lov Grover related to the search of an unsorted database. It is well known how difficult it is to find a needle in a haystack. This proverbially corresponds to the problem of the search for a marked object in a large database. A considerable speed-up in solving this problem by using a quantum computer provided another major impetus for the interest in quantum computing.

16.1 How to Factorize *N*?

We first address the question how to find prime factors of a number N, i.e., if we have a number $N = pq$ where p and q are primes then how to find p and q. This can be done as follows.

We can adopt the obvious course: divide N by all the prime numbers starting with the lowest up to $\sqrt{N}$ and see if N is divisible by any of them. If so, then that prime number would be a factor. There should be at least one prime factor of N less than or equal to $\sqrt{N}$. For example, if we want to find the factors of 21583, we first list all the primes less than $\sqrt{21583} \approx 147$. There are 34 such primes and they are:

Quantum Mechanics for Beginners: With Applications to Quantum Communication and Quantum Computing. M. Suhail Zubairy.
© M. Suhail Zubairy 2020. Published in 2020 by Oxford University Press. DOI: 10.1093/oso/9780198854227.001.0001

$$2, 3, 5, 7, 11, 13, 17, 19, 23, 29, 31, 37, 41, 43, 47, 53, 59, 61, 67,$$
$$71, 73, 79, 83, 89, 97, 101, 103, 107, 109, 113, 127, 131, 137, 139$$

Next, we divide $N = 21583$ by each prime, one by one, and find that this number is exactly divisible by 113. From $21583/113 = 191$, it follows that the prime factors of 21583 are 113 and 191.

This procedure may be suitable for a relatively small number N for which we may know all the prime numbers smaller than N. However, for a large number such as a 256- digit number, the number of primes is prohibitively large and this procedure is not practical. In the late seventeenth century, famous mathematicians Gauss and Legendre conjectured that, for a large number N, the number of primes less than N is approximately equal to $N/\ln N$, which can be extremely large. We therefore need to have a formal procedure that should be able to provide the prime factors via a well-defined algorithm. Here we describe such a procedure or algorithm.

Before formally stating the algorithm, we motivate it with a simple example by finding the prime factors of $N = 15$. In order to do so, first, select an integer x that is co-prime with N, i.e., there are no common factors between x and N and the great common divisor (gcd) is equal to 1.[1] In our case we choose $x = 2$. We can verify that 2 and 15 have no common factors. The next step is to find the sequence formed by the function

$$f(a) = x^a \bmod N. \tag{16.1}$$

Here recall the definition of the modulo function: a statement

$$b = a \bmod N \tag{16.2}$$

means that b is the remainder when a is divided by N. Thus $3 = 21 \bmod 6$ and $5 = 65 \bmod 12$ etc. We encountered the modulo function when discussing the RSA algorithm in Chapter 13.

It follows on substituting $x = 2$ and $N = 15$ in Eq. (16.1), we obtain the following sequence:

$$f(0) = 2^0 \bmod 15 = 1$$
$$f(1) = 2^1 \bmod 15 = 2$$
$$f(2) = 2^2 \bmod 15 = 4$$
$$f(3) = 2^3 \bmod 15 = 8$$
$$f(4) = 2^4 \bmod 15 = 1$$
$$f(5) = 2^5 \bmod 15 = 2$$
$$f(6) = 2^6 \bmod 15 = 4$$
$$f(7) = 2^7 \bmod 15 = 8$$
$$f(8) = 2^8 \bmod 15 = 1$$
$$f(9) = 2^9 \bmod 15 = 2$$

and so on. We observe that the sequence is periodic with periodicity $r = 4$, i.e., for any value of a

---

[1] A Euclidean algorithm can be used to find the gcd of two numbers a and b (with $a > b$). First, divide a by b and let the remainder be r_1. Next we divide b by r_1 and let the remainder be r_2. Next divide r_1 by r_2 and continue the process until the remainder is 0 and we stop. The final non-zero remainder is the $\gcd(a, b)$. For example, if we want to find $\gcd(45, 30)$, we divide 45 by 30 and the remainder is 15. We then divide 30 by 15 and the remainder is 0. Therefore $\gcd(45, 30) = 15$.

$$f(a) = f(a + r).$$ (16.3)

For example, $f(1) = f(1 + 4) = f(5)$ etc. We also note that $f(r) = f(4) = 1$. Thus according to Eq. (16.1),

$$1 = x^r \bmod N.$$ (16.4)

In the present case, when $x = 2$, $N = 15$, and the periodicity $r = 4$, we have

$$1 = 2^4 \bmod 15.$$

This equation can be rewritten as

$$0 = (2^4 - 1) \bmod 15,$$

i.e., $(2^4 - 1)$ is divisible by 15. However $(2^4 - 1) = (2^2 - 1)(2^2 + 1) = 3 \cdot 5$. Therefore 3 and 5 should be the factors of 15. We confirm this by dividing 15 by 5 and 3 and find that these are indeed factors of 15.

It is interesting to note that the crucial point of this procedure is to find the periodicity of the function $f(a)$. For an even value of r ($r = 4$ in the present case), we find smaller factors $(2^{r/2} - 1) = (2^2 - 1)$ and $(2^{r/2} + 1) = (2^2 + 1)$ that contain the prime factors of $N = 15$. This is the essence of the method to finding the prime factors of N.

In order to illustrate this method further, we consider another slightly more complicated example. We try to find the prime factors of $N = 91$ and choose $x = 3$. The sequence

$$f(a) = x^a \bmod N$$

is given by

$$f(0) = 3^0 \bmod 91 = 1$$
$$f(1) = 3^1 \bmod 91 = 3$$
$$f(2) = 3^2 \bmod 91 = 9$$
$$f(3) = 3^3 \bmod 91 = 27$$
$$f(4) = 3^4 \bmod 91 = 81$$
$$f(5) = 3^5 \bmod 91 = 61$$
$$f(6) = 3^6 \bmod 91 = 1$$
$$f(7) = 3^7 \bmod 91 = 3$$
$$f(8) = 3^8 \bmod 91 = 9$$
$$f(9) = 3^9 \bmod 91 = 27$$

and so on. The period, r, of the function $f(a)$ is 6, i.e., for any value a, $f(a) = f(a+6)$. It follows from

$$f(r) = 1 = x^r \bmod N,$$

that, for $x = 3$ and $N = 91$, we have

$$1 = 3^6 \bmod 91.$$

This equation can be rewritten as

$$0 = (3^3 - 1)(3^3 + 1) \bmod 91.$$

Thus $(3^3 - 1) = 26 = 13 \times 2$ and $(3^3 + 1) = 28 = 7 \times 4$ have the prime factors of 91. We can verify that these factors are 13 and 7.

These examples clearly show that the main objective of an algorithm to find the prime factors of N is to find the period r of the function $f(a) = x^a \pmod N$. We can now formulate the general procedure of finding the prime factors p and q of a number N as follows.

Choose a random number x that has no common factors with N and find the period r of the sequence

$$f(a) = x^a \bmod N$$

If the period r is even, we proceed. If it is odd then we start over by choosing a different x and finding the period of the sequence $f(a)$ with the new value of x. If the period r is even then

$$1 = x^r \bmod N$$
$$0 = (x^{r/2} - 1)(x^{r/2} + 1)\bmod N$$

Thus, $(x^{r/2} - 1)$ or $(x^{r/2} + 1)$ must have common factors with N. We factorize the relatively small numbers $(x^{r/2} - 1)$ and $(x^{r/2} + 1)$ and verify which of these prime numbers are factors of N by finding whether N is divisible by these numbers.

The crucial step in finding the prime factors is therefore determining the period r of the function $f(a) = x^a \pmod N$.

16.2 Discrete Quantum Fourier Transform

How can quantum mechanics help in finding the period r of the function $f(a)$? It turns out that a discrete "quantum" Fourier transform helps in this objective. This is a beautiful example where the basic features of quantum mechanics can be employed in solving a very practical problem. The principles of superposition, quantum interference, quantum entanglement, and the probabilistic nature of measurement all come together to achieve this goal.

A quantum Fourier transform is given by the following transformation on $|a\rangle$:

$$|a\rangle \rightarrow \frac{1}{\sqrt{q}} \sum_{c=0}^{q-1} e^{2\pi i \frac{ac}{q}} |c\rangle, \tag{16.5}$$

i.e., a state $|a\rangle$ is prepared in a coherent superposition of all possible states. As a simple example, we assume $q = 8$. Then, for $a = 0$ and 1, we obtain the following transformations:

$$|0\rangle \rightarrow \frac{1}{\sqrt{8}} [|0\rangle + |1\rangle + |2\rangle + |3\rangle + |4\rangle + |5\rangle + |6\rangle + |7\rangle]$$

$$|1\rangle \rightarrow \frac{1}{\sqrt{8}} \left[e^{i0}|0\rangle + e^{i\frac{\pi}{4}}|1\rangle + e^{i\frac{\pi}{2}}|2\rangle + e^{i\frac{3\pi}{4}}|3\rangle + e^{i\pi}|4\rangle + e^{i\frac{5\pi}{4}}|5\rangle + e^{i\frac{3\pi}{2}}|6\rangle + e^{i\frac{7\pi}{4}}|7\rangle \right]$$

In a similar manner, the transformation of the states $|2\rangle \dots |7\rangle$ can be obtained. The important thing is to realize that each amplitude has a well-defined phase. We now address the question how this transformation of the quantum states helps in finding the periodicity of the function $f(a)$.

For this purpose, consider the quantum Fourier transform of an entangled state $|\phi(A, B)\rangle$ between two systems A and B as follows:

$$| \phi(A, B)\rangle = \frac{1}{\sqrt{q}} \sum_{a=0}^{q-1} | a\, f(a)\rangle \rightarrow |\psi(A.B)>= \frac{1}{q} \sum_{a=0}^{q-1}\sum_{c=0}^{q-1} e^{2\pi i \frac{ac}{q}} | c, f(a)\rangle. \qquad (16.6)$$

This is a double sum that contains q^2 terms.

Again, for simplicity's sake, consider the case: $q = 8$. We then obtain

$$| \psi(A, B)\rangle = \frac{1}{8} \sum_{a=0}^{7}\sum_{c=0}^{7} e^{2\pi i \frac{ac}{8}} | c, f(a)\rangle. \qquad (16.7)$$

This double sum can be written explicitly as follows:

$$
\begin{aligned}
| \psi(A, B) = \tfrac{1}{8} | 0\rangle &\big[|f(0)\rangle + |f(1)\rangle + |f(2)\rangle + |f(3)\rangle + |f(4)\rangle + |f(5)\rangle + |f(6)\rangle + |f(7)\rangle\big]\\
+ \tfrac{1}{8} | 1\rangle &\big[e^{i0}|f(0)\rangle + e^{i\frac{\pi}{4}}|f(1)\rangle + e^{i\frac{2\pi}{4}}|f(2)\rangle + e^{i\frac{3\pi}{4}}|f(3)\rangle + e^{i\frac{4\pi}{4}}|f(4)\rangle + e^{i\frac{5\pi}{4}}|f(5)\rangle + e^{i\frac{6\pi}{4}}|f(6)\rangle + e^{i\frac{7\pi}{4}}|f(7)\rangle\big]\\
+ \tfrac{1}{8} | 2\rangle &\big[e^{i0}|f(0)\rangle + e^{i\frac{2\pi}{4}}|f(1)\rangle + e^{i\frac{4\pi}{4}}|f(2)\rangle + e^{i\frac{6\pi}{4}}|f(3)\rangle + e^{i0}|f(4)\rangle + e^{i\frac{2\pi}{4}}|f(5)\rangle + e^{i\frac{4\pi}{4}}|f(6)\rangle + e^{i\frac{6\pi}{4}}|f(7)\rangle\big]\\
+ \tfrac{1}{8} | 3\rangle &\big[e^{i0}|f(0)\rangle + e^{i\frac{3\pi}{4}}|f(1)\rangle + e^{i\frac{6\pi}{4}}|f(2)\rangle + e^{i\frac{\pi}{4}}|f(3)\rangle + e^{i\frac{4\pi}{4}}|f(4)\rangle + e^{i\frac{7\pi}{4}}|f(5)\rangle + e^{i\frac{2\pi}{4}}|f(6)\rangle + e^{i\frac{5\pi}{4}}|f(7)\rangle\big]\\
+ \tfrac{1}{8} | 4\rangle &\big[e^{i0}|f(0)\rangle + e^{i\frac{4\pi}{4}}|f(1)\rangle + e^{i0}|f(2)\rangle + e^{i\frac{4\pi}{4}}|f(3)\rangle + e^{i0}|f(4)\rangle + e^{i\frac{4\pi}{4}}|f(5)\rangle + e^{i0}|f(6)\rangle + e^{i\frac{4\pi}{4}}|f(7)\rangle\big]\\
+ \tfrac{1}{8} | 5\rangle &\big[e^{i0}|f(0)\rangle + e^{i\frac{5\pi}{4}}|f(1)\rangle + e^{i\frac{2\pi}{4}}|f(2)\rangle + e^{i\frac{7\pi}{4}}|f(3)\rangle + e^{i\frac{4\pi}{4}}|f(4)\rangle + e^{i\frac{\pi}{4}}|f(5)\rangle + e^{i\frac{6\pi}{4}}|f(6)\rangle + e^{i\frac{3\pi}{4}}|f(7)\rangle\big]\\
+ \tfrac{1}{8} | 6\rangle &\big[e^{i0}|f(0)\rangle + e^{i\frac{6\pi}{4}}|f(1)\rangle + e^{i\frac{4\pi}{4}}|f(2)\rangle + e^{i\frac{2\pi}{4}}|f(3)\rangle + e^{i0}|f(4)\rangle + e^{i\frac{6\pi}{4}}|f(5)\rangle + e^{i\frac{4\pi}{4}}|f(6)\rangle + e^{i\frac{2\pi}{4}}|f(7)\rangle\big]\\
+ \tfrac{1}{8} | 7\rangle &\big[e^{i0}|f(0)\rangle + e^{i\frac{7\pi}{4}}|f(1)\rangle + e^{i\frac{6\pi}{4}}|f(2)\rangle + e^{i\frac{5\pi}{4}}|f(3)\rangle + e^{i\frac{4\pi}{4}}|f(4)\rangle + e^{i\frac{3\pi}{4}}|f(5)\rangle + e^{i\frac{2\pi}{4}}|f(6)\rangle + e^{i\frac{\pi}{4}}|f(7)\rangle\big],
\end{aligned}
$$
$$(16.8)$$

where we used

$$e^{i(2\pi n+\theta)} = e^{2\pi in}e^{i\theta} = e^{i\theta}\ (n = 0, \pm 1, \pm 2, \cdots). \qquad (16.9)$$

There are $8 \times 8 = 64$ terms in the double summation representing $|\psi(A, B)\rangle$. In Fig. 16.1, the phases of all these 64 terms are plotted. The intersection of the ith horizontal and the jth vertical lines gives the phase contribution of the state $| c = j, f(i)\rangle$ in the sum.

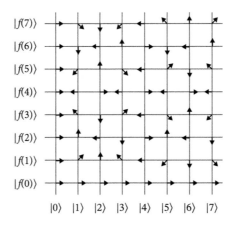

Fig. 16.1 The arrow at each intersection corresponds to the phase of the corresponding term $|c, f(a)\rangle$.

Next assume that the period r of the function $f(a)$ is equal to 2, i.e., $f(a) = f(a+2)$. In this case

$$f(0) = f(2) = f(4)) = f(6),$$
$$f(1) = f(3) = f(5) = f(7).$$

The summation (16.8) can then be rewritten as

$$| \psi(A, B)\rangle = \frac{1}{8} |0\rangle \left[|f(0)\rangle(1 + e^{i0} + e^{i0} + e^{i0}) + |f(1)\rangle(1 + e^{i0} + e^{i0} + e^{i0}) \right]$$

$$+ \frac{1}{8} |1\rangle \left[|f(0)\rangle(1 + e^{i\frac{2\pi}{4}} + e^{i\frac{4\pi}{4}} + e^{i\frac{6\pi}{4}}) + |f(1)\rangle(e^{i\frac{\pi}{4}} + e^{i\frac{3\pi}{4}} + e^{i\frac{5\pi}{4}} + e^{i\frac{7\pi}{4}}) \right]$$

$$+ \frac{1}{8} |2\rangle \left[|f(0)\rangle(1 + e^{i\frac{4\pi}{4}} + e^{i0} + e^{i\frac{4\pi}{4}}) + |f(1)\rangle(e^{i\frac{2\pi}{4}} + e^{i\frac{6\pi}{4}} + e^{i\frac{2\pi}{4}} + e^{i\frac{6\pi}{4}}) \right]$$

$$+ \frac{1}{8} |3\rangle \left[|f(0)\rangle(1 + e^{i\frac{6\pi}{4}} + e^{i\frac{4\pi}{4}} + e^{i\frac{2\pi}{4}}) + |f(1)\rangle(e^{i\frac{3\pi}{4}} + e^{i\frac{\pi}{4}} + e^{i\frac{7\pi}{4}} + e^{i\frac{5\pi}{4}}) \right]$$

$$+ \frac{1}{8} |4\rangle \left[|f(0)\rangle(1 + e^{i0} + e^{i0} + e^{i0}) + |f(1)\rangle(e^{i\frac{4\pi}{4}} + e^{i\frac{4\pi}{4}} + e^{i\frac{4\pi}{4}} + e^{i\frac{4\pi}{4}}) \right]$$

$$+ \frac{1}{8} |5\rangle \left[|f(0)\rangle(1 + e^{i\frac{2\pi}{4}} + e^{i\frac{4\pi}{4}} + e^{i\frac{6\pi}{4}}) + |f(1)\rangle(e^{i\frac{5\pi}{4}} + e^{i\frac{7\pi}{4}} + e^{i\frac{\pi}{4}} + e^{i\frac{3\pi}{4}}) \right]$$

$$+ \frac{1}{8} |6\rangle \left[|f(0)\rangle(1 + e^{i\frac{4\pi}{4}} + e^{i0} + e^{i\frac{4\pi}{4}}) + |f(1)\rangle(e^{i\frac{6\pi}{4}} + e^{i\frac{2\pi}{4}} + e^{i\frac{6\pi}{4}} + e^{i\frac{2\pi}{4}}) \right]$$

$$+ \frac{1}{8} |7\rangle \left[|f(0)\rangle(1 + e^{i\frac{6\pi}{4}} + e^{i0} + e^{i\frac{2\pi}{4}}) + |f(1)\rangle(e^{i\frac{7\pi}{4}} + e^{i\frac{5\pi}{4}} + e^{i\frac{3\pi}{4}} + e^{i\frac{\pi}{4}}) \right].$$

$$(16.10)$$

It follows from using the relation

$$e^{i(\pi+\theta)} = e^{i\pi} e^{i\theta} = -e^{i\theta} \tag{16.11}$$

that all the terms involving states $|1\rangle, |2\rangle, |3\rangle, |5\rangle, |6\rangle$, and $|7\rangle$ vanish. Only the terms involving $|0\rangle$ and $|4\rangle$ survive and the end result is that, for $f(a) = f(a+2)$,

$$| \psi(A, B)\rangle = \frac{1}{2}(|0, f(0)\rangle + |0, f(1)\rangle + |4, f(0)\rangle - |4, f(1)\rangle). \tag{16.12}$$

Thus the outcome of state A is $|0\rangle$ or $|4\rangle$ with equal probability.[2] A clever application of quantum Fourier transform led to a cancellation of most of the terms; only few survived. We can see pictorially from Fig. 16.1 that only the states $|0\rangle$ or $|4\rangle$ survive when $f(a) = f(a+2)$.

Next we ask the obvious question: What is the significance of the states $|0\rangle$ and $|4\rangle$ that survived? What have these states to do with finding the period r?

---

[2] Following the discussion in Section 2.4, the probability for the outcome $|0\rangle$ is

$$P(0) = |\langle 0, f(0)|\psi(A, B)\rangle|^2 + |\langle 0, f(1)|\psi(A, B)\rangle|^2 = 1/2.$$

Similarly the probability for the outcome $|4\rangle$ is

$$P(4) = |\langle 4, f(0)|\psi(A, B)\rangle|^2 + |\langle 4, f(1)|\psi(A, B)\rangle|^2 = 1/2.$$

In general, for $f(a) = f(a + r)$, the only $|c\rangle$ states that survive are

$$|c_0\rangle = |0\rangle, |q/r\rangle, |2q/r\rangle, \cdots, |(r-1)q/r\rangle.. \tag{16.13}$$

All the others vanish. In our example, $q = 8$ and $r = 2$, therefore only surviving states are $|0\rangle$ and $|4\rangle$. Due to the probabilistic nature of quantum mechanics, these outcomes are equally probable.[3]

This is an amazing result and is a consequence of quantum interference – some amplitudes interfere destructively and get cancelled and the others interfere constructively and survive. Thus, if the state $|\psi(A, B)\rangle$ of the form (16.6) is prepared such that $f(a)$ is periodic with an unknown period r, the outcome of the measurement of the state $|c\rangle$ carries the information about r. Even for very large q, if we make a measurement of the state $|c\rangle$ in identically prepared systems, we should get the outcomes $|0\rangle, |q/r\rangle, |2q/r\rangle, ..., |(r-1)q/r\rangle$ with equal probability. Thus, after a modest number of measurements, we should be able to determine the value of r. This, as shown below, is the essence of Shor's algorithm.

16.3 Shor's Algorithm

Armed with a mathematical method for finding the prime factors of a number N and the ability of the quantum Fourier transform to find the period of a function, a simplified form of Shor's algorithm to factorize a number quantum mechanically with a much faster speed-up can be described.

The quantum computer is prepared in two registers A and B. The registers are essentially two separate sets of qubits. As an example, if our qubits are two-level atoms, then register A would correspond to one set of atoms and register B to another set of atoms. Each register consists of qubits initialized to values "0", i.e., all qubits in both registers are initially in states $|0\rangle$. Thus if we have n qubits in register A and m qubits in register B, then the initial state is

$$|A_0; B_0\rangle = |0_1, 0_2 \cdots 0_n; 0_1, 0_2 \cdots 0_m\rangle. \tag{16.14}$$

For the sake of simplicity we denote it by $|0; 0\rangle$, i.e.,

$$|A_0; B_0\rangle = |0; 0\rangle. \tag{16.15}$$

Register A is used to hold the arguments of function $f(a)$ whose unknown period is sought whereas register B is used to store the value of $f(a)$.

Now consider the sequence

$$f(0), f(1), f(2), ..., f(q-1)$$

---

[3] This result is strictly true if q is divisible by r. In case this condition is not satisfied, the outcome of the measurement of the state $|c\rangle$ may have contributions other than those listed in Eq. (16.13). However it can be shown that the probability of getting those states is significantly lower. In order to keep things simple, we assume in the following that the condition, q is divisible by r, is satisfied.

where $q = 2^k$. Register A stores k qubits so that it can store all the values of a, from 0 to 2^k. Here we recall that k qubits can store 2^k numbers. Register B should have enough qubits to hold the largest value contained in the above sequence.

The Shor algorithm requires the following four steps.

1. On each qubit of register A, perform the Hadamard transformation H, i.e.,

$$H\,|0\rangle = \frac{1}{\sqrt{2}}(|0\rangle + |1\rangle). \tag{16.16}$$

The resulting state of register A is the superposition of all possible 2^k states:

$$|0,0\rangle \Rightarrow \frac{1}{\sqrt{q}} \sum_{a=0}^{q-1} |a,0\rangle. \tag{16.17}$$

Thus, we get every possible bit of length k in register A.

2. Map the state $|a, 0\rangle$ to the state $|a, f(a)\rangle$ for any input. The number of qubits required for register B must be at least sufficient to store the longest result $f(a)$ for any of these computations:

$$\frac{1}{\sqrt{q}} \sum_{a=0}^{q-1} |a,0\rangle \Rightarrow \frac{1}{\sqrt{q}} \sum_{a=0}^{q-1} |a,f(a)\rangle. \tag{16.18}$$

This is a highly entangled state!!!

In order to illustrate these two steps, consider the case when $N = 91$ is factorized. In the factorization algorithm, discussed above, choose $x = 3$. If only 8 terms are retained, then only three qubits are needed in register A. The Hadamard transformation on each qubit leads to

$$|0,0,0\rangle \Rightarrow \frac{1}{\sqrt{2}}(|0\rangle + |1\rangle)\frac{1}{\sqrt{2}}(|0\rangle + |1\rangle)\frac{1}{\sqrt{2}}(|0\rangle + |1\rangle),$$

$$= \frac{1}{\sqrt{8}}(|\,0,0,0\rangle + |\,0,0,1\rangle + |\,0,1,0\rangle + |\,0,1,1\rangle$$

$$+ |\,1,0,0\rangle + |\,1,0,1\rangle + |\,1,1,0\rangle + |\,1,1,1\rangle$$

$$= \frac{1}{\sqrt{8}}\sum_{a=0}^{7} |\,a\rangle. \tag{16.19}$$

Here, the qubit binary states can be translated into the decimal number states. For example, $|0,0,0\rangle \equiv |0\rangle, |0,0,1\rangle \equiv |1\rangle, |0,1,0\rangle \equiv |2\rangle, |0,1,1\rangle \equiv |3\rangle$, and so on.

In the next step, the entangled state

$$\frac{1}{\sqrt{8}}\sum_{a=0}^{7} |\,a,f(a)\rangle = \frac{1}{\sqrt{8}}(|0,1\,\rangle + |\,1,3\rangle + |\,2,9\,\rangle + |\,3,27\rangle + |\,4,81\,\rangle$$

$$+ |\,5,61\rangle + |\,6,1\,\rangle + |\,7,3\rangle \tag{16.20}$$

is prepared. As the maximum value of $f(a)$ is 81, we need minimum seven qubits to store the values for $f(a)$ for all values of a. The decimal states in Eq. (16.20) can be converted to binary states: $|0,1\rangle \equiv |000; 0000001\rangle, |1,3\rangle \equiv |001; 0000011\rangle, |2,9\rangle \equiv |010; 0001001\rangle$, and so on. Thus the joint state for the entire A and B registers consist of an entanglement of 10 qubits:

$$\frac{1}{\sqrt{8}} \sum_{a=0}^{7} |a,f(a)\rangle = \frac{1}{\sqrt{8}}(|000;0000001\rangle + |001;0000011\rangle$$
$$+ |010;0001001\rangle + |011;0001011\rangle \qquad (16.21)$$
$$+ |100,1010001\rangle + |101,0111101\rangle$$
$$+ |110,0000001\rangle + |111,0000011\rangle).$$

Here we see that, even for a small number like 91, the requirement for an entangled state is very severe. This step is a big stumbling block for the practical realization of the Shor algorithm for factorizing a realistically large number.

3. Perform a discrete "quantum" Fourier transform on the first register

$$|a\rangle \Rightarrow \frac{1}{\sqrt{q}} \sum_{c=0}^{q-1} e^{2\pi i \frac{ac}{q}} |c\rangle \qquad (16.22)$$

so that

$$\frac{1}{\sqrt{q}} \sum_{a=0}^{q-1} |a,f(a)\rangle \Rightarrow \frac{1}{q} \sum_{a=0}^{q-1} \sum_{c=0}^{q-1} e^{2\pi i \frac{ac}{q}} |c,f(a)\rangle. \qquad (16.23)$$

4. Finally, carry out the measurement. We retrieve the output from the quantum computer by measuring the state of all qubits in register A. As seen in earlier discussion of the quantum Fourier transform, if $f(a) = f(a + r)$, the sum over a yields constructive interference from coefficients

$$e^{2\pi i \frac{ac}{q}}$$

only when c/q is a multiple of $1/r$. All other values of c/q produce destructive interference.

Each measurement will give an outcome for one of the allowed values for c, e.g.,

$$c = \frac{q}{r}n, n = 0, 1, 2, \cdots, \qquad (16.24)$$

i.e., a measurement of c gives an outcome which is either 0 or q/r or $2q/r$ and so on. If we make a large number of measurements on the first register on identically prepared systems, the histogram for c/q as shown in Fig. 16.2 is created. After a sufficient number of measurements, we should be able to infer the value of the period r. This completes Shor's algorithm. It is interesting to note that a quantum Fourier transform played a key role in finding the periodicity of the function $f(a)$.

Despite the mathematical simplicity, it is still a very difficult problem to experimentally realize the Shor algorithm to factorize a realistically large number. In addition to the usual

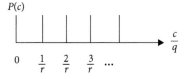

Fig. 16.2 The probability of the outcome $|c\rangle$ in the first register is non-vanishing only for those c that satisfy the condition $c = qn/r$ where n is an integer.

problem associated with decoherence discussed in the last chapter, there are issues coming up with possible schemes to generate the massive entangled states of the form in Eq. (16.23) for very large q. Nevertheless Shor's algorithm has opened up the possibility of factoring large numbers in a considerably shorter time than previously. We recall that the most efficient presently known algorithm can factorize a 1000-digit number in a time equal to almost the age of the universe. Shor's algorithm, when experimentally feasible, would be able to accomplish this task only in few million steps.

16.4 Quantum Shell Game

Another important application of a quantum computing algorithm is to find an object in an unsorted database. This algorithm is presented in the next section. Here a shell game that we may have played in our childhood is described and see how a quantum computer can help in playing this game and winning each time in an almost unbelievable way.

Consider a shell game in which four inverted cups or nutshells are moved about with only one of the shells hiding a pea underneath. The contestants must spot the shell with the pea underneath (Fig. 16.3). The search process requires inverting each shell one by one to find the pea. If we are lucky, we may find the pea under the first shell and win. The probability of this happening is only 25%. If we are not lucky, we may have to flip all four before we find the shell with the pea. On the average it requires more than two searches to find the pea.

Can we find the pea in only one try with certainty every time? Conventional wisdom tells us that this is simply not possible. However the situation is different in the quantum shell game. In a quantum shell game, where the shell is replaced by a quantum state and the pea is replaced by an "inverted" target state, we can find the target state with certainty only with one measurement.

For four shells, we need two qubits with $|0_1, 0_2\rangle \equiv |0\rangle, |0_1, 1_2\rangle \equiv |1\rangle, |1_1, 0_2\rangle \equiv |2\rangle, |1_1, 1_2\rangle \equiv |3\rangle$. More formally the whole database can be represented as a state of the form

$$|S\rangle = \frac{1}{2} [|0_1, 0_2\rangle + |0_1, 1_2\rangle + |1_1, 0_2\rangle + |1_1, 1_2\rangle].$$ (16.25)

This state can be depicted in a pictorial form as follows:

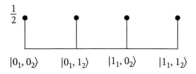

$|0_1, 0_2\rangle \qquad |0_1, 1_2\rangle \qquad |1_1, 0_2\rangle \qquad |1_1, 1_2\rangle$

Fig. 16.3 In the shell game, contestants must spot the shell with a pea underneath.

Here the horizontal axis represents the two-qubit states $| n, m \rangle$ $(n, m = 1, 2)$. The vertical axis represents the amplitudes. In the case of the state $|S\rangle$, all the amplitudes are equal to 1/2. Each term in $|S\rangle$ represents a shell.

Suppose, without us knowing, someone flips the sign of one the states (say $| 1_1, 0_2 \rangle$) such that the resulting state is

$$|F\rangle = \frac{1}{2} \left[|0_1, 0_2\rangle + |0_1, 1_2\rangle - |1_1, 0_2\rangle + |1_1, 1_2\rangle \right]. \tag{16.26}$$

Pictorially it is represented as:

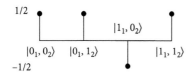

The question is whether we can find this flipped target state with certainty in just one measurement?

The answer remarkably is YES and is the basis of one of the most famous quantum computing algorithms – the so-called Grover's algorithm. Here, we show how the quantum shell game is played.

There are three steps: In the preparation step, starting with two qubits in the $|0, 0\rangle$ state, the state $|S\rangle$ is prepared; in the flipping stage, a target state such as $|10\rangle$ acquires a phase shift and the new state is $|F\rangle$; and finally, in the measurement stage we find in just one measurement of the two qubits the target state. All these steps can be carried out using the quantum gates discussed in the last chapter.

Initially, the two qubits are prepared in the state $|0_1, 0_2\rangle$. If we apply the Hadamard gates H_1 and H_2 to the two qubits, we obtain

$$H_1 H_2 |0_1, 0_2\rangle = H_1 | 0_1 \rangle H_2 | 0_2 \rangle = \frac{1}{\sqrt{2}}(|0_1\rangle + |1_1\rangle)\frac{1}{\sqrt{2}}(|0_2\rangle + |1_2\rangle) \tag{16.27}$$

This step is equivalent to preparing the data basis.

In the second step, a π-phase shift is implemented for a target state which can be any one of the four states: $|0_1, 0_2\rangle$, $|0_1, 1_2\rangle$, $|1_1, 0_2\rangle$, or $|1_1, 1_2\rangle$. This can be done by first applying the quantum phase gate, Q_π, on $|S\rangle$. We recall that the quantum phase gate leaves all the states undisturbed except $|1_1, 1_2\rangle$, for which $Q_\pi | 1_1, 1_2 \rangle = - | 1_1, 1_2\rangle$, Thus

$$|C_{11}\rangle = Q_\pi |S\rangle = \frac{1}{2} \left[|0_1, 0_2\rangle + |0_1, 1_2\rangle + |1_1, 0_2\rangle - |1_1, 1_2\rangle \right]. \tag{16.28}$$

If the target state is $|1_1, 1_2\rangle$, then the desired state is ready. If however the target state is (say) $|1_1, 0_2\rangle$, then we apply the X_2-gate on the second qubit making the transformation $| 1_2 \rangle \leftrightarrow | 0_2 \rangle$. The resulting state is:

$$|C_{10}\rangle = X_2 Q_\pi |S\rangle = \frac{1}{2} \left[|0_1, 0_2\rangle + |0_1, 1_2\rangle - |1_1, 0_2\rangle + |1_1, 1_2\rangle \right]. \tag{16.29}$$

Similarly,

$$|C_{01}\rangle = X_1 Q_\pi |S\rangle = \frac{1}{2} \left[|0_1, 0_2\rangle - |0_1, 1_2\rangle + |1_1, 0_2\rangle + |1_1, 1_2\rangle \right], \tag{16.30}$$

$$|C_{00}\rangle = X_1 X_2 Q_\pi \, |S\rangle = \frac{1}{2} \left[-|0_1, 0_2\rangle + |0_1, 1_2\rangle + |1_1, 0_2\rangle + |1_1, 1_2\rangle \right]. \tag{16.31}$$

Thus, any desired state can be flipped by a combination of operators X_i and Q_π. The resulting states are C_{ij} with $i, j = 0, 1$.

Now the challenge!! Can we do just one measurement on these qubits to find out which state is flipped? This operation is accomplished by inverting about the mean as shown in Fig. 16.4. The rule of the game is that we do not know the target state. So the operation should be blind to the state that is flipped.

Inversion about the mean operation (denoted by arrows) takes the marked state $|1_1, 0_2\rangle$ to unit amplitude, while the other states are taken to zero amplitude. It turns out that an application of the following sequence of logic gates

$$N = H_1 H_2 Q_\pi H_1 H_2 X_1 X_2 \tag{16.32}$$

can accomplish the task. We can show that

$$N \, |C_{11}\rangle = |\, 1_1, 1_2\rangle, \tag{16.33}$$

$$N \, |C_{01}\rangle = |\, 0_1, 1_2\rangle, \tag{16.34}$$

$$N \, |C_{10}\rangle = |\, 1_1, 0_2\rangle, \tag{16.35}$$

$$N \, |C_{00}\rangle = |\, 0_1, 0_2\rangle. \tag{16.36}$$

Thus, by measuring the output qubits we can determine uniquely which qubit is flipped. For example if we find the two qubits in the state $|1_1, 0_2\rangle$, we know that the target state was $|C_{01}\rangle$ as given by Eq. (16.30).

A full circuit diagram for the implementation of the shell game is shown in Fig. 16.5. The two qubits are initially prepared in the state $|0_1, 0_2\rangle$. The database is prepared by applying the

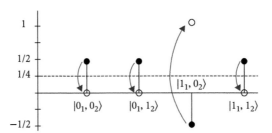

Fig. 16.4 Inversion about the mean. The inversion of all the states about the mean value of the amplitude 1/4 takes the amplitude of all the states to zero except the flipped state $|1_1, 0_2\rangle$ whose amplitude becomes 1.

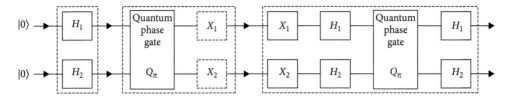

Fig. 16.5 Circuit diagram to implement the quantum shell game.

Hadamard gates H_1 and H_2 to the two qubits as shown in the first dotted box. Next, the target state is flipped as shown in the second dotted box by applying the quantum phase gate Q_π and a combination of Hadamard gates H_1 and H_2 as discussed above. The quantum gates in the third dotted box represent the operation of inversion about the mean as given by the operator N in Eq. (16.32). Finally the two qubits are measured.

16.5 Searching an Unsorted Database

We now discuss Grover's algorithm, for the search of an object in an unsorted database. The searching problem can be stated as follows: Suppose there is a database consisting of N items out of which just one item satisfies a given condition. We would like to retrieve that item.

There are two kinds of databases, a sorted database and an unsorted database. Consider a telephone book with names listed in alphabetical order with their telephone numbers. If we are given a name and want to find the telephone number, we can search alphabetically to recover the telephone number. This is an example of sorted database. However, what if we are given the telephone number and we would like to know the name of the person it belongs to? This is an example of an unsorted database as the telephone numbers in the book are listed completely randomly. In the search for an item in an unsorted database of N items, we would look at each item one by one. If we are lucky, the first item can be the searched item. However, there is a possibility that the searched item is the last on our search list, thus requiring N searches. On average, it may require $N/2$ searches before we find the desired item. In the example of the telephone book, if we are searching for the name of the person with a certain telephone number, we may require, on average, a search through half the telephone book before we see the name matching the telephone number. This is the essence of what we call the classical search.

Next, we ask the question: Can quantum mechanics provide a speed-up? Can we do better than $N/2$? Here, we show that, for a certain class of search problems, the searching process of an unsorted database in a quantum computer may require only $\sqrt{N}$ searches instead of $N/2$ for large databases.

In the previous section (16.4) on the quantum shell game, we saw an example of a database of four items described by quantum states $|0_1, 0_2\rangle \equiv |0\rangle, |0_1, 1_2\rangle \equiv |1\rangle, |1_1, 0_2\rangle \equiv |2\rangle, |1_1, 1_2\rangle \equiv |3\rangle$. One of the items had an inverted phase and we searched that item in just one measurement. To illustrate how we may require only about $\sqrt{N}$ searches instead of $N/2$ for large databases, consider the example of sixteen items, $N = 16$, and follow the same steps as in the shell game.

(i) In the first step, a database is prepared as a superposition of all 16 states. This can be done by applying the Hadamard gate H to each of the four qubits initially in states $|0_1\rangle, |0_2\rangle, |0_3\rangle$, and $|0_4\rangle$, leading to

$$S = H_1 \,|\, 0_1 \,\rangle H_2 \,|0_2\rangle \, H_3 \,|0_3\,\rangle H_4 \,|0_4\rangle$$

$$= \frac{1}{\sqrt{2}}(|0_1\rangle + |1_1\rangle)\frac{1}{\sqrt{2}}(|0_2\rangle + |1_2\rangle)\frac{1}{\sqrt{2}}(|0_3\rangle + |1_3\rangle)\frac{1}{\sqrt{2}}(|0_4\rangle + |1_4\rangle)$$

$$= \frac{1}{4}\left[|0000\rangle + |0001\rangle + \cdots\cdots\cdots + |1111\rangle\right], \tag{16.37}$$

where the notation $|\,0000\,\rangle \equiv |\,0_1\rangle\,|\,0_2\,\rangle|\,0_3\rangle\,|\,0_4\rangle$ etc. is used. A more convenient notation is to represent this superposition of all the 16 possible states as

$$\frac{1}{4}(1,1,1,1,1,1,1,1,1,1,1,1,1,1,1,1). \tag{16.38}$$

Here each number in the bracket represents the amplitude of the corresponding state. As an example

$$(c_0,c_1,c_2,c_3) \equiv c_0\,|00\,\rangle + c_1\,|01\rangle + c_2\,|10\,\rangle + c_3\,|11\rangle.$$

Since, the coefficients of all the states, $|0000\rangle$, $|0001\rangle$, $\cdots$, are equal to 1/4 for the superposition state (16.37), it is represented by Eq. (16.38) in the simplified notation.

In the next step, an *oracle* inverts the amplitude of the target state (say) $|0011\rangle$ thus changing it from $+1$ to -1. This inversion operation leads to

$$\frac{1}{4}(1,1,1,-1,1,1,1,1,1,1,1,1,1,1,1,1). \tag{16.39}$$

Note that if we make a measurement on the qubits at this stage, there is equal probability for each state, including the target state $|0011\rangle$, equal to $1/16 = 0.0625$. Our objective is to find the target state $|0011\rangle$.

Also note that the mean amplitude is the sum of all the amplitudes of states $|\,0000\rangle$, $|\,0000\rangle$, $\cdots\cdots|\,1111\rangle$ divided by 16, i.e.,

$$x_0 = \frac{1}{16}\left(\frac{1}{4}+\frac{1}{4}+\frac{1}{4}-\frac{1}{4}+\frac{1}{4}+\frac{1}{4}+\frac{1}{4}+\frac{1}{4}+\frac{1}{4}+\frac{1}{4}+\frac{1}{4}+\frac{1}{4}+\frac{1}{4}+\frac{1}{4}+\frac{1}{4}+\frac{1}{4}\right)$$

$$= \frac{1}{16}\left(\frac{14}{4}\right) = \frac{7}{32}. \tag{16.40}$$

Our measurement, as in the shell game, is an inversion about the mean. If the probability amplitude of a particular state is x, then an inversion about the mean x_0, leads to new amplitude

$$x_0 - (x - x_0) = 2x_0 - x. \tag{16.41}$$

Thus

$$\frac{1}{4} \rightarrow 2\cdot\frac{7}{32} - \frac{1}{4} = \frac{6}{32} = \frac{3}{16}, \tag{16.42}$$

$$-\frac{1}{4} \rightarrow 2\cdot\frac{7}{32} + \frac{1}{4} = \frac{22}{32} = \frac{11}{16}. \tag{16.43}$$

The resulting state is

$$\frac{1}{16}(3,3,3,11,3,3,3,3,3,3,3,3,3,3,3,3). \tag{16.44}$$

If, at this point, a measurement is made, the probability of finding $|0011\rangle$ is $(11/16)^2 = 0.39$. This is a substantial increase from the original probability of $(1/4)^2 = 0.0625$. However, this is still far from a unit probability. Therefore refrain from making a measurement here and repeat the process once more.

(ii) In the next cycle, our starting state is (16.44). As before, the *oracle* inverts the amplitude of the target state $|0011\rangle$ from 11/16 to $-11/16$. The resulting state is

$$\frac{1}{16}(3, 3, 3, -11, 3, 3, 3, 3, 3, 3, 3, 3, 3, 3, 3, 3). \tag{16.45}$$

The mean amplitude is

$$\frac{1}{(16)^2}[3 \times 15 - 11] = \frac{17}{128}. \tag{16.46}$$

The operation corresponding to the inversion about the mean leads to

$$\frac{3}{16} \rightarrow 2 \cdot \frac{17}{128} - \frac{3}{16} = \frac{5}{64}, \tag{16.47}$$

$$-\frac{11}{16} \rightarrow 2 \cdot \frac{17}{128} + \frac{11}{16} = \frac{61}{64}, \tag{16.48}$$

and the final state after this step is

$$\frac{1}{64}(5, 5, 5, 61, 5, 5, 5, 5, 5, 5, 5, 5, 5, 5, 5, 5). \tag{16.49}$$

The probability of finding the target state $|0011\rangle$ has now gone up to $(61/64)^2 = 0.91$. This is a very large probability. Can we do better? To find out we repeat the process again.

(iii) The starting state is (16.49). The same steps (*oracle* inverts the target state $|0011\rangle$ and we invert about the mean) lead to the new state

$$\frac{1}{256}(-13, -13, -13, 251, -13, -13, -13, -13, -13, -13, -13, -13, -13, -13, -13, -13). \tag{16.50}$$

The probability of finding $|0011\rangle$ has increased to $(251/256)^2 = 0.96$. Can we improve further?

(iv) The same steps as above lead to

$$\frac{1}{1024}(-171, -171, -171, 781, -171, -171, -171, -171,$$
$$- 171, -171, -171, -171, -171, -171, -171, -171). \tag{16.51}$$

The probability of finding the target state $|0011\rangle$ has now gone down to $(781/1024)^2 = 0.58$. This indicates that the best strategy was to stop after 3 steps when the probability of detection of the target state was 96 percent.

But how do we know when to stop? A careful analysis indicates that the number of measurements required for the search of a target state in N states is the integer closest to $\pi\sqrt{N}/4$. In our example of $N = 16$, this number is 3. We should therefore stop after 3 steps.

Thus the Grover search algorithm allows the search for a target state in a superposition of N states in the order of $\sqrt{N}$ steps.

Problems

16.1 Find the period of the function

$$f(a) = 5^a \bmod 4069.$$

Use this result to find the factors of 4069.

16.2 Consider the state

$$|\psi(A, B)\rangle = \frac{1}{8} \sum_{a=0}^{7} \sum_{c=0}^{7} e^{2\pi i \frac{ac}{8}} \, | c, f(a)\rangle.$$

Show that, for a function $f(a)$ with period $r = 4$, the only allowed values of c are 0, 2, 4, and 6.

16.3 Consider the state

$$|\psi(A, B)\rangle = \frac{1}{16} \sum_{a=0}^{15} \sum_{c=0}^{15} e^{2\pi i \frac{ac}{16}} \, | c, f(a)\rangle.$$

Show that, for a function $f(a)$ with period $r = 4$, the only allowed values of c are 0, 4, 8, and 16.

16.4 Discuss Shor's algorithm to factorize 15 by choosing $x = 11$ in Eq. (16.1). What is the period of the function $f(a)$? What is the minimum number of qubits required for the implementation of the algorithm?

16.5 Show explicitly that

$$N \, | C_{11}\rangle = |1_1, 1_2\rangle,$$
$$N \, | C_{01}\rangle = | 0_1, 1_2\rangle,$$
$$N \, | C_{10}\rangle = |1_1, 0_2\rangle,$$
$$N \, | C_{00}\rangle = | 0_1, 0_2\rangle,$$

where the states $| C_{ij}\rangle (i, j = 0, 1)$ are given by Eqs. (16.28)–(16.31) and

$$N = H_1 H_2 Q_\pi H_1 H_2 X_1 X_2.$$

16.6 How many steps are required for the implementation of Grover's algorithm for $N = 32$? What is the final success probability?

BIBLIOGRAPHY

P. Shor, in Proceedings of the 35th Annual Symposium on Foundations of Computer Science, Santa Fe, NM, Edited by S. Goldwasser, (IEEE Computer Society Press, New York, 1994), p. 124.

L. K. Grover, *Quantum mechanics helps in searching for a needle in a haystack*, Physical Review Letters **79**, 325 (1997).

E. Farhi and S. Gutmann, *Analog analogue of a digital quantum computation*, Physical Review A 57, 2403 (1998).

M. O. Scully and M. S. Zubairy, *Quantum optical implementation of Grover's algorithm*, Proceedings of the National Academy of Sciences (USA) **98**, 9490 (2001).

A. Muthukrishnan, M. Jones, M. O. Scully, and M. S. Zubairy, *Quantum shell game: finding the hidden pea in a single attempt*, Journal of Modern Optics **16**, 2351 (2004).

M. A. Nielson and I. L. Chuang, *Quantum Computation and Quantum Information* (Cambridge University Press 2000).

S. M. Barnett, *Quantum Information* (Oxford University Press 2009).

S. Stenholm and K.-A. Suominen, *Quantum Approach to Informatics* (John Wiley 2005).

P. Lambropoulos and D. Petrosyan, *Fundamentals of Quantum Optics and Quantum Information* (Springer 2007).

The Schrödinger Equation

17

The Schrödinger Equation

Felix Bloch, the first Ph.D. student of Werner Heisenberg and the winner of the 1952 Nobel Prize, was a student at the Swiss Federal Institute of Technology (ETH) in Zurich in 1925. In his reminiscences on the occasion of the 50th anniversary of quantum mechanics, he described the birth of the Schrödinger equation in these words (Physics Today, December 1976):

> Once at the end of a colloquium I heard Debye saying something like: "Schrödinger, you are not working right now on very important problems anyway. Why don't you tell us some time about that thesis of de Broglie, which seems to have attracted some attention." So, in one of the next colloquia, Schrödinger gave a beautifully clear account of how de Broglie associated a wave with a particle and how he could obtain the quantization rules of Niels Bohr . . . . by demanding that an integer number of waves should be fitted along a stationary orbit. When he had finished, Debye casually remarked that he thought this way of talking was rather childish . . . . to deal properly with waves, one had to have a wave equation. It sounded quite trivial and did not seem to make a great impression, but Schrödinger evidently thought a bit more about the idea afterwards. Just a few weeks later he gave another talk in the colloquium which he started by saying: "My colleague Debye suggested that one should have a wave equation; well, I have found one!"

The equation Schrödinger found is what we know as the "Schrödinger equation." This is one of the most important equations of physics, and perhaps science, of all time. This is an equation on a par with Newton's equation $F = ma$ in terms of importance and impact. The Schrödinger equation provides the tools to tackle most problems in physics.

In this chapter, the Schrödinger equation is derived from the de Broglie wave description of matter. A solution of this equation is presented for some of the simplest problems and illustrates some novel and highly counterintuitive effects based on these solutions.

But first a warning! Unlike the previous 16 chapters, this chapter assumes a knowledge of basic calculus. This involves a knowledge of the differentiation and integration of simple functions. In some places we present the results without explicit calculation. The physical meaning of the results should however be clear, even if the mathematical derivation is not.

17.1 The Schrödinger Equation in One Dimension

As seen in Chapter 7, de Broglie waves describe particles in terms of matter waves. However, it is not clear from de Broglie's formalism how these waves are produced and how they evolve. Schrödinger derived the governing wave equation for these waves in close analogy with a description of classical mechanics as formulated by the nineteenth century Irish mathematician

Quantum Mechanics for Beginners: With Applications to Quantum Communication and Quantum Computing. M. Suhail Zubairy.
© M. Suhail Zubairy 2020. Published in 2020 by Oxford University Press. DOI: 10.1093/oso/9780198854227.001.0001

William Hamilton. Through a simple approach, the Schrödinger equation is "derived" directly from the de Broglie postulate.

So, how can we derive it from the de Broglie wave description of particles? First, the discussion is restricted to one-dimensional problems and then generalized to three dimensions. Central to this equation is the concept of the wavefunction $\psi(x,t)$ that contains the information about the position and the momentum of the particle at time t. The object is to derive a dynamical equation for $\psi(x,t)$.

From the discussion of waves in Chapter 4, recall that a wave of frequency ν and wavevector k propagating in the x-direction is described by

$$\psi(x, t) = Ae^{i(kx-\nu t)}, \tag{17.1}$$

where A is an amplitude. The function $\psi(x,t)$ can be differentiated with respect to x and t as follows:

$$\frac{\partial \psi(x, t)}{\partial x} = ikAe^{i(kx-\nu t)}, \tag{17.2}$$

$$\frac{\partial^2 \psi(x, t)}{\partial x^2} = -k^2 Ae^{i(kx-\nu t)}, \tag{17.3}$$

$$\frac{\partial \psi(x, t)}{\partial t} = -i\nu Ae^{i(kx-\nu t)}, \tag{17.4}$$

$$\frac{\partial^2 \psi(x, t)}{\partial t^2} = -\nu^2 Ae^{i(kx-\nu t)}. \tag{17.5}$$

If $k = \nu/c$, as we learned in Chapter 4, we obtain a wave equation of the form

$$\frac{\partial^2 \psi(x, t)}{\partial x^2} - \frac{1}{c^2}\frac{\partial^2 \psi(x, t)}{\partial t^2} = 0. \tag{17.6}$$

This is the wave equation for the propagation of light waves in free space at the speed of light c. The important step in the derivation of this equation is the relation

$$\nu = ck. \tag{17.7}$$

Such a relation between the frequency and the wave number is called the dispersion relation.

For a particle like an electron, the corresponding relation between frequency ν and wavenumber k is very different. According to de Broglie's description, the momentum of the particle is described in terms of the de Broglie wavelength, λ, and the corresponding wavenumber k as

$$p = \frac{h}{\lambda} = \hbar k. \tag{17.8}$$

Here we used $k = 2\pi/\lambda$ and $\hbar = h/2\pi$. The kinetic energy, E, is related to the momentum, p, via the classical equation, $E = p^2/2m$. However, Einstein showed that the energy can be related to frequency ν via $E = \hbar\nu$. Therefore the wave–particle duality implies

$$E = \hbar\nu = \frac{p^2}{2m}. \tag{17.9}$$

Combining Eqs. (17.8) and (17.9), we obtain

$$v = \frac{\hbar k^2}{2m}.$$
(17.10)

With these expressions, we can write the exponent in Eq. (17.1) as follows:

$$i(kx - vt) = \frac{i}{\hbar}(\hbar kx - \hbar vt) = \frac{i}{\hbar}(px - Et) = \frac{i}{\hbar}\left(px - \frac{p^2}{2m}t\right).$$
(17.11)

The wavefunction of the de Broglie wave is therefore given by

$$\psi(x, t) = Ae^{\frac{i}{\hbar}\left(px - \frac{p^2}{2m}t\right)}.$$
(17.12)

The function $\psi(x,t)$ can be differentiated with respect to x and t as before and we obtain:

$$\frac{\partial \psi(x, t)}{\partial x} = \frac{i}{\hbar}pAe^{\frac{i}{\hbar}\left(px - \frac{p^2}{2m}t\right)},$$
(17.13)

$$\frac{\partial^2 \psi(x, t)}{\partial x^2} = -\frac{p^2}{\hbar^2}Ae^{\frac{i}{\hbar}\left(px - \frac{p^2}{2m}t\right)},$$
(17.14)

$$\frac{\partial \psi(x, t)}{\partial t} = -i\frac{p^2}{2m\hbar}Ae^{\frac{i}{\hbar}\left(px - \frac{p^2}{2m}t\right)}.$$
(17.15)

The resulting wave equation for $\psi(x,t)$ is

$$-\frac{\hbar^2}{2m}\frac{\partial^2 \psi(x, t)}{\partial x^2} = i\hbar\frac{\partial \psi(x, t)}{\partial t}.$$
(17.16)

This is the Schrödinger equation for a free particle like an electron (or a baseball...) moving along the x-axis.

Recall that the corresponding equation for a free particle (force $F = 0$) is $ma = 0$ where $a = d^2x/dt^2$ is the acceleration. Therefore the equation of motion of a free particle in Newtonian mechanics is

$$m\frac{d^2x}{dt^2} = 0.$$
(17.17)

This equation has no resemblance to the Schrödinger equation (17.16) even when they are describing an identical system. How the results of Newtonian mechanics can be recovered from the Schrödinger equation is discussed in the next section.

Next, we try to understand the Schrödinger equation (17.16) as the law of conservation of energy. Following a departure from the classical concept of physical quantities as numbers, we define position, momentum, and energy as differential operators

$$\hat{x} = x,$$
(17.18)

$$\hat{p} = -i\hbar\frac{\partial}{\partial x},$$
(17.19)

$$\hat{E} = i\hbar\frac{\partial}{\partial t}.$$
(17.20)

With these substitutions, the Schrödinger equation (17.16) reads

$$\frac{\hat{p}^2}{2m}\psi(x,t) = \hat{E}\psi(x,t).$$ (17.21)

This looks like an equation for a free particle ($E = p^2/2m$) except that the "observable" quantities, such as momentum and energy, are differential operators operating on the wavefunction. This novel feature has no analog in classical mechanics where the position and momentum are always some real numbers.

In the presence of potential energy, $V(x)$, this equation can be completed as follows. The total energy is the sum of the kinetic energy and the potential energy, i.e.,

$$E = \frac{p^2}{2m} + V(x).$$ (17.22)

If E and p are replaced by operators operating on the wavefunction, we obtain

$$\left(\frac{\hat{p}^2}{2m} + V(x)\right)\psi(x,t) = \hat{E}\psi(x,t).$$ (17.23)

It then follows from Eqs. (17.18)–(17.20) that

$$\left(-\frac{\hbar^2}{2m}\frac{\partial^2}{\partial x^2} + V(x)\right)\psi(x,t) = i\hbar\frac{\partial\psi(x,t)}{\partial t}.$$ (17.24)

This is the complete Schrödinger equation for the one-dimensional problems. Thus if a particle is moving under the action of a force, the correct equation is no longer Newton's equation $F = ma$ but the Schrödinger equation. Equation (17.24) is called the time-dependent Schrödinger equation.

In many problems, we are often interested in the steady state after the system settles down to a state with a characteristic energy E. In those situations, the wavefunction $\psi(x,t)$ can be expressed as

$$\psi(x,t) = \phi(x)e^{-i(E/\hbar)t}.$$ (17.25)

Then

$$\frac{\partial\psi(x,t)}{\partial t} = -i\frac{E}{\hbar}\psi(x,t).$$ (17.26)

The Schrödinger equation reduces to the so-called stationary-state Schrödinger equation, which is of the form

$$\left(-\frac{\hbar^2}{2m}\frac{\partial^2}{\partial x^2} + V(x)\right)\psi = E\psi.$$ (17.27)

In the following sections, both forms of the Schrödinger equation (Eqs. (17.24) and (17.27)) are considered along with solutions for some simple problems.

An immediate question of interest is: What does it all mean? Both forms of the Schrödinger equation look rather abstract. Even if we can solve these equations mathematically, how do we find the quantities of interest such as the position, the momentum, and the energy of the particle.

The central quantity in the Schrödinger equation is the wavefunction $\psi(x,t)$ and all the physical quantities can be determined when we know it. From a knowledge of $\psi(x,t)$, we can immediately calculate $|\psi(x,t)|^2$. According to the interpretation of Max Born, $|\psi(x,t)|^2$ is the probability density of finding the particle at position x at time t. Since the particle exists at some point in space,

$$\int |\psi(x,t)|^2 dx = 1. \tag{17.28}$$

This is called the normalization condition and is satisfied by all the wavefunctions.

The Schrödinger equation does not give us a well-defined or deterministic answer about the location and the momentum of the particle. However, in view of Born's interpretation of $|\psi(x,t)|^2 = \psi^*(x,t)\,\psi(x,t)$ as being the probability density, we can calculate the average or expectation value of any observable O via

$$\langle O \rangle = \int \psi^*(x,t)\,\hat{O}\,\psi(x,t)\,dx, \tag{17.29}$$

where $\psi^*(x,t)$ is the complex conjugate of $\psi(x,t)$ and $\hat{O}$ is the operator corresponding to the observable quantity O.

For example, the mean value of the position and the momentum can be calculated as follows:

$$\langle x \rangle = \int \psi^*(x,t)\,\hat{x}\,\psi(x,t)\,dx = \int \psi^*(x,t)\,x\,\psi(x,t)\,dx, \tag{17.30}$$

$$\langle p \rangle = \int \psi^*(x,t)\,\hat{p}\,\psi(x,t)\,dx = \int \psi^*(x,t)\left(-i\hbar\frac{\partial}{\partial x}\right)\psi(x,t)\,dx. \tag{17.31}$$

Similarly mean square position and mean square momentum of the particle are calculated from

$$\langle x^2 \rangle = \int \psi^*(x,t)\,\hat{x}^2\,\psi(x,t)\,dx = \int \psi^*(x,t)\,x^2\,\psi(x,t)\,dx, \tag{17.32}$$

$$\langle p^2 \rangle = \int \psi^*(x,t)\,\hat{p}^2\,\psi(x,t)\,dx = \int \psi^*(x,t)\left(-i\hbar\frac{\partial}{\partial x}\right)^2\psi(x,t)\,dx. \tag{17.33}$$

The root-mean-square deviations or the uncertainties of the position and momentum are

$$\Delta x = \sqrt{\langle x^2 \rangle - \langle x \rangle^2}, \tag{17.34}$$

$$\Delta p = \sqrt{\langle p^2 \rangle - \langle p \rangle^2}. \tag{17.35}$$

It can be shown that these uncertainties satisfy the Heisenberg uncertainty relation

$$\Delta x \Delta p \geq \frac{\hbar}{2}. \tag{17.36}$$

The allowed energy values for a given system can be obtained by solving the steady-state Schrödinger equation (17.27). A remarkable result is that a solution of the three-dimensional version of the Schrödinger equation (17.27) yields the correct energy values for the allowed energy levels for the hydrogen atom. This is discussed in Section 17.5. There is no need for a quantum postulate, as was done by Bohr, to explain the spectrum of light emitted by a gas of hydrogen atoms. What is more, we can solve practically any problem of particle dynamics using this version of the Schrödinger equation. In general, it is difficult to analytically solve the Schrödinger equation for a realistic problem and we usually have to resort to the numerical calculations.

Finally, we note that the Schrödinger equation is valid for all problems within what we call the non-relativistic limit, i.e., for those problems where the particle velocity is much less than the speed of light. For relativistic particles, Paul Dirac derived another more general equation in 1928, called the Dirac equation.

17.2 Kinematics in Classical and Quantum Mechanics–Newton vs. Schrödinger

The dynamical equations for a particle in quantum mechanics and classical mechanics are very different and there is no resemblance between the Schrödinger equation and Newton's second law of motion. So how do we understand the connection between the two? We do not see the probabilistic nature that is central to quantum mechanics in our everyday life. Newtonian mechanics gives very accurate and precise results. How do we reconcile these observations?

In order to answer these questions, consider the simplest problem of particle dynamics, solve it in both Newtonian and quantum mechanics, and see where the fundamental differences come in the two approaches. This simple problem also helps to illustrate that the fundamental theory is quantum mechanics and classical mechanics is an approximation—a remarkably good approximation—when considering macroscopic objects.

The problem of the motion of a free particle of mass m along the x-axis is considered. Assume that the particle is at position x_i at time $t = 0$ moving with momentum p_i. There is no force acting on it. The question posed is: What is the location of the particle at a later time, t?

The answer to this question in Newtonian mechanics is simple and straightforward. According to Newton's second law of motion, $F = ma$, where F is the applied force and a is the acceleration. Since the external force F is equal to zero, we have $ma = 0$, i.e.,

$$ma = m\frac{d^2x}{dt^2} = 0,\tag{17.37}$$

where we have used the definition of acceleration, $a = d^2x/dt^2$. This equation can be integrated to obtain the momentum at the time t as follows:

$$p(t) = m\frac{dx}{dt} = p_i.\tag{17.38}$$

Another integration yields the position at time t:

$$x(t) = x_i + \frac{p_i}{m}t.\tag{17.39}$$

That this is, indeed, the solution of the Newtonian equation can be verified by differentiating it twice with respect to t and seeing that Eq. (17.37) is satisfied. Also, it is clear from Eqs. (17.39) and (17.38) that the position and momentum at $t = 0$ are x_i and p_i, respectively. Thus, knowing the position and momentum at the initial time t_i, we can predict the position and momentum at the later time precisely. This is the hallmark of Newtonian mechanics.

What about the solution of the same problem within the framework of quantum mechanics? The corresponding equation for the wavefunction $\psi(x,t)$ for a free particle is the Schrödinger equation (with $V(x) = 0$)

$$-\frac{\hbar^2}{2m}\frac{\partial^2\psi(x,t)}{\partial x^2} = i\hbar\frac{\partial\psi(x,t)}{\partial t}. \tag{17.40}$$

The first issue is: How do we describe the initial position and momentum of the particle? The well-defined values x_i and p_i cannot be identified for the initial position and momentum as that would violate the Heisenberg uncertainty relation. The Heisenberg uncertainty relation

$$\Delta x \Delta p \geq \frac{\hbar}{2}$$

does not allow us to assign precise values for position and momentum. We cannot therefore think of the particle as a point object precisely located at x_i and moving with a well-defined momentum p_i. Instead we have to describe the particle as a wavefunction $\psi(x,0)$ at $t = 0$, the modulus square of the wavefunction $|\psi(x,0)|^2$ describing the probability density of finding the particle at position x. The particle is thus described by a wave packet centered at the position $x = x_i$. For a discussion of the wave packet description we refer to Section 7.2.

The wave packet is created by combining many wavefunctions with different wavelengths, where the wavelength is determined by the particle momentum (remember the de Broglie relation $p = h/\lambda$). This wave packet provides information about both the location and momentum of the object. However, it cannot give us exact values for either quantity. The wavefunction leaves us with an uncertainty in position and an uncertainty in momentum for the particle in accordance with the Heisenberg uncertainty relation (17.36). This wave packet approach is used to look at how knowledge of an electron's position and momentum evolves with time. For this, we need to solve the Schrödinger equation (17.40).

A simple wave packet describing the particle at time $t = 0$ is the so-called Gaussian wave packet of the form

$$\psi(x,0) = \frac{1}{\sqrt{\sqrt{2\pi}\sigma_i}}e^{-\frac{(x-x_i)^2}{4\sigma_i^2}}e^{i\frac{p_i}{\hbar}(x-x_i)}. \tag{17.41}$$

The probability of finding the electron in this wave packet is given by

$$|\psi(x,0)|^2 = \frac{1}{\sqrt{2\pi}\sigma_i}e^{-\frac{(x-x_i)^2}{2\sigma_i^2}}. \tag{17.42}$$

This wave packet is centered at the position $x = x_i$ and is symmetric about this position as shown in Fig. 17.1. Physically, this means that the particle, instead of being localized at the position $x = x_i$, is given by the probability density which is maximum at $x = x_i$ and decreases as we go away from this point. The probability distribution is narrow if σ_i is small. Thus σ_i is a measure of how "localized" a particle is. The important point to note is that the particle can exist at any value of x with certain probability. This is, as we mentioned earlier, contrary to the classical picture of the particle.

Using the integral

$$\int_{-\infty}^{\infty} e^{-u^2} du = \sqrt{\pi},$$ (17.43)

the wavefunction (17.41) is normalized:

$$\int_{-\infty}^{\infty} |\psi(x,0)|^2 dx = 1.$$ (17.44)

Also, the average position of the particle is

$$\langle x(0) \rangle = \int_{-\infty}^{\infty} x|\psi(x,0)|^2 dx = x_i.$$ (17.45)

The mean-square position is given by

$$\langle x^2(0) \rangle = \int_{-\infty}^{\infty} x^2|\psi(x,0)|^2 dx = x_i^2 + \sigma_i^2.$$ (17.46)

The root-mean-square deviation or the spread of the wave packet is therefore equal to

$$\Delta x = \sqrt{\langle x^2(0) \rangle - \langle x(0) \rangle^2} = \sigma_i.$$ (17.47)

This is shown in Fig. 17.1.

Next consider the evolution of this wave packet according to the Schrödinger equation (17.40). The solution is given by

$$\psi(x,t) = \frac{1}{\sqrt{\sigma_i[1 + i\hbar t/2m\sigma_i^2]}\sqrt{2\pi}} e^{-\left\{ \frac{(x-x_i)^2}{4\sigma_i^2[1+i(\hbar t/2m\sigma_i^2)]} - \frac{i}{\hbar} \frac{p_i(x-x_i)-(p_i^2/2m)t}{[1+i(\hbar t/2m\sigma_i^2)]} \right\}}$$ (17.48)

A formal derivation of this result is quite complicated and is not produced here. However, it can be verified that this is indeed a solution of Eq. (17.40) by differentiating $\psi(x,t)$ twice with respect to x and differentiating it once with respect to t and substituting the resulting expressions in Eq. (17.40). In addition we note that this expression of $\psi(x,t)$ reduces to the expression of $\psi(x,0)$ as given in Eq. (17.41) for $t = 0$.

The probability density of the particle at time t is given by the modulus square of the wavefunction:

$$|\psi(x,t)|^2 = \frac{1}{\sigma_i\sqrt{1 + (\hbar t/2m\sigma_i^2)^2}\sqrt{2\pi}} e^{-\frac{(x-x_i-(p_i/m)t)^2}{2\sigma_i^2\left(1+(\hbar t/2m\sigma_i^2)^2\right)}}.$$ (17.49)

Fig. 17.1 A Gaussian wave packet according to Eq. (17.42).

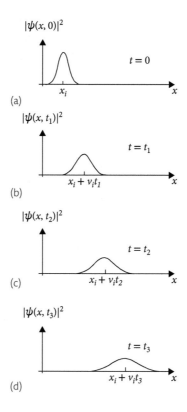

Fig. 17.2 Spreading of a one-dimensional Gaussian wave packet as it propagates.

This quantity can be used to find the probability of locating the particle at a later time t. Note that the wave packet (17.49) at time t has the same form as the initial wave packet (17.42) but with a displaced average position location and enhanced fluctuations. As we see in Fig. 17.2, the wave packet spreads as it propagates.

First, note that the center of the wave packet moves from $x = x_i$ to

$$x = x_i + \frac{p_i}{m}t$$

at time t. This is the same equation as obtained by classical mechanics (Eq. (17.39)). Here this is the equation, not for the location of the particle, but for the center of a spreading wave packet. The Gaussian wave packet is symmetric about the center of the wave packet. Therefore the center of the wave packet is identical to the mean position of the particle. This can be seen by substituting for $|\psi(x,t)|^2$ from Eq. (17.49) in the definition (17.30) and integrating using Eq. (17.43). The result is

$$\langle x(t) \rangle = \int_{-\infty}^{\infty} x|\psi(x, t)|^2 dx = x_i + \frac{p_i}{m}t. \tag{17.50}$$

Thus a more general result is obtained: the mean position of the wave packet obeys the same dynamics as the classical particle.

We have thus shown that the mean position obeys the same equation as derived from the Newtonian equations of motion for the simple case of uniform motion in the absence of any external force. However this result can be shown to be quite general and is valid even in the presence of any potential $V(x)$. The result that the average location follows the same trajectory as predicted by the Newton's equation of motion is often referred as the Ehrenfest theorem and forms the bridge between the Newtonian and the Schrödinger descriptions.

From this result, it is seen that a very narrow wave packet that hardly spreads with time should be a good description for the particle dynamics in Newtonian mechanics. A narrow wave packet corresponds to a well-localized particle and a lack of significant spreading of the wave packet corresponds to a reasonably well defined trajectory.

Next we study the spread of the wave packet as it propagates. The spreading of the wave packet is given by the root-mean-square width of the wave packet at time t, $\sigma_f(t)$. From Eq. (17.49),

$$\sigma_f(t) = \Delta x = \sqrt{\langle x^2(t)\rangle - \langle x(t)\rangle^2} = \sigma_i \sqrt{1 + \left(\frac{\hbar t}{2m\sigma_i^2}\right)^2}. \tag{17.51}$$

The wave packet representing the particle spreads in time. The spreading can be very slow for a massive particle. The wave packet spreads to twice the initial value ($\sigma_f(t) = 2\sigma_i$) when

$$1 + \left(\frac{\hbar t}{2m\sigma_i^2}\right)^2 = 4 \tag{17.52}$$

or

$$t = \frac{2\sqrt{3}m\sigma_i^2}{\hbar}. \tag{17.53}$$

For example, if we describe a marble of mass one gram ($= 10^{-3}$ kg) by a wave packet corresponding to its size equal to 1 mm ($= 10^{-3}$ m), it spreads to twice the value (2 mm) in a time

$$t_{marble} = \frac{2\sqrt{3}m\sigma_i^2}{\hbar} = \frac{2\sqrt{3}\times 10^{-3}\times 10^{-6}}{1.1\times 10^{-34}} \text{ sec} \approx 10^{25} \text{ sec} \approx 10^{15}\text{years}, \tag{17.54}$$

which is about 10 000 times the known age of the universe.

As a second example, we consider an electron of mass $m = 9.1 \times 10^{-31}$ kg confined within 1 μm, the spread time for the wave packet is

$$t_{electron} = \frac{2\sqrt{3}m\sigma_i^2}{\hbar} = \frac{2\sqrt{3}\times 9.11 \times 10^{-31}\times 10^{-12}}{1.1\times 10^{-34}} \text{ sec} \approx 10^{-8} \text{ sec}. \tag{17.55}$$

Thus, a fast spreading of the wave packet is observed for electrons.

Next, we ask the question: Where is the particle at time t? We cannot give a deterministic answer to this question (as we can do in Newtonian mechanics). We can only say that the particle is located at position x at time t with a probability distribution given by Eq. (17.49). The most probable location of the particle is the same as predicted by Newton's law of motion but it can also be found far from the classical trajectory, even if with very small probability.

As the particle moves, the corresponding wave packet spreads and the region where it can be found increases.

The spreading of the wave packet can be understood by invoking the Heisenberg uncertainty relation. According to the uncertainty relation, a wave packet spread in space has also a corresponding spread in momentum. Thus the wave packet contains waves of different momenta and hence different velocities. This causes the wave packet to spread. In the case where the wave packet is narrow in space, the momentum distribution is wider and the result is that the spreading is faster. This agrees with our conclusions based on rigorous calculations.

The spreading of the wave packet can be directly derived from the Heisenberg uncertainty relation without resorting to the Schrödinger equation as follows. First we note that each point inside the initial wave packet follows the path defined by

$$x = x_i + \frac{p_i}{m}t. \tag{17.56}$$

The mean location then follows the trajectory

$$\langle x \rangle = \langle x_i \rangle + \frac{\langle p_i \rangle}{m}t. \tag{17.57}$$

Next we find the mean square location

$$\langle x^2 \rangle = \left\langle \left(x_i + \frac{p_i}{m}t\right)^2 \right\rangle = \langle x_i^2 \rangle + 2\langle x_i \rangle \frac{\langle p_i \rangle}{m}t + \frac{\langle p_i^2 \rangle}{m^2}t^2. \tag{17.58}$$

The position spread is defined as

$$\sigma_f(t) = \Delta x = \sqrt{\langle x^2(t) \rangle - \langle x(t) \rangle^2} = \sqrt{(\Delta x_i)^2 + \frac{(\Delta p_i)^2}{m^2}t^2}, \tag{17.59}$$

where we substituted for $\langle x \rangle$ and $\langle x^2(t) \rangle$ from Eqs. (17.57) and (17.58), respectively. According to the Heisenberg uncertainty relation, under the most optimal conditions,

$$\Delta p_i = \frac{\hbar}{2\Delta x_i}. \tag{17.60}$$

On substituting this expression for Δp_i in Eq. (17.59), we recover the wave packet spreading relation (17.51) with the substitution $\sigma_i \equiv \Delta x_i$. This simple derivation brings out the intimate relation between the uncertainty relation and the spreading of the wave packet, and consequently an understanding of the origin of quantum dynamics.

17.3 Particle Inside a Box

One of the simplest problems in quantum mechanics is a single free particle confined inside a one-dimensional box of length L. The potential energy is zero inside the box but infinite at the boundaries $x = 0$ and $x = L$. Inside the box, Schrödinger equation

$$\left[-\frac{\hbar^2}{2m}\frac{d^2}{dx^2} + V(x)\right]\psi = E\psi \tag{17.61}$$

reduces to

$$-\frac{\hbar^2}{2m}\frac{d^2\psi}{dx^2} = E\psi. \tag{17.62}$$

The particle always remains inside the box because of the infinite potential barrier at the walls. So the probability of finding the particle outside the box is zero, i.e., $\psi = 0$ outside the box. The wavefunction must be continuous at the boundaries of the potential well at $x = 0$ and $x = L$. Therefore the wavefunction satisfies the boundary conditions

$$\psi(0) = \psi(L) = 0. \tag{17.63}$$

The Schrödinger equation (17.62) is solved subject to these boundary conditions.

Equation (17.62) can be rewritten in the form

$$\frac{d^2\psi}{dx^2} + k^2\psi = 0, \tag{17.64}$$

where

$$k = \sqrt{\frac{2mE}{\hbar^2}}. \tag{17.65}$$

This equation has the familiar form of the equation for a harmonic oscillator. The general solution of this equation is given by

$$\psi(x) = A \sin kx + B \cos kx, \tag{17.66}$$

where A and B are constants that can be determined from the boundary conditions as well as normalization.

The condition that $\psi(0) = 0$ leads to $B = 0$. Therefore

$$\psi(x) = A \sin kx. \tag{17.67}$$

The condition that $\psi(L) = 0$ implies that

$$A \sin kL = 0. \tag{17.68}$$

This equation can be satisfied in two different ways. First is to choose $A = 0$. However, this leads to $\psi(x) = 0$, which is not possible as it implies that particle is not inside the box. The other possibility is that

$$kL = n\pi, \tag{17.69}$$

where $n = 1, 2, \cdots$ are integers. Thus the solutions for the wavefunction are

$$\psi_n(x) = A \sin\left(\frac{n\pi}{L}x\right). \tag{17.70}$$

The constant A can be found from the condition that the particle is somewhere inside the box, i.e.,

$$\int_0^L |\psi_n(x)|^2 dx = 1. \tag{17.71}$$

We thus have

$$A^2 \int_0^L \sin^2\left(\frac{n\pi}{L}x\right) = A^2 \frac{L}{2} = 1$$

The normalization constant A is equal to $\sqrt{2/L}$. The wavefunction is

$$\psi_n(x) = \sqrt{\frac{2}{L}}\, \sin\left(\frac{2\pi}{\lambda_n}x\right).$$ (17.72)

In Fig. 17.3, the wavefunctions $\psi_n(x)$ are plotted for different values of n. The corresponding probability densities are given by

$$|\psi_n|^2 = \frac{2}{L}\sin^2\left(\frac{n\pi}{L}x\right)$$ (17.73)

and are plotted in Fig. 17.4.

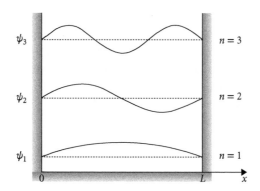

Fig. 17.3 The wavefunctions for a particle inside a one-dimensional box for $n = 1, 2, 3$.

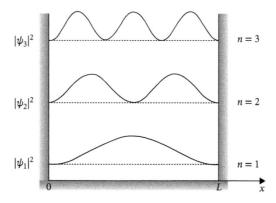

Fig. 17.4 The probability density of the particle in a one-dimensional box, $|\psi_n|^2$, for $n = 1, 2, 3$.

What about the energy of the particle when it exists in the quantum state $\psi_n(x)$? If we substitute the value of k from Eq. (17.65) into Eq. (17.69), we obtain

$$E_n = \frac{n^2\pi^2\hbar^2}{2mL^2} \tag{17.74}$$

($n = 1, 2 \cdots$). We thus have a very counterintuitive result: A particle can have only quantized energies as given by Eq. (17.74). Even more surprising is that the location of the particle can be given only probabilistically with the corresponding probability density given by Eq. (17.73). As seen in Fig. 17.4, for a given energy, there are locations where the particle cannot exist.

Why do we not see such strange behavior in our everyday life? What we observe is that the particles (like a tennis ball) can have any energy and can be found anywhere between the two walls. To explain this contradictory behavior, we note that the quantum behavior as depicted by Eq. (17.73) for the wavefunction and Eq. (17.74) for energy can be observed only for small values of the quantum number n. This happens when

$$\frac{2mL^2E_n}{\pi^2} \sim \hbar^2. \tag{17.75}$$

For example, an electron of mass $m = 9.1 \times 10^{-31}$ kg and an energy of (say) 7 eV $= 7 \times 1.6 \times 10^{-19}$J confined in a region $L = 4 \times 10^{-10}$ m would be in quantum states with $n \sim 1 - 2$. Such a situation approximately corresponds to a free electron moving inside a metal. A metal consists of atoms occupying positions in a lattice. The free electron experiences a repulsive force due to the electronic cloud surrounding the atom. For a sufficiently large repulsion, the situation is similar to an electron confined between two walls.

As the energy and/or the mass of the particle increases, the quantum number n increases to very high values and the energy levels get closer. A tennis ball of mass 50 grams moving with a velocity of 5 m/s has the energy $(1/2)\, mv^2 = 0.625$ J. If it is confined between two walls 10 m apart, then

$$n = \frac{\sqrt{2mL^2E_n}}{\pi\hbar} \sim 10^{34}, \tag{17.76}$$

i.e., there are 10^{33} nodes of the wavefunction within a distance of 1 m (the spacing between the two nodes being 10^{-33} m), forming essentially a continuum. This continuum means the particle is free, can be anywhere between the walls, and can have continuous values of allowable energies. The quantum mechanical model is no longer necessary and the classical model gives accurate results.

17.4 Tunneling Through a Barrier

In this section, the problem of a particle crossing a barrier is addressed. An example of such a barrier is a wall of height h_0. According to Newtonian mechanics, only those particles of mass m moving with velocity v whose total energy (sum of kinetic and potential energies) is larger than the energy mgh_0 are able to surmount the barrier and get to the other side of the wall. In the absence of any frictional forces, such particles should always be able to cross the barrier. However, if the total energy of the particle E is less than the required energy, i.e.,

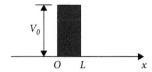

Fig. 17.5 A potential barrier of width L and height V_0.

$$E < mgh_0, \tag{17.77}$$

then there is no way that the particle can cross over to the other side. This is a common sense result and can be observed around us all the time.

What about the corresponding result in quantum mechanics? Consider a one-dimensional potential barrier of width L and height V_0, i.e.,

$$V(x) = \begin{cases} 0, & x < 0 \\ V_0, 0 \leq x \leq L \\ 0, & x > 0 \end{cases} \tag{17.78}$$

Such a barrier is shown in Fig. 17.5. The question we ask is: What happens when a particle with energy

$$E < V_0 \tag{17.79}$$

is incident on the barrier? Do we get the classical result that it definitely bounces back (Fig. 17.6a) or is there some possibility that it can "tunnel" through the barrier and is found on the other side (Fig. 17.6b)? The surprising answer is that there is a definite probability that the particle tunnels through the barrier. In the following, we show how to calculate the tunneling probability by solving the Schrödinger equation

$$\left(-\frac{\hbar^2}{2m}\frac{d^2}{dx^2} + V(x)\right)\psi = E\psi. \tag{17.80}$$

The method works as follows. The wavefunction $\psi(x)$ of the particle should be spread over the entire one-dimensional space from $x = -\infty$ to $x = \infty$. The wavefunction should be uniform with no kinks or discontinuities throughout the space. In order to find such a wavefunction, we first write down and solve three separate Schrödinger equations for the three regions: before the barrier ($x < 0$) with wavefunction $\psi_I(x)$, inside the barrier ($0 \leq x \leq L$) with wavefunction $\psi_{II}(x)$, and beyond the barrier ($x > 0$) with wavefunction $\psi_{III}(x)$. In order to ensure that the total wavefunction is smooth in the entire one-dimensional space, we equate the wavefunctions and their derivatives at the boundaries of the potential. For example, matching the wavefunction and ensuring its smoothness at the boundary $x = 0$ requires

$$\psi_I(0) = \psi_{II}(0), \tag{17.81}$$

$$\left.\frac{d\psi_I(x)}{dx}\right|_{x=0} = \left.\frac{d\psi_{II}(x)}{dx}\right|_{x=0}. \tag{17.82}$$

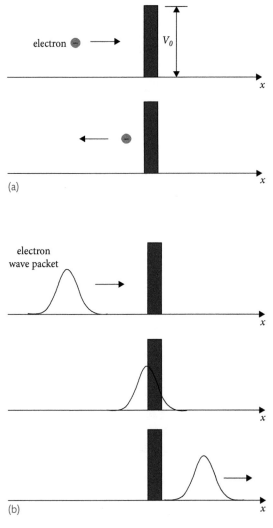

Fig. 17.6 (*a*) In the classical picture, a particle whose energy *E* is less than that of the height of the potential barrier V_0 is completely reflected with no chance of crossing over to the right side. (*b*) In the quantum picture, the same particle treated as a wave packet has a finite probability of tunneling through the barrier.

Similar conditions are obtained at the boundary located at $x = L$, i.e.,

$$\psi_{II}(L) = \psi_{III}(L), \tag{17.83}$$

$$\left.\frac{d\psi_{II}(x)}{dx}\right|_{x=L} = \left.\frac{d\psi_{III}(x)}{dx}\right|_{x=L}. \tag{17.84}$$

These boundary conditions along with the solutions of the Schrödinger equations in the three regions give the complete information, and the probabilities of transmission and reflection of the particle can be calculated.

In the region I $(-\infty < x < 0)$, where the potential $V(x)$ is zero, the Schrödinger equation for the wavefunction $\psi_I(x)$ is given by

$$-\frac{\hbar^2}{2m}\frac{d^2\psi_I(x)}{dx^2} = E\psi_I(x). \tag{17.85}$$

In the region II $(0 < x < L)$, the potential is $V(x) = V_0$ and the Schrödinger equation for the wavefunction ψ_{II} is

$$\left(-\frac{\hbar^2}{2m}\frac{d^2}{dx^2} + V_0\right)\psi_{II}(x) = E\psi_{II}(x). \tag{17.86}$$

Similarly, in the region III $(L \leq x < \infty)$, the potential is zero and the equation for the wavefunction ψ_{III} is

$$-\frac{\hbar^2}{2m}\frac{d^2\psi_{III}(x)}{dx^2} = E\psi_{III}(x). \tag{17.87}$$

Next we find a general solution of these equations. First we find the solution of Eq. (17.85) which can be rewritten as

$$\frac{d^2\psi_I(x)}{dx^2} + k^2\psi_I(x) = 0, \tag{17.88}$$

where

$$k = \sqrt{\frac{2mE}{\hbar^2}}. \tag{17.89}$$

A general solution of this equation is

$$\psi_I(x) = Ae^{ikx} + Be^{-ikx}. \tag{17.90}$$

Here A and B are constants that can be obtained from the boundary conditions discussed above.

In a similar manner Eq. (17.87) for $\psi_{III}(x)$ can be solved. The result is

$$\psi_{III}(x) = Fe^{ikx} + Ge^{-ikx}. \tag{17.91}$$

Again F and G are constants.

Finally Eq. (17.86) for $\psi_{II}(x)$ can be rewritten as

$$\frac{d^2\psi_{II}(x)}{dx^2} - \beta^2\psi_{II}(x) = 0, \tag{17.92}$$

where

$$\beta = \sqrt{\frac{2m(V_0 - E)}{\hbar^2}}. \tag{17.93}$$

The solution of this equation is given by

$$\psi_{II}(x) = Ce^{\beta x} + De^{-\beta x}, \tag{17.94}$$

where the constants C and D are determined from the boundary conditions.

Before proceeding to use the boundary conditions, first consider the physical nature of these solutions. The wavefunction in the region 1, $\psi_I(x)$, consists of two terms: The first term, $A \exp (ikx)$, corresponds to a wave traveling to the right, and A is the amplitude of this incident wave. The second term, $B \exp (-ikx)$, corresponds to a wave traveling to the left, and B is the amplitude of this reflected wave. The ratio B/A is therefore the reflection amplitude and

$$R = \left| \frac{B}{A} \right|^2 \tag{17.95}$$

is the probability that the particle is reflected back. If the particle behaves classically we would expect $R = 1$.

Next look at the solution in the region 3, $\psi_{III}(x)$, as given by Eq. (17.91). Here, the first term, $F \exp(ikx)$, corresponds to a wave traveling to the right, and F is the amplitude of this transmitted wave. The second term, $G \exp(-ikx)$, corresponds to a wave traveling to the left. However, in region 3, the particle can only move in the $+x$-direction, there is no reflection. Therefore G should be equal to zero. The transmission coefficient can be defined as

$$T = \left| \frac{F}{A} \right|^2. \tag{17.96}$$

Here T is the probability that the particle somehow tunneled through the barrier. The sum of the reflection and transmission coefficient should be equal to unity, i.e.,

$$R + T = 1. \tag{17.97}$$

Next we substitute the general solutions (17.90), (17.91), and (17.94) in the boundary conditions (17.81)–(17.84). The resulting equations (with $G = 0$) are

$$A + B = C + D, \tag{17.98}$$

$$ik(A - B) = \beta(C - D), \tag{17.99}$$

$$Ce^{\beta L} + De^{-\beta L} = Fe^{ikL}, \tag{17.100}$$

$$\beta(Ce^{\beta L} - De^{-\beta L}) = ikFe^{ikL}. \tag{17.101}$$

These are four equations in five unknowns A, B, C, D, and F. However the number of unknowns can be reduced to four if the unknowns are defined as B/A, C/A, D/A, and F/A. Since we are only interested in the reflection and transmission coefficients, we are interested in finding B/A and F/A.

After a lengthy but straightforward calculation (which we do not reproduce here), we find

$$\frac{B}{A} = \frac{-i \left(k^2 + \beta^2\right) \sinh (\beta L)}{2k\beta \cosh (\beta L) + i (\beta^2 - k^2) \sinh (\beta L)}, \tag{17.102}$$

$$\frac{F}{A} = \frac{2k\beta e^{-ikL}}{2k\beta \cosh (\beta L) + i (\beta^2 - k^2) \sinh (\beta L)}. \tag{17.103}$$

The reflection coefficient R and the transmission coefficient T are then given by

$$R = \left| \frac{B}{A} \right|^2 = \frac{\left(k^2 + \beta^2\right)^2 \sinh^2 (\beta L)}{4k^2 \beta^2 + (k^2 + \beta^2)^2 \sinh^2(\beta L)}, \tag{17.104}$$

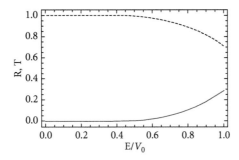

Fig. 17.7 The reflection coefficient, R (solid curve), and the transmission coefficient, T (dashed curve), of a particle for a potential barrier of width $L = 0.5\,\lambda$ ($kL = \pi$) where λ is the de Broglie wavelength as a function of the ratio of the energy E of the incoming particle to the height V_0 of the barrier.

$$T = \left|\frac{F}{A}\right|^2 = \frac{4k^2\beta^2}{4k^2\beta^2 + (k^2 + \beta^2)^2 \sinh^2(\beta L)}. \tag{17.105}$$

In Fig. 17.7, we plot the reflection coefficient R and the transmission coefficient, T of a particle for a potential barrier of width $L = 0.5\,\lambda$ ($kL = \pi$) where λ is the de Broglie wavelength. The simplified expressions of R and T are

$$R = \frac{V_0^2 \sinh^2\left(\sqrt{(V_0/E) - 1}\,\pi\right)}{4E\,(V_0 - E) + V_0^2 \sinh^2\left(\sqrt{(V_0/E) - 1}\,\pi\right)},$$

$$T = \frac{4E\,(V_0 - E)}{4E\,(V_0 - E) + V_0^2 \sinh^2\left(\sqrt{(V_0/E) - 1}\,\pi\right)}.$$

It can be seen that, when the energy of the incoming particle, E, is much less than the height of the potential barrier, V_0, the reflection coefficient, R, is equal to 1 and the transmission coefficient, T, is equal to zero. This is the same result we expect from the classical mechanics. However, the quantum features, completely contrary to our classical intuition, start to appear when the ratio $E/V_0 > 0.5$. For these values, the energy of the incident particle is still substantially less than the height of the potential barrier V_0 but the transmission coefficient, T, is non-zero. Therefore the particle has a non-vanishing probability of "tunneling" through the barrier. This remarkable result has been used in many quantum devices such as transistors and high precision microscopes.

17.5 The Schrödinger Equation in Three Dimensions and the Hydrogen Atom

Historically, the most important success of quantum mechanics was the derivation of the energy levels in a hydrogen atom. This was indeed the test of the validity of a full theory. Schrödinger was able to solve his equation for the hydrogen atom and recover analytically all the known results at that time regarding the spectra of hydrogen atoms. This went far beyond

Bohr's model where the hydrogen atom spectrum was derived by using a postulate with no justification except that it gave an answer which agreed with experimental results.

We begin by generalizing the one-dimensional stationary Schrödinger equation (17.27) to the three-dimensional Schrödinger equation

$$\left[-\frac{\hbar^2}{2m} \left(\frac{\partial^2}{\partial x^2} + \frac{\partial^2}{\partial y^2} + \frac{\partial^2}{\partial z^2} \right) + V(x, y, z) \right] \psi = E\psi. \tag{17.106}$$

In simplified notation, Eq. (17.106) can be rewritten as

$$\left[-\frac{\hbar^2}{2m} \nabla^2 + V(r) \right] \psi = E\psi, \tag{17.107}$$

where $r \equiv (x, y, z)$ and

$$\nabla^2 = \frac{\partial^2}{\partial x^2} + \frac{\partial^2}{\partial y^2} + \frac{\partial^2}{\partial z^2}. \tag{17.108}$$

The differential operator ∇^2 is called Laplacian.

For many problems of interest with spherical symmetry, it is more convenient to use spherical coordinates (r, ϕ, θ) instead of rectangular coordinates (x, y, z), as shown in Fig. 17.8. They transform into each other via the relations

$$x = r \sin \theta \cos \phi, \tag{17.109}$$

$$y = r \sin \theta \sin \phi, \tag{17.110}$$

$$z = r \cos \theta. \tag{17.111}$$

This transformation can be verified from Fig. 17.8.

The Laplacian operator in the spherical coordinates is given by

$$\nabla^2 \psi = \frac{1}{r^2} \frac{\partial}{\partial r} \left(r^2 \frac{\partial \psi}{\partial r} \right) + \frac{1}{r^2 \sin \theta} \frac{\partial}{\partial \theta} \left(\sin \theta \frac{\partial \psi}{\partial \theta} \right) + \frac{1}{r^2 \sin^2 \theta} \frac{\partial^2 \psi}{\partial \phi^2}. \tag{17.112}$$

Therefore the Schrödinger equation in the spherical coordinates can be written as

$$\left[-\frac{\hbar^2}{2m} \left(\frac{1}{r^2} \frac{\partial}{\partial r} \left(r^2 \frac{\partial}{\partial r} \right) + \frac{1}{r^2 \sin \theta} \frac{\partial}{\partial \theta} \left(\sin \theta \frac{\partial}{\partial \theta} \right) + \frac{1}{r^2 \sin^2 \theta} \frac{\partial^2}{\partial \phi^2} \right) \right.$$
$$\left. + V(r, \phi, \theta) \right] \psi = E\psi. \tag{17.113}$$

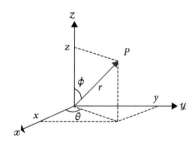

Fig. 17.8 Transformation from the rectangular coordinates (x, y, z) to (r, ϕ, θ).

Next we turn to the problem of the hydrogen atom and solve the Schrödinger equation for energy values and the corresponding wavefunctions.

A hydrogen atom consists of one proton and one electron. In the Rutherford and Bohr models, the proton formed the nucleus and the electron revolved around it. Our aim is thus to find the energy and the wavefunction of the electron. The potential energy of the proton–electron system is given by

$$V(r) = -\frac{e^2}{4\pi\varepsilon_0 r}, \tag{17.114}$$

where r is the distance of the electron from the proton in the nucleus. This problem has spherical symmetry as the potential energy only depends on the spherical coordinate r. We therefore use the Schrödinger equation in the spherical coordinates. It follows, on substituting the potential energy (17.114) in the Schrödinger equation (17.113),

$$\left[-\frac{\hbar^2}{2m} \left(\frac{1}{r^2} \frac{\partial}{\partial r} \left(r^2 \frac{\partial}{\partial r} \right) + \frac{1}{r^2 \sin\theta} \frac{\partial}{\partial \theta} \left(\sin\theta \frac{\partial}{\partial \theta} \right) + \frac{1}{r^2 \sin^2\theta} \frac{\partial^2}{\partial \phi^2} \right) \right.$$
$$\left. -\frac{e^2}{4\pi\varepsilon_0 r} \right] \psi = E\psi. \tag{17.115}$$

This equation can be solved exactly for both the allowed values of energy E and the corresponding wavefunction ψ. The derivation is rather involved and it is not reproduced here. We only give the salient features of the solution.

A most remarkable result in the early development of quantum mechanics was that Eq. (17.115) could be solved and the only allowed values of the energy E were

$$E_n = -\frac{me^4}{2(4\pi\varepsilon_0)^2 \hbar^2} \frac{1}{n^2}, \tag{17.116}$$

where $n = 1, 2, 3 \cdots$ is called the principal quantum number. These values of electron energy are in perfect agreement with experimental observation of the hydrogen spectrum.

As an example, consider the case, $n = 1$. It turns out that the corresponding wavefunction is given by

$$\psi_1 = \frac{1}{\sqrt{\pi}} \left(\frac{1}{a_B} \right)^{3/2} e^{-(r/a_B)}, \tag{17.117}$$

where

$$a_B = \frac{\hbar^2}{m} \left(\frac{4\pi\varepsilon_0}{e^2} \right) \tag{17.118}$$

is the Bohr radius as given by Eq. (6.20). It is easy to verify that ψ_1 is the solution of the Schrödinger equation when the energy is given by

$$E_1 = -\frac{me^4}{2(4\pi\varepsilon_0)^2 \hbar^2}. \tag{17.119}$$

First, we note that the wavefunction ψ_1 depends only on radial coordinate r and does not depend on ϕ and θ. Therefore the simplified Schrödinger equation is obtained by neglecting the ϕ and θ dependent terms in Eq. (17.115). The resulting equation is

$$-\frac{\hbar^2}{2m}\left(\frac{1}{r^2}\frac{\partial}{\partial r}\left(r^2\frac{\partial}{\partial r}\right)\right)\psi_1 - \frac{e^2}{4\pi\varepsilon_0 r}\psi_1 = E_1\psi_1. \tag{17.120}$$

This equation can be verified by substituting for E_1 and ψ_1 from Eqs. (17.119) and (17.117), respectively. We note that

$$\frac{1}{r^2}\frac{\partial}{\partial r}\left(r^2\frac{\partial\psi_1}{\partial r}\right) = \frac{\partial^2\psi_1}{\partial r^2} + \frac{2}{r}\frac{\partial\psi_1}{\partial r}.$$

It follows from Eq. (17.117) that

$$\frac{\partial\psi_1}{\partial r} = -\frac{1}{\sqrt{\pi}}\left(\frac{1}{a_B}\right)^{5/2}e^{-(r/a_B)}, \tag{17.121}$$

$$\frac{\partial^2\psi_1}{\partial r^2} = \frac{1}{\sqrt{\pi}}\left(\frac{1}{a_B}\right)^{7/2}e^{-(r/a_B)}, \tag{17.122}$$

where the Bohr radius, a_B, is given by Eq. (17.119). On substituting from Eqs. (17.121) and (17.122) into Eq. (17.120), it follows that

$$-\frac{\hbar^2}{2m}\left(\frac{1}{r^2}\frac{\partial}{\partial r}\left(r^2\frac{\partial}{\partial r}\right)\right)\psi_1 - \frac{e^2}{4\pi\varepsilon_0 r}\psi_1 = -\frac{me^4}{2(4\pi\varepsilon_0)^2\hbar^2}\psi_1$$

Thus, the energy, E_1, is given by Eq. (17.119).

It turns out that, for $n = 1$, there is only one allowed energy level described by the wavefunction ψ_1. For $n > 1$, there is more than one energy level for each value of n and the number of allowed energy levels increases as n increases. Here, we concentrate only on the case $n = 1$. We have seen that the Schrödinger equation can be solved and the correct value of the electron energy is obtained.

But what about the location of the electron? Bohr's theory predicted a well-defined orbit of radius r equal to the Bohr radius a_B. The full quantum mechanical picture is dramatically different.

In Fig. 17.9, we plot

$$|\psi_1(r)|^2 = \frac{1}{\pi}\left(\frac{1}{a_B}\right)^3 e^{-(2r/a_B)}. \tag{17.123}$$

This equation represents the probability density of finding the electron inside the atom. This distribution has no resemblance to an orbit. The picture of the hydrogen atom is therefore remarkably different from what we expect from classical or quasi-classical (Bohr) pictures. In

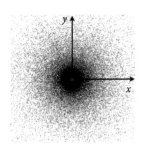

Fig. 17.9 The probability density $|\psi_1(r)|^2$ of the electron in the lowest atomic level ($n = 1$). The nucleus is located at the origin ($r = 0$).

the quantum mechanical picture of the hydrogen atom in its lowest energy state, the proton forms the nucleus and the electron is described, not by a definite location or an orbit, but by a cloud that represents the probability of finding the electron. This probability of finding the electron is higher close to the nucleus and decreases as the distance increases from the nucleus. The shape of the probability cloud becomes more complex for higher values of n but the basic idea remains the same.

Problems

17.1 Verify that

$$\psi(x, t) = \frac{1}{\sqrt{\sigma_i \left[1 + i\hbar t/2m\sigma_i^2\right]}\sqrt{2\pi}} e^{-\left\{\frac{(x-x_i)^2}{4\sigma_i^2\left[1+i\left(\hbar t/2m\sigma_i^2\right)\right]} - \frac{i}{\hbar}\frac{p_i(x-x_i)-\left(p_i^2/2m\right)t}{\left[1+i\left(\hbar t/2m\sigma_i^2\right)\right]}\right\}}$$

is the solution of the Schrödinger equation

$$-\frac{\hbar^2}{2m}\frac{\partial^2\psi(x, t)}{\partial x^2} = i\hbar\frac{\partial\psi(x, t)}{\partial t}$$

subject to the initial condition

$$\psi(x, 0) = \frac{1}{\sqrt{\sqrt{2\pi}\sigma_i}} e^{-\frac{(x-x_i)^2}{4\sigma_i^2}} e^{i\frac{p_i}{\hbar}(x-x_i)}.$$

17.2 Consider an electron of energy E incident on a potential barrier of width L and height V_0 as discussed in Section 17.5. Show that, when $E > V_0$, the reflectivity is not zero as would be the case in classical mechanics.

17.3 Show that the wavefunction

$$\psi_2 = \frac{1}{4\sqrt{2\pi}}\left(\frac{1}{a_B}\right)^{3/2}\left(2 - \frac{r}{a_B}\right)e^{-(r/2a_B)}$$

satisfies the Schrödinger equation for the hydrogen atom (Eq. (17.115)) when the energy is given by

$$E_2 = -\frac{me^4}{2(4\pi\varepsilon_0)^2\hbar^2}\frac{1}{4}.$$

BIBLIOGRAPHY

D. J. Griffiths and D. F. Schroeter, *Introduction to Quantum Mechanics* (Cambridge University Press 2018).

M. G. Raymer, *Quantum Mechanics: What Everyone Needs to Know* (Oxford University Press 2018).

L. Susskind, *Quantum Mechanics: The Theoretical Minimum* (Basic Books 2015).

Index

acceleration 32, 33, 34, 35,
    267, 270
AND gate, truth table
    of 233, 234
Airy pattern 66
algebra, fundamental theorem
    of 15
algebraic equation 13
Ampere's law 46, 84
Andromeda galaxy 68
angular acceleration 42
    displacement 40
    motion 39
    resolution 66
        of an eye 67
    velocity 39
ASCII code 202, 203
atom, Dalton's theory of 90, 91
    history of 93
    Rutherford model of 92–4
    Thomson's plum-pudding
        model of 91, 92

Balmer series 80, 93, 99
beam splitter 133, 134, 150,
    219, 220
    for a single photon state 149,
        150, 216, 217
    input–output relation for 148,
        149, 150, 218
    lossless 148
    polarization 150, 151, 152,
        155, 157, 158, 183, 193.
        208, 235
Bell basis states 158, 161, 162,
    165, 166, 167, 169, 236,
    237, 240, 243
    Bells's inequality 187–90, 192
    CHSH inequality 193–7
Bennett-92 (B-92)
        protocol 211–14, 216
    Brassard 84 (BB-84)
        protocol 207–11, 214
Bessel function 64, 65
binary numbers 202, 207, 215,
    233, 234

binomial expansion 117, 223
bit 8, 208
    flip gate 235
blackbody radiation 4, 73, 81, 85,
    86, 87
    spectrum 74, 86, 87
Bohr's model of atom 4, 94–8,
    102, 286
Bohr's principle of
    complementarity 4, 6, 7,
        76, 77, 121, 128, 158, 172,
        182, 208
    Bohr's reply 182, 186
    quantization condition 74, 99,
        102, 103
    radius 96, 286
Bose–Einstein condensation 6,
    108–10, 119

centripetal force 42, 96
circular motion 40, 41, 43
classical mechanics 1, 2, 3, 4,
    5, 7, 11, 32, 34, 35, 73,
    74, 76, 78, 80, 158, 183,
    265, 268, 270, 273,
    283, 287
    deterministic nature of 34
    trajectory 7, 32–5
cloning machine 174, 176
    of quantum state 7, 172,
        174, 175
CNOT gate 236, 237, 238, 240,
    242, 243
coherent superposition 78,
    79, 230
    of states 62, 154, 229
collision 37
    elastic 39, 49
    inelastic 39
    one-dimensional 37
complex conjugate 15,
    19, 30
    numbers 13, 16, 20, 30
        addition 16
        division 16
        multiplication 16

polar coordinates 20
    subtraction 16
complex plane 15, 19
Compton scattering 5,
    114–19
    wavelength 119
computer technology 5
concurrence 161, 169
correlation function 195
Coulomb force 38, 45, 49, 92
counterfactual
        communication 217,
        218, 223–6
cross product of two vectors 24,
    25, 30
cryptography 8, 201, 203
    quantum 202

Davisson–Germer experiment 5,
    103, 104, 105
de Broglie's conjecture 5, 102, 114
    wave 6, 100–5, 108, 110, 265,
        266, 267
    wavelength 75, 101, 102,
        105, 108, 111, 117, 119,
        129, 135, 136, 266,
        271, 283
decoherence 231, 232, 233
delayed-choice experiment 6,
    130, 131
De Moivre's theorem 19
dense coding, quatum
        241, 242
Deutsch problem 237–9
diffraction 61–6, 79, 80
    from a circular aperture 64
    from a slit 61, 62
Dirac's ket-bra notation 142, 145
dispersion relation 117
Di Vincenzo criteria 232
dot product of two vectors 23, 24,
    30, 143

eavesdropper 8, 208, 210,
    211, 213
e-commerce 8, 9, 207

electron, charge-to-mass
        ratio 48
energy, conservation of 115,
        116, 117
Einstein–Bohr debate 7
    on complementarity 127–30
Einstein–Podolsky–Rosen (EPR)
        paradox 182–6, 188
Einstein's theory of relativity 1,
        101, 116, 173, 175, 184
electric field 45, 47, 48
    force 2
electromagnetic wave 84,
        137, 138
electron microscope 105
entangled state 8, 9, 79, 158, 159,
        160, 161, 164, 167, 169,
        170, 173, 174, 192, 196
entanglement 78, 80, 154, 158,
        159, 169, 229, 230
    measure of 161
    swapping 167, 168, 169
Euclidean algorithm 246
Euler's formula 19
exponent 18

factoring a number 9, 206, 245–8
    algorithm 207
fidelity of a quantum state
        176–9, 181
frequency 51
    angular 52
    of standing wave 56

geocentric model 2
gravity, force of 33
Grover's algorithm 245, 255, 257,
        259, 260

Hadamard gate 235, 237, 242,
        255, 257
    transformation 239, 240,
        242, 252
half adder, circuit of 234
Heisenberg microscope 110–114
    uncertainty relation 4, 6, 7, 76,
        77, 100, 110, 111, 114, 121,
        128, 129, 130, 172, 231,
        269, 271, 275
heliocentric model 2
hidden variables 182, 186, 187,
        188, 189, 195
Hubble telescope 68
Huygen's principle 61, 82, 83

hydrogen atom 3, 10, 74, 76, 80,
        90, 95, 98, 284, 285
    spectrum 93, 94, 97

imaginary number 14
interaction-free
        measurement 220, 221
interference 5, 61
    constructive 54, 55, 57, 116
    destructive 54, 58, 62, 63,
        104, 121

kinetic energy 37, 38, 39, 49
Kirchhoff function 86

Laplacian 284
light, corpuscular nature
        of 82, 83
    extramission theory of 81
    history of 81–4
    wave nature of 82, 121
locality 7, 183, 184, 187–92, 195
logic gates 232–7, 241, 242, 256
Lorenz formula 46

Mach–Zehnder
        interferometer 218–26
magnetic field 45–8
    force 2, 52
Malus' law 139, 141, 142, 144
    for a single photon 142, 146
Maxwell's equations 3
modulo function 205, 215, 246
moment of inertia 41, 42, 44
momentum 36–9
    angular 44, 49, 77, 96
    conservation of 36, 37, 39,
        115, 127
    linear 35–7, 44

needle in a haystack 9
Newton's corpuscular theory of
        light 2, 3
    equation 3, 274
    laws of motion 6, 32
    second law of motion 33, 38,
        41, 270
    third law of motion 13
Newtonian mechanics 3, 6, 76,
        187, 270
no-cloning theorem 8, 173, 175,
        176, 214
nucleus 9
number theory 204

orthogonality relation 145, 146

particle inside a box 275–8
permittivity 38, 49, 96
photoelectric effect 4, 5, 7, 74, 75,
        81, 87, 88, 89, 90, 99, 100,
        114, 116
photon energy 95, 101
    momentum 101, 111
Planck's constant 77, 86, 87
    distribution function 86
    quantum postulate 3
Pockel cell 150, 151, 152, 153,
        155, 193, 196, 209, 210
Poisson spot 83
polarization of a single
        photon 144, 145, 147,
        155, 172, 194
    light 137–42, 147
    state 145, 146, 147, 155, 156,
        157, 174, 185, 209, 210,
        211, 213, 214
polarizer 138, 139, 140, 142,
        146, 147, 152, 153, 159,
        188, 196
potential energy 9, 16, 17, 37, 38,
        97, 268, 275, 285
probability 25, 27
    distribution 27
    joint 27, 28, 29
    theory 13, 25–30
projectile motion 35
psychic communication 8
public key distribution 8,
        203, 204
Pythagoras theorem 16, 22

quadratic equation 13, 14
quantization condition
        74, 87
    of energy 3, 73, 74
quantum communication 5,
        143, 169
    computer 8, 9, 229, 230, 231,
        232, 233, 237, 238, 245,
        251, 253, 254, 257
    computing 5, 9, 73, 80, 154,
        158, 207, 245, 254, 255
    introduction to 229–33
    copier 176–80
    cryptography 8, 207, 208
    entanglement
        (see entanglement)
    eraser 6, 75, 131–5

Fourier transform,
    discrete 248–51, 253
logic gates 233–7
mechanics, foundation of 6
    probabilistic nature of 72,
        73, 231
    vacuum in 231
money 214, 215
phase gate 235, 236
shell game 254–7
swapping 237
teleportation 8, 164, 165, 166,
    167, 170, 237, 240, 241
tunneling 9
qubit 9, 229–33, 251–8, 260

Rayleigh criterion 66–68,
    113, 114
Jeans ultraviolet catastrophe 86
law 86, 98
limit 68
reality 7, 135, 163, 183, 187, 189,
    190, 191, 195
    Einstein's concept of 164, 182,
        184, 185, 186, 188
real numbers 13
root-mean-square (rms)
    deviation 27
RSA algorithm 8, 9, 202, 203–7,
    215, 245, 246
Rutherford gold-foil
    experiment 92
Rydberg constant 97
    formula 80, 93

scalar product of two
    vectors 23
    quantity 20

Schrödinger equation 3, 4, 9, 10,
    73, 74, 76, 98, 100, 265,
    271, 272, 275, 280
    in one dimension 265–70, 275,
        276, 279, 281
    in spherical coordinates 284
    in three dimensions 283–7
    Schrödinger's cat paradox 162,
        163, 164
secure communication 8
shell game 9
Shor's algorithm 9, 207, 245,
    251–4, 260
simple harmonic motion 38
Star Trek 8
Stern-Gerlach experiment
    78, 79
state vector 143, 144, 159, 176
superluminal
    communication 167, 172,
        173, 174
superposition, principle
    of 53, 106

teleportation 8
thermodynamics 2
Thomson experiment 45, 46
threshold frequency 74,
    88, 99
torque 41, 42
transistor 8, 9, 283
trigonometry 13, 16–19
tunneling through a
    barrier 278–83

vector, adding 21
    analysis 13
    quantity 20

wavelength, definition of 51
wavefunction 79, 98, 105, 106,
    108, 121, 126, 154, 266,
    267, 268, 269, 271, 272
wave, longitudinal 50, 51, 82, 138
    motion 50–6
    number 52
    phase of a 54
    running wave 54
    standing 54, 55, 56, 102
    transverse 50, 51, 52, 137
    vector 56
wave packet 7, 105–8, 271–5, 280
    Gaussian 108, 271, 273
wave–particle description of
    light 4, 75
    duality 6, 50, 74, 75, 77, 90,
        100, 105, 114, 121, 126,
        127, 130, 131, 266
which-path information 6, 125,
    126, 129, 131, 132, 133,
    134, 135
Wien's displacement law 85, 98
work 37
    function of a metal 88

X gate 235
XOR gate, truth table of
    233, 234

Young's double-slit experiment 2,
    5, 6, 57–63, 68, 75, 82, 121,
    129, 130, 131, 135
    with bullets 123, 125
    with waves 122, 123
    with electrons 121–7

Z gate 235